Solid State Fermentation
for Foods and Beverages

FERMENTED FOODS AND BEVERAGES SERIES

Series Editors

M.J.R. Nout and Prabir Kumar Sarkar

Solid State Fermentation for Foods and Beverages (2013)
Editors: Jian Chen and Yang Zhu

Valorization of Food Processing By-Products (2013)
Editor: M. Chandrasekaran

Fermented Foods and Beverages Series

Solid State
Fermentation
for Foods and Beverages

Edited by **Jian Chen** • **Yang Zhu**

CRC Press
Taylor & Francis Group
Boca Raton   London   New York

CRC Press is an imprint of the
Taylor & Francis Group, an **informa** business

CRC Press
Taylor & Francis Group
6000 Broken Sound Parkway NW, Suite 300
Boca Raton, FL 33487-2742

First issued in paperback 2016

© 2014 by Taylor & Francis Group, LLC
CRC Press is an imprint of Taylor & Francis Group, an Informa business

No claim to original U.S. Government works

ISBN 13: 978-1-138-19932-3 (pbk)
ISBN 13: 978-1-4398-4496-0 (hbk)

Library of Congress Cataloging-in-Publication Data

Solid state fermentation for foods and beverages / editors, Jian Chen, Yang Zhu.
    pages cm -- (Fermented foods and beverages series ; 2)
    Summary: "This is the first book about fermented food and beverage to focus on solid-state fermentation. The text describes the various reactors used in solid-state fermentation, including static state reactors, dynamic reactors, and other new types of reactors, and provides a detailed introduction to various solid-state fermented foods and beverages, including product category, characteristics, functionalities, safety issues and consumer perception. The authors address such engineering issues as mass and heat transfer and energy equation calculation of solid-state fermentation, dynamic modeling of solid-state fermentation, and process control of solid-state fermentation. "-- Provided by publisher.
    Includes bibliographical references and index.
    ISBN 978-1-4398-4496-0 (hardback)
    1. Solid-state fermentation. I. Chen, Jian (Chemist) II. Zhu, Yang, 1955-

TP248.25.S64S57 2013
660'.28449--dc23
    2013038027

Visit the Taylor & Francis Web site at
http://www.taylorandfrancis.com

and the CRC Press Web site at
http://www.crcpress.com

Contents

V

Series Preface

Natural fermentation precedes human history, and since ancient times humans have been controlling the fermentation process. Fermentation, the anaerobic way of life, has attained a wider meaning in the biotransformations resulting in a wide variety of fermented foods and beverages.

Fermented products made with uncontrolled natural fermentations or with defined starter cultures, achieve their characteristic flavor, taste, consistency and nutritional properties through the combined effects of microbial assimilation and metabolite production, as well as from enzyme activities derived from food ingredients.

Fermented foods and beverages span a wide diversity range of starchy root crops, cereals, pulses, vegetables, nuts and fruits, as well as animal products such as meats, fish, seafood, and dairy.

The science of chemical, microbiological and technological factors and changes associated with manufacture, quality and safety is progressing and is aimed at achieving higher levels of control of quality, safety, and profitability of food manufacture.

Both producer and consumer benefit from scientific, technological and consumer-oriented research. Small-scale production needs to be better controlled and safeguarded. Traditional products need to be characterized and described to establish, maintain and protect their authenticity. Medium- and large-scale food fermentation required selected, tailor-made or improved processes that provide sustainable solutions for the future conservation of energy and water, and responsible utilization of resources and disposal of by-products in the environment.

The scope of the CRC book series on "Fermented Foods and Beverages" shall include (i) globally known foods and beverages of plant and animal origin (such as dairy, meat, fish, vegetables, cereals, root crops, soybeans, legumes, pickles, cocoa and coffee, wines, beers, spirits, starter cultures and probiotic cultures), their manufacture, chemical and microbiological composition, processing, compositional and functional modifications taking place as a result of microbial and enzymic effects, their safety, legislation, development of novel products, and opportunities for industrialization, (ii) indigenous commodities from Africa, Asia (South, East, and South-East), Europe, Latin America, and Middle East, their traditional and industrialized processes and their contribution to livelihood, and (iii) several aspects of general interest such as valorization of food-processing by-products, biotechnology, engineering of solid-state processes, modern chemical and biological analytical approaches (genomics, transcriptomics, metabolomics and other ~omics), safety, health and consumer perception.

The second book, born in the series, is *Solid State Fermentation for Foods and Beverages*. This treatise, edited by Professor Jian Chen and Dr. Yang Zhu, deals with fermentation in the absence of free flowing water, thus called "solid-state fermentation or, SSF." SSF is environment-friendly, requiring a minimum of water and energy; the technique has been used traditionally in Asia and presently develops into industrial-scale. As such, the engineering aspects require attention and form the basis of this book. In addition major SSF processes for a range of foods, beverages and condiments are outlined by the contributors of this book, who are experts in their area.

Preface

Solid-state fermentation (SSF) is one of the oldest microbial technologies that have been applied in food processing, although neglected for a long time. SSF technology originated in China, dating back thousands of years. For the past 10 years, SSF has again attracted much attention, due to the energy crisis and environmental pollution. Scientists are trying to explore the mechanisms in the fermentation process through multidisciplinary and different research tools. Based on a series of established mathematical models, new design concepts for SSF bioreactors and process control strategies have been proposed that are taking SSF technology to new levels.

Food and beverage production is an important application field of SSF technology that has received widespread interest. This book systematically describes the production of solid-state fermented food and beverages in terms of the history and development of SSF technology and SSF foods, bioreactor design, fermentation process, various substrate origins, and sustainable development. More emphasis has been placed on the description of Oriental traditional foods produced by SSF in Chapters 6, 8, and 9, such as sufu, vinegar, soy sauce, Chinese distilled spirits, and rice wine. This book makes a unique contribution to the understanding of Oriental traditional foods. In addition, it has comprehensively incorporated the latest developments and achievements in the field of SSF.

The authors of this book are from Wageningen University (Netherlands), Jiangnan University (China), and China Agricultural University, all well known for their achievements in the field of food science and technology in general and food fermentations in particular. We are particularly grateful to all authors for their indispensable contributions.

The Authors

Yang Zhu, PhD, studied biochemical engineering at East China University of Science and Technology, where he worked also as a faculty member in the Department of Food Science and Technology, and the Department of Biochemical Engineering after his graduation. He also was a visiting scholar at the Institute of Microbiology of the University of Münster, Germany. Since 1993, he has been at TNO Nutrition and Food Research, the Netherlands and Wageningen University, the Netherlands. He received his PhD in bioprocess engineering in 1997 at Wageningen University. His main research focus covers food fermentation, bioprocess engineering, food nutrition, and pharmacokinetics.

The author/coauthor of dozens of scientific publications and books, currently Dr. Yang Zhu is an independent consultant in agro-food and biotechnology in the Netherlands, as well as a guest scientist at Wageningen University, the Netherlands.

Jian Chen, PhD, is a professor of food biotechnology at Jiangnan University. Professor Chen earned his PhD from Jiangnan University (former Wuxi University of Light Industry) in 1990. His research interests include stress tolerance and the response of food microorganisms, the production of food additives by biotechnology, and food safety issues in fermented foods. Using lactic bacteria and yeast as models, he investigated the roles of key genes and antioxidants on tolerance and response of food microorganisms to cold, acidic, oxidative, and osmotic stresses. He has successfully achieved the industrial-scale production of more than 30 different food additives by metabolic engineering and process optimization. He also screened and improved a series of food enzymes, which have been widely applied in juice processing and bakery areas. He is working on mechanisms inside the accumulation of nitrogen-containing small molecule harmful components during the production of fermented foods, such as ethyl carbamate and bioamines.

Dr. Chen has published over 250 research articles and 16 invited reviews. He currently serves on the board of editors for eight peer-review journals, including the *Journal of Agricultural and Food Chemistry* and *Process Biochemistry*. Professor Chen was named as a fellow of the International Academy of Food Science & Technology (IAFoST) in 2012.

Contributing Authors

Jian Chen
Key Laboratory of Industrial
    Biotechnology
Ministry of Education, School
    of Biotechnology
Jiangnan University
Wuxi, China

Yingjia Chen
State Key Laboratory of Food
    Science and Technology
School of Food Science and
    Technology
Jiangnan University
Wuxi, China

Yongqiang Cheng
College of Food Science &
    Nutritional Engineering
China Agricultural University
Beijing, China

Wenlai Fan
State Key Laboratory of Food
    Science and Technology
School of Biotechnology
Jiangnan University
Wuxi, China

Fang Fang
Key Laboratory of Industrial
    Biotechnology
Ministry of Education and
    School of Biotechnology
Jiangnan University
Wuxi, China

Beizhong Han
College of Food Science &
    Nutritional Engineering
China Agricultural University
Beijing, China

Tiancheng Li
State Key Laboratory of Food
  Science and Technology
School of Food Science and
  Technology
Jiangnan University
Wuxi, China

Dasong Liu
State Key Laboratory of Food
  Science and Technology
School of Food Science and
  Technology
Jiangnan University
Wuxi, China

Long Liu
State Key Laboratory of Food
  Science and Technology
School of Food Science and
  Technology
Jiangnan University
Wuxi, China

Song Liu
Key Laboratory of Industrial
  Biotechnology
Ministry of Education, School
  of Biotechnology
Jiangnan University
Wuxi, China

Xiaoming Liu
State Key Laboratory of Food
  Science and Technology
School of Food Science and
  Technology
Jiangnan University
Wuxi, China

Xiaoqing Mu
State Key Laboratory of Food
  Science and Technology
School of Biotechnology
Jiangnan University
Wuxi, China

Dong Wang
State Key Laboratory of Food
  Science and Technology
School of Biotechnology
Jiangnan University
Wuxi, China

Haiyan Wang
State Key Laboratory of Food
  Science and Technology
School of Biotechnology
Jiangnan University
Wuxi, China

Qun Wu
State Key Laboratory of Food
  Science and Technology
School of Biotechnology
Jiangnan University
Wuxi, China

Ganrong Xu
Key Laboratory of Industrial
  Biotechnology
Ministry of Education, School
  of Biotechnology
Jiangnan University
Wuxi, China

Yan Xu
State Key Laboratory of Food
    Science and Technology
School of Biotechnology
Jiangnan University
Wuxi, China

Bobo Zhang
Key Laboratory of Industrial
    Biotechnology
Ministry of Education, School
    of Biotechnology
Jiangnan University
Wuxi, China

Dongxu Zhang
Key Laboratory of Industrial
    Biotechnology
Ministry of Education, School
    of Biotechnology
Jiangnan University
Wuxi, China

Peng Zhou
State Key Laboratory of Food
    Science and Technology
School of Food Science and
    Technology
Jiangnan University
Wuxi, China

Yang Zhu
Department of Biosciences
TNO Quality of Life
Zeist, The Netherlands

1

HISTORY AND DEVELOPMENT OF SOLID-STATE FERMENTATION

SONG LIU, DONGXU ZHANG, JIAN CHEN

Key Laboratory of Industrial Biotechnology, Ministry of Education, School of Biotechnology, Jiangnan University, Wuxi, China

YANG ZHU

Department of Biosciences, TNO Quality of Life, Ziest, The Netherlands

Contents

1.1 Introduction

1.1.1 Definition

Solid-state fermentation (SSF) may be defined as a technique of cultivating microorganisms on and inside humidified particles (solid substrate). The liquid content, bound with this solid matrix, is maintained at the level corresponding to the water activity, assuring correct growth and metabolism of cells, but should not exceed the maximum water-holding capacity of the solid matrix (Singhania et al., 2009). The solid matrix is either the source of nutrients or simply a support impregnated by the proper nutrients. SSF has many advantages over liquid-state fermentation (LSF): lower energy requirements, producing

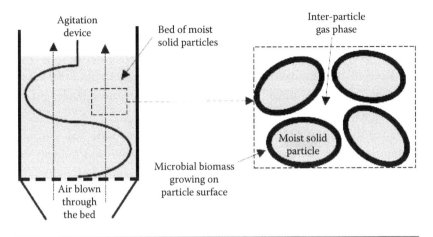

Figure 1.1   A typical SSF process in a bioreactor.

lesser wastewater, environmentally friendly process, etc. Nearly all the fermentation products in ancient times were derived from SSF.

The aim of SSF is to bring the cultured microorganisms in close contact with the insoluble substrates and to achieve the highest nutrient concentration from the substrates for fermentation (Krishna, 2005). Growth of microorganisms, especially filamentous fungi and yeasts during SSF is more close to natural conditions than LSF, and some enzymes and metabolites are easier to be produced. For these reasons, SSF has been paid great attention. Because SSF is able to stimulate the growth of microorganisms in nature on moist solids, it has widely been used in fermentation in last decades, and has considerable economical potential in producing products for the food, feed, pharmaceutical and agricultural industries. A typical solid-state fermentation process in a bioreactor is shown in Figure 1.1 (Mitchell et al., 2005).

1.1.2 Categories of SSF

According to the types of microorganisms involved, SSF processes are divided into two major categories: natural (indigenous) SSF and pure culture SSF (Pandey, 1992). Natural SSF is a process that involves natural microflora, and does not need inoculum. Typical examples are the fermented foods (such as Chinese wine, soy sauce, and vinegar) made since ancient times (Chen, 1992), and composting/ensiling SSF using a pure culture. SSF is a fermentation using a single strain or a mixed culture to improve the control of the substrates utilization and end-product formation, but in these processes, contamination by undesired microorganisms must be avoided (Pandey et al., 2001). To that effect, substrates must be sterilized to kill unwanted species prior to inoculation with the desired microorganisms.

Depending upon the nature of the solid phase used, SSF can be distinguished by two systems. The most commonly used (and most often described) system involves cultivation on a natural material, which is also named *solid substrate fermentation*. The substrates of solid substrate fermentation are used as a carbon source and a solid phase. Frequently used substrates are wheat bran, potato, foxtail millet, soybean cake, and so on. However, natural substrates usually cause problems when fermentation goes on for a period of time, the geometric and physical characteristics of the medium changing (Oriol et al., 1988). Consequently, heat and mass transfer decrease (Barrios-Gonzalez and Mejia, 1996). These problems can be resolved by the second system, which involves cultivation on an inert support impregnated with a liquid medium. This provides a homogenous aerobic condition throughout the fermentor, and also does not contribute impurities to the fermentation product. It facilitates maximal recovery of the leachate with low viscosity and high specificity for the target product (Gautam et al., 2002). Hemp, perlite, polyurethane foam, sugarcane bagasse, and vermiculite are commonly used as inserts.

1.1.3 Difference between Solid-State Fermentation and Liquid Fermentation

The essential difference between SSF and LSF is whether a gas phase or liquid phase is used as a continuous phase. In addition, moisture content is a prime parameter. In SSF, microorganisms grow among moist solid particles where nutrition is obtained. Their moisture content can effectively be controlled from 12% to 80%, but most by 60%. Contrary to SSF, there are only 5% of solutes during LSF, and moisture content is about 95%. So far, LSF has been the dominant technique in the fermentation industry. The detailed differences between SSF and LSF are summarized in Table 1.1 (Chen and Xu, 2004).

Table 1.1   Difference between SSF and LSF

SSF	LSF
There is no flow of water in medium; water content is low	There is always flow of water; water content is high
SSF stimulates the growth of microorganisms in nature; some enzymes and metabolites can be produced that are not secreted in LSF, such as mycomycin and conidiospore	Microorganisms grow in artificial liquid medium
Water activity of the medium is lower than 0.99; between 0.93 and 0.98 is fit for microorganism growth; application is restricted	Most microorganisms can be used in LSF
System makes for microbial distribution, especially filamentous fungi	Microbial dissociation is restrained by liquid surroundings, which is not good for manufacture of secondary metabolite
Microorganisms absorb nutrition from solid matrix; there is a concentration gradient in medium	Microorganisms absorb nutrition from water; there is no concentration gradient in medium
Culture system contains gas phase, solid phase and liquid phase. Only gas phase is continuous phase	Mostly culture includes gas phase and liquid phase. Only liquid is continuous phase
Large inoculum amount, $\geq 10\%$	Small inoculum amount, $\leq 10\%$
A small quantity of atmosphere is required, which can provide oxygen needed by microorganisms. Energy consumption is low	Large energy is consumed on dissolving oxygen in water, which can be utilized by microorganisms
Microorganisms grow on the surface or in the pores of matrix	Microorganisms have a homogeneous distribution in water
When fermentation is finished, the medium is in moist material state, and product concentration is high	When fermentation is finished, the medium is in liquid state and product concentration is low
Because concentrated medium and small SSF bioreactors are used, production rate is high, but yield and growth rate are low	Because of using diluted medium and large LSF bioreactors, production rate is low
High substrate concentration can bring high product concentration	High substrate concentration led to phenomenon of non-Newtonian fluid; even feeding system is needed
Aeration pressure is low on account of low systematic pressure	System requests high bleed pressure in order to overcome head of liquid and pressure from gas phase to liquid phase
Metabolic heat is hard to be removed, and it is easily overheating	Fermentation temperature is controlled easily, but equipment is bulky
Nonuniform fermentation, especially in microbial growth, absorption of nutrient substance and secretion of metabolites	Homogeneous fermentation

continued

Table 1.1 (continued)   Difference between SSF and LSF

SSF	LSF
It is difficult to provide online measurements	Course of fermentation is measured online
Product extraction process is easy, but substrates in the extracts are hard to remove	Product extraction process is complicated, and expensive, and large volume of organic wastewater of high concentration is discharged. There is high purity in products
Because solid raw materials are utilized in SSF, saccharification and fermentation can be preceded simultaneously. Operation sequence is reduced, and energy consumption is low	Complex working procedures are required for solid raw materials, and energy consumption is high

1.2 History of Solid-State Fermentation

Although our ancestors did not know the science of SSF, they applied SSF in daily life for many centuries, evidence even dating from thousands of years ago.

The earliest SSF product is the natural wine that is produced by SSF with fruits, honey, animal milk, etc. SSF was mainly referred to as *distiller fermentation* in these cases. For example, *yuanjiu*, a Chinese old natural wine, was the first drink fermented in a natural environment without any saccharifying agents. Wine-making equipments and wine-making sets have been found in Chinese Longshan Culture Period 5000 years ago. With the appearance of the use of the yeast ingredient *qubo* 3000 years ago, people in China began to make hard liquor with yeast. At almost the same time, the ancient Egyptians were developing beer and wine brewing, which were illustrated in a fresco of a Theban pharaoh's pyramid.

In prehistoric times, another type of the SSF product was wheaten food. Paste fermentation, a typical SSF technology, is applied in the initial period of the production of wheat-fermented food. Over 2500 years ago, the ancient Egyptians mastered bread-making techniques using microorganism fermentation. There are also records describing wheat-fermented food processing in *Shihchi*, a famous history book in ancient China 2600 years ago.

After the paste fermentation, many food products produced by SSF came into the market, such as vinegar, soy sauce, fermented bean curd, and soya beans. China has the earliest historical documents on cereal vinegar in the world. The book *Ceremony Notes* (about 800 BC; Xiao,

2000) showed that vinegar played an important role in celebrations during the time of the China Zhou dynasty (from 1000 BC to 256 BC). There were also similar records of vinegar in another famous Chinese book *Analects of Confucius* (450 BC; Xiao, 2000). In the Chinese *Techniques of Qinese* (533–544 AD), there were 23 methods for vinegar preparation described in detail, among which the SSF method with cereals, a repeated successive fermentation process, was a unique technique for vinegar preparation (Bao, 1985). Until the late Qing Dynasty (1644–1911 AD), when industrial-scale vinegar production appeared (Zhang, 2000), Chinese vinegar was only domestically produced on a small scale. It is generally believed that the production process of Shanxi old mature vinegar was invented by Wang Laifu during the Qing Dynasty (1644 and 1661 AD; Yan and Xiong, 1997).

As vinegar fermentation evolved, the kojic acid process was developed in China about 550 BC years ago, and then this technique was taken to Japan by the Buddhists during the 7th century, growing into Japanese *koji*. Apple vinegar fermented by apple pomace emerged during the 18th century (Pandey, 1992). Another typical SSF product is Chinese fermented bean curd, which is dated from approximately 220 AD to 265 AD. It became a popular condiment and expanded to many types of products as SSF technology developed. *Miso* is a soft Japanese food and generally used in vegetarian cooking; it is also made from soybeans through SSF. *Tempeh* is yet another solid-state fermented food used in Indonesia, New Guinea, and Surinam.

In summary, SSF products in the early stage were composed of different kinds of fermented foods. Our ancestors left us with precious experiences in SSF technology, which is a solid foundation for SSF technology in modern times. As SSF technology proceeds, SSF products have been extended from food products into many kinds of nonfood products. In the early 40 years of the 20th century, SSF technology was developed to produce fungal enzymes, gluconic acid and citric acid, etc. During World War II in the 1940s, SSF developed rapidly because of the increased demand for antibiotics, extending the product range of SSF from foods to herbals. Soon after that, SSF technology became used for steroid transformation by fungal cultures during the 1950s. In the following 20 years, SSF technology was applied in the production of mycotoxins and protein-enriched feed. Today, the products based on SSF cover bioactive products, enzymes, organic acids, biopesticides, biofuel, aroma compounds, and miscellaneous compounds.

1.3 Development of Solid-State Fermentation

1.3.1 General Considerations

1.3.1.1 Industrial Strains   Selection of a suitable microorganism is one of the most important tasks in SSF. Filamentous fungi and yeasts are generally termed as predominant and ideal types of microorganisms suitable for SSF. For example, *Aspergillus niger* can produce as many as 19 types of enzymes through SSF (Pandey, 1992). The reason why they have advantages over unicellular microorganisms in the colonization of solid substrates and utilization of available nutrients is probably that their hyphal can spread into a solid-state medium and is resistant to low water activity (*Aw*) and high osmotic pressure conditions (Raimbault, 1998). The cultivation of the filamentous fungi on solid substrates has been widely used for different purposes on a laboratory and industry scale. *Phycomycetes* (*Mucor* and *Rhizopus*), *Ascomycetes*, and *Basidiomycetes* (white-rot fungi) are filamentous fungi that are most commonly used in SSF processes. Some bacteria are also successfully adopted in SSF for production of enzymes, and one successful case is the production of inulinase by *Staphylococcus* sp. (Selvakumar and Pandey, 1999). The productivity of a strain is a key factor affecting its commercial application through SSF, but a great quantity of wild-type microorganisms cannot achieve commercially acceptable yields. For this reason, wild strains will usually be improved before being used in industry. Several chemical or physical mutagenesis methods, including nitrosoguanidine (NTG), diethyl sulfate and nitrous acid, UV-irradiation, γ-ray, fast neutron irradiation, and neodymium-doped yttrium aluminum garnet (Nd:YAG) laser, have been successfully applied to breed microorganisms with improved productivity. In addition, gene engineering, such as genome shuffling, has also been used in strain improvement for the SSF process. However, gene engineering was usually less efficient in strain improvement than traditional chemical and physical mutagenesis methods.

1.3.1.2 Nutritional Requirements from a Medium and Solid Particles
Selection of substrates for SSF depends on physical parameters such
as particles size, cost, intraparticle spacing, and nutrient composition
requirements. If the particle size is too small, aeration of microorganisms is restricted and cell growth is poor. Although large particle size
(when void space between particles is big) is favorable for aeration, the
specific surface area which microorganisms can attach on is limited.
Thus, selecting a right particle size of substrates is essential for SSF
process. Agro-industrial residues are considered as the best substrates
for SSF as they supply the nutrients as well as spaces for the growth of
microbes (Xu, 2006). The substrates include sugarcane bagasse, wheat
bran, rice, maize and grain brans, straws of wheat and rice, coconut
coir pith, banana waste, tea and coffee wastes, cassava waste, palm
oil mill waste, sugar beet pulp, sorghum pulp, pomace of apple and
grape, peanut meal, rapeseed cake, coconut oil cake, soybean cake,
peanut cake, mustard oil cake, wheat flour, cassava flour and vegetable
wastes, hemp, and perlite. When these substrates are crushed prop-
erly, nutrient utilization rate and product yields are usually improved.
Filling materials, which are beneficial to aeration, are added in order
to increase the void space of substrates during SSF. In this respect,
raw materials for SSF can be divided into two types, including the
materials for offering nutrition and the materials utilized for aeration. Additionally, a suitable C/N ratio also should be considered,
which has significant impact on microorganism growth and secretion
of products. Excessive nitrogen often induces overgrowth of microorganisms, which is usually unfavorable for products formation. When
there is a shortage of carbon and nitrogen, it will lead to poor growth,
aging, and self-digestion of thalli. In general, the optimal C/N ratio
determined for specific products ranges from 1/10 to 1/100.

1.3.1.3 Moisture Content of a Medium The moisture content of a
medium is a key factor that determines the process stability and yield
of products during SSF. Moisture content can affect the physical state
of substrates, diffusion and utilization of nutrient substance, exchange
between oxygen and carbon dioxide, and heat/mass transfer. High
moisture content usually leads to particle agglomeration, gas transfer
limitation, bad aeration, and bacterial contamination. Low moisture

content result in substrate swelling and reduce nutrient diffusion, cell growth, enzyme stability, and product yields. Thus, an optimum moisture level can be determined according to the properties of raw materials (fineness and water-binding capacity), types of microorganisms (anaerobic, amphimicrobian, and aerobic), and culture conditions (temperature, humidity, and ventilation status).

In SSF, the moisture content of the medium should reach the level that is favorable for the growth of filamentous fungi and yeasts (low moisture content) but unsuitable for bacterial growth (high moisture content). The optimum moisture content for growth of filamentous fungi and yeasts and substrates utilization is between 40% and 70% according to the type of microorganisms and substrates (Raimbault, 1998). Moisture content may decrease because of water evaporation, metabolic activity, and aeration during the SSF process. Additional water should be added at this time, which is usually accomplished by feeding humid air.

1.3.1.4 pH   Each microorganism possesses a specific pH range for its growth and activity. Filamentous fungi have reasonably good growth over a broad range of pH, 2 to 9, with an optimal range of 3.8 to 6.0. Yeasts have an optimum pH between 4 and 5 and can grow over a wide pH range of 2.5 to 8.5. This typical pH versatility of fungi can be beneficially exploited to prevent or minimize bacterial contamination, especially when a low pH condition is applied (Krishna, 2005). A pH of SSF substrates is usually changed in response to metabolic activities. Organic acids produced by the strains often decrease the pH. Nevertheless, the assimilation of organic acids in some media may lead to an increase in pH (Raimbault, 1998). The nitrogen source plays a significant role in pH adjustment. For example, ammonium salts cause acidification of culture medium, while urea leads to alkalization. The pH control is usually integrated with moisture control in the SSF system that employs evaporative cooling for heat removal (Ryoo et al., 1991; Lu et al., 1998). The acid or alkali substances are dissolved in water at the desired concentrations, and then these solutions are sprayed on the surface of the SSF medium. It is easy to scale-up this pH control strategy in a large-scale SSF process.

1.3.1.5 Oxygen Transfer in a Medium   Oxygen transfer aims directly at aerobic SSF. Microbial growth results in the formation of biomass film on the solid surface, and the metabolized substrates agglomerate and become viscous. Otherwise, high moisture and slimy substrates will also affect oxygen transfer in the medium. In large-scale applications of SSF, aeration and agitation are utilized in order to avoid the deficiency of oxygen in medium and promote microbial growth.

Aeration mainly fulfills four functions in SSF: (1) maintaining aerobic conditions; (2) desorbing carbon dioxide; (3) regulating the temperature of substrates; and (4) regulating moisture content. Aeration volume is dependent on the volume of substrates and production state of strains, and moisture evaporation from the substrates should be also considered at the same time. A typical aeration rate ranges from 0.05×10^{-3} to 0.2×10^{-3} $m^3Kg^{-1}min^{-1}$ (Chisti, 1999). Some substrates with strong water-absorbing ability, such as lignocellulose, are added to the medium so as to avoid excessive water loss caused by evaporation.

Agitation can strengthen oxygen transfer and reduce compacting and mycelial binding of the substrate particles. However, excessive agitation is not favorable for exposure of the substrates to hyphae and even breaks the hyphae. The frequency of agitation is generally based on experience. Agitation is often used in conjunction with aeration to enhance oxygen transfer in many cases. The other approaches that can improve oxygen transfer are as follows: (1) reducing the depth of the substrate bed; (2) using polyporate, coarse and crumbly materials as substrate; (3) using a foraminiferous culture plate; and (4) employing materials with low moisture content.

1.3.1.6 Incubation Temperature and Heat Transfer   At the early stage of the fermentation, the temperature of the solid mass increases quickly as microorganisms grow rapidly. If these superabundant calories cannot be removed in time, the enhanced temperature may inhibit the growth and metabolism of strains and eventually cause mass mortality of hyphae. As a result of the relative low moisture and poor thermal conductivity of substrates, it is difficult to control temperature and heat transfer during the SSF process.

One of the most effective methods for dealing with heat accumulation is forced aeration, which plays multiple roles in SSF. Temperature is controlled by adjusting the aeration rate during evaporative cooling in SSF. Besides aeration, agitating the substrates regularly is another effective method of reducing cultural temperature. In summer it is more difficult to reduce the temperature because of high environmental temperature and humidity. In this case, refrigeration by liquid ammonia and air conditioning may be employed according to different conditions.

1.3.1.7 Biomass Measurement   Biomass is a fundamental parameter in characterization of microbial growth, and measurement of biomass is helpful to analyze and control the progress of SSF. Because the fungal hyphae usually penetrate it, it is difficult to directly measure the biomass in SSF in the substrates and tightly bind it. Currently SSF biomass is analyzed in an indirect way: (1) determination of physical parameters, including temperature, composition of effluent gas, and light reflectance; (2) determination of biological activities such as ATP levels, enzymatic activities, DNA content, respiration rate (oxygen consumption and carbon dioxide release), immunological activity, and nutrient consumption; and (3) analysis of cell constituents such as chitin, glucosamine, nucleic acids, ergosterol, and protein. Each determination method has drawbacks and is effective for a specific SSF.

1.3.2 SSF Bioreactors

SSF bioreactors differ according to the type of SSF products. On the basis of how they are mixed and aerated, SSF bioreactors can be divided into four groups (Ali and Zulkali, 2011):

Group 1: unforced aeration bioreactors without mixing
Group 2: forcefully-aerated bioreactors without mixing
Group 3: unforced aeration bioreactors with continuous or intermittent mixing
Group 4: forcefully-aerated bioreactors with continuous or intermittent mixing

1.3.2.1 Unforced Aeration Bioreactors without Mixing   Unforced aeration bioreactors without mixing usually are tray bioreactors. A tray bioreactor is a type of static bioreactor that is composed of a closed room with many removable trays. The fermentation temperature of the substrates is regulated by circulating warm or cool air as necessary. Its relative humidity can be controlled in the same way with humid or dry air. Because there is no continuous mixing, microbial cells, especially the hyphae of filamentous fungi, will not be damaged by the enormous stress of agitation. The structural simplicity of the tray bioreactor makes it relative easy to scale-up SSF using tray bioreactors. However, a tray bioreactor requires much space and manual labor in contrast to other SSF bioreactors. Thus, it may be appropriate for a new process if the product is not produced in very large quantities, or if labor is relatively cheap. The major disadvantage of tray bioreactors is that steep gaseous concentration gradients develop within the substrate bed, owing to mass transfer resistances, which may adversely affect the bioreactor performance (Gowthaman et al., 1993). Ito et al. (2011) recently constructed a nonairflow box with a moisture-permeable fluoropolymer membrane, making it possible to control and maintain uniform and optimal conditions in the substrate. Arturo et al. (2011) improved the heat transfer of the substrate in trays through forced air circulation. Recently, a novel deep-bed solid-state bioreactor was designed and fabricated for cellulolytic enzymes production using mixed fungal cultures. Better temperature and moisture control was achieved through a unique bioreactor design comprising an outer wire-mesh frame with internal air distribution along with near-saturation conditions within the cabinet (Brijwani et al., 2011).

1.3.2.2 Forcefully-Aerated Bioreactors without Mixing These bioreactors are typically represented by packed-bed bioreactors. The packed-bed bioreactor is another kind of static bioreactor in which substrates are packed closely as a bed and forced aeration is used. Packed-bed bioreactors are also appropriate for those SSF processes in which it is not desirable to mix the substrate bed at all during the fermentation due to deleterious effects on either microbial growth or the physical structure of the final product. Because air is blown through a sieve at the bottom of trays or columns, the fermentation control of packed-bed bioreactors is easier than that of tray bioreactors. However, the temperature and the O_2 concentration of the air that flows within the bed increase along the bed towards the outlet, and the increased temperature may slow growth or causes death, causing low product yields (Ali and Mahmoodzadeh, 2009). Thus, the main challenge in scaling-up packed-bed bioreactors is to reduce temperatures at the outlet end of the bioreactor. Lu et al. (1998) designed a multilayer packed-bed reactor in which the bed was divided into several layers. This division can enhance heat and mass transfer in comparison with a bioreactor in which the same total amount of substrate was loaded on a single perforated base plate. The multilayer packed bed allows a new operation strategy in which mass and heat transfer can be regulated by changing the order of the layers during SSF. Alexander et al. (2010) studied the effect of order of layer on SSF and pointed out that the operation of multilayer packed beds in the continuous plug-flow mode can reduce metabolic heat significantly.

1.3.2.3 Unforced Aeration Bioreactors with Continuous or Intermittent Mixing
The solid-state cultures in this kind of bioreactor are mixed either continuously or intermittently with a frequency from minutes to hours, and the air is circulated through the head space of the bed. The advantage is that the mixing event prevents the pressure drop from becoming too high within the bed and that water can be added to the bed in a reasonably uniform manner during the mixing event. The choice between continuous and intermittent mixing will depend on both the sensitivity of the organism to shear effects during mixing and the properties of the substrate particles, such as their mechanical strength and stickiness. Although forced aeration is not applied, the mixing can efficiently accelerate the mass and heat transfer during the fermentation. Based on the difference of agitation mechanism, this kind of bioreactor can be divided into the "rotating-drum bioreactors" and the "stirred-drum bioreactors." A rotating-drum bioreactor is generally composed of a matrix bed, gaseous phase flux space, and rotating-drum wall which are the mixing component. Under natural aeration, the ratio of length to diameter of the rotating-drum seems to be an important parameter that affects oxygen supply in a rotating-drum bioreactor (Rodríguez-Meza et al., 2010). Unlike the rotating-drum bioreactor, the stirred-drum bioreactor is equipped with paddles or scrapers to achieve agitation, though its global structure is similar to that of a rotating-drum bioreactor. Due to the application of continuous or intermittent mixing, the growth of the microorganism in stirred-drum

bioreactors or stirred-drum bioreactors is considered more uniform than in the tray fermentation or bed packed fermentation. Researchers from Wageningen University designed a continuous mixed horizontal stir bioreactor in which temperature and humidity are well controlled during the SSF process (Liao and Zheng, 2005).

1.3.2.4 Forcefully-Aerated Bioreactors with Continuous or Intermittent Mixing
In this kind of SSF bioreactor, air blown forcefully through the substrate bed is mixed continuously or intermittently. The continuous mixing mode is usually not favorable for growth of microorganisms, and the intermittent mixing mode is commonly used for its unduly deleterious effects on microorganism cells. There are many different ways in which bioreactors can be mixed, and thus bioreactors of this group usually have quite different structures to achieve mixing.

Among these bioreactors, the rotary-type automatic koji-making equipment have been widely used in Asian countries (Durand, 2003). In this bioreactor, the treated substrate is heaped up on a rotary disc with a layer of maximum thickness 50 cm, and the agitation is achieved by a rotating perforated table. This nonsterile reactor operates with a microcomputer that controls all the parameters (temperature of the air-inlet, air flow rate, and agitation periods). One of the major weaknesses of this bioreactor is the need to prepare and inoculate the substrate in other equipment before filling the reactor.

The fluidized-bed bioreactor is a typical SSF bioreactor with forceful aeration and continuous mixing. In a fluidized-bed bioreactor, the air flows into the reactor through a porous distributor at the bottom of the column, and the flow rate is sufficient for achieving the fluidized state of a solid (Liao and Zheng, 2005). The substrate bed is well mixed without additional agitation components because of the special aeration mechanism. In contrast to the packed-bed bioreactor, the fluidized-bed bioreactor has several advantages, such as the high fluency, low pressure, and favorable homogeneity of the culture. The difficulty of operating a fluidized bed is dependent on the size and distribution of the cultural particles (Moreira et al., 1996). The small particles with narrow size distribution are generally favorable for keeping the bed in a fluidized state. In contrast, the particles tendency towards aggregation makes it often difficult to maintain a fluidized state due to collisions among them.

The bioreactor is the core of the fermentation process. An ideal SSF bioreactor should have several features: (1) The material used to produce the bioreactor should be strong, anticorrosive, and nontoxic to the microorganisms. (2) The bioreactor can prevent the microbial contamination from external environment and restrict the release of organic materials produced by the fermentation. It is particularly hard to satisfy the former requirement because the solid-state substrates cannot be delivered by pumps as the liquid state substrate does in LSF. (3) The bioreactor can precisely control the operation parameters, such as temperature, water activity, and oxygen concentration in the air. (4) The bioreactor can maintain the homogeneity of the matrix bed. It is also critical to minimize the thermal gradient, which is an important factor affecting the SSF process. (5) The bioreactor should contain all the unit operations of the fermentation process, including medium preparation and sterilization, and product recovery.

1.3.3 Modeling in SSF

In the past decades, bioreactor design and process control developed concurrently. Modeling in SSF is an important method to push both of them forward. The objective of SSF modeling is to obtain mathematical expressions that represent the system under consideration, and subsequently resolve the relationships among different variables that characterize SSF (Singhania et al., 2009).

1.3.3.1 Modeling of Heat and Mass Transfer   One of the major problems to be overcome in large-scale SSF is heat accumulation and heterogeneous distribution in the complex gas–liquid–solid multiphase bioreactor system. The interaction between the heat and mass transfer phenomena usually leads to the formation of steep concentrations and temperature gradients, resulting in inhomogeneous conditions in the SSF bioreactor. Subsequently, efficiency of the bioreactor capacity is reduced.

Since agitation is not applied, the problems of the heat and mass transfer are especially noticeable in SSF using packed-bed bioreactors. Von Meien and Mitchell (2002) developed a two-phase dynamic model that describes heat and mass transfer in intermittently-mixed SSF fermentation bioreactors. The model predicts that in the regions of the bed near the air inlet there can be significant differences in the air and solid temperatures, while in the remainder of the bed the gas and solid phases are much closer to equilibrium. Sangsurasak and Mitchell (1998) developed a two-dimensional heat transfer model in which both radial temperature gradients and axial temperature gradients of the bed were well described during SSF of *A. niger*. A similar mathematical model is also developed in the case of a rotating-drum bioreactor under an anaerobic condition (Wang et al., 2010). These models are obviously useful tools that can guide scale-up and test control strategies of temperatures.

Oxygen limitation in SSF has become an important topic of modeling studies. Oostra et al. (2001) studied the oxygen transfer limitation at the particle level by modeling investigation. The penetration depth of oxygen could adequately be described using a reaction-diffusion model based on the zero-order intrinsic oxygen uptake kinetics of *Rhizopus oligosporus*. As indicated by the model, the thickness of the liquid layer and the available gas–liquid interface are identified as key parameters in optimizing oxygen transfer to the microorganism in solid-state fermentation. As mentioned before, wheat bran, rice husk, and bagasse that form solid particles are usually used as the substrates in SSF. To understand the transfer of the particles, a method was developed to characterize mixing in a rotating-drum bioreactor based on the use of wheat bran particles dyed with Rhodamine-WT as tracer particles (Marsh et al., 2000). Recently, particle collision intensity in a SSF bioreactor has been estimated directly by means of a hard-sphere model. This method is entirely general and can be used to assess bioreactors designed for slurry bioprocesses (Jin et al., 2012).

1.3.3.2 Modeling of Cell Growth   In SSF, the fungal growth on the solid substrate particles induces the formation of the cell film around the particles. The size of the biofilm subsequently leads to the decrease in the porosity of the substrate bed and diffusivity of oxygen in the bed. Moreover, the steric hindrance limitation for the cell growth is also worth mentioning. A cell growth model in which microorganism film thickness around the substrate particles and steric hindrance was taken into account was thus developed (Rajagopalan and Modak, 1995). Based on this model, the upper limit on cell density in a tray bioreactor could be accurately predicted (Rajagopalan and Modak, 1995). For static cultivation, the fungus *Aspergillus oryzae* can also form a mycelial network that binds substrate particles, thereby forming agglomerates. These agglomerates are problematic because they restrict the transport of O_2 to the surfaces of the particles within the agglomerates. Recently, a three-dimensional discrete lattice-based system for modeling the growth of aerial hyphae of filamentous fungi on solid surfaces was established (Coradin et al., 2011). In this model, the space above the solid surface is divided into a 3-dimensional array of cubes, and the elongation of hyphae was represented by successively adding the microcubes, with random numbers being used to choose the direction of growth. In comparison with the continuous model (Nopharatana et al., 1998), this discrete model and its modified versions are able to accurately describe the hyphal densities as a function of height above the surface in the later periods of growth. It therefore represents a useful tool for investigating phenomena that occur at the microscale in SSF systems.

In many cases, the period after cell growth is frequently ignored in SSF. However, the latter period may be important especially during the secondary metabolite production. In addition, biomass in different regions of the bed is not necessarily in the same growth phase because of the inhomogeneous growth conditions. Modeling of the postgrowth stage is also important for the SSF process. Smits et al. (1999) developed four models describing inactivation of biomass of *A. niger* after long fermentation periods under nonisothermal conditions. These models indicated that heat transfer in the fermentation bed determines the biomass of the fermentation to a large extent.

It is usually very difficult to calibrate the kinetic models that describe cell growth under SSF conditions because there are too many parameters needing to be estimated and violation of standard conditions of numerical regularity. Recently, the advanced nonlinear programming techniques have been used to calibrate a complex kinetic model describing growth of *Gibberella fujikuroi* on an inert solid support in glass columns (Araya et al., 2007). To be noted, a special purpose optimization code was used during the data processing to solve the resultant large-scale nonlinear program. The faster convergence and higher accuracy can be achieved in comparison with the standard sequential solution (Araya et al., 2007).

In most cases, the remarkable gradients existed in static beds, and partial differential equations have to be used to set up submodels at macroscale. It is difficult to model the intraparticle diffusion processes of heat and mass in this situation. Simple empirical equations were thus adopted to construct growth kinetic models of SSF bioreactor, as it did not describe the effect of substrate concentration on cell growth. A series of empirical equations have been used to depict the cell growth kinetic models under SSF conditions, including linear, exponential, logistic, and monod equations (Viccini et al., 2001; Iyer and Singhal, 2010). The relative simplicity of parameter estimation is the major advantage of these kinds of empirical equations, among which the logistic model was the most frequently used due to its description of a final concentration of biomass (Spier et al., 2009; Iyer and Singhal, 2010). However, application of these empirical equations to simulate the cell growth has its natural limitation. For example, the logistic model is symmetric around the inflection point, and rates of acceleration are equal to that of deceleration; in a real SSF system, growth profile showed a short acceleration period followed by a long period of slow deceleration of growth, resulting in a systematic deviation of the mathematic models by using a logistic equation (Mitchell et al., 2004). Based on the concept of active and nonactive hyphal segments, Ikasari and Mitchell (2000) established a two-phase model of the growth kinetics of *R. oligosporus* in a membrane culture, in which a series of parameters with biological significance was proposed, such as a biomass in active segments (μ_g) and mass death of tips at the switchover from exponential to deceleration phase (L). Both exponential phase and the following deceleration phase could be accurately predicted by this two-phase model. After that, a new two-phase model consisting of exponential and logistic models with two different temperature-dependent specific growth rates for each phase was proposed by Hamidi-Esfahani et al. (2007). Since the maximum specific growth rate in the logistic model was related to the environmental temperature, this model can predict the entire growth rate profile at various temperatures. As mentioned by Viccini et al (2001), incorporation of parameters with biological significance and environmental parameters was an important strategy to improve the accuracy and application of those models based on empirical equations in the future.

Recently, an empirical model based on an artificial neural network was developed to predict the cell growth rate in SSF (Mazutti et al., 2010). After enough runs of data training and validation procedures, the model could accurately describe the dependence of cell growth on the fermentation time, initial total reducing sugar concentration, and inlet and outlet air temperatures. Obviously, these models combined with the balance/transport submodel could be a more robust and reliable tool for dynamic simulations of the SSF processes, and shall be an important supplement to the traditional empirical models.

1.3.4 SSF Products

Based on the analysis of literature, SSF offers potential advantages for bioprocessing and production of various value-added products. It is well established that in many cases such as enzymes, bioactive compounds, etc., the productivity in SSF is many-fold higher than in LSF (Singhania et al., 2009). The cost of production in LSF is relatively high, and it is uneconomical to use many enzymes, in contrast to SSF. In addition to enzymes and bioactive compounds, SSF has been used to produce many other products such as organic acids, biopesticides, biofuel, biosurfactants, food flavor compounds, miscellaneous compounds, etc. (Table 1.2).

1.4 Future Trends

1.4.1 Intensive Study of the SSF Mechanism

In SSF, microbial growth, substrate consumption, and product formation are the key targets for the research of SSF mechanism, as well as the mathematical model of mass transfer and heat transfer. Understanding these aspects contributes to the scale-up of SSF and its commercialization, making the SSF technology economically feasible.

1.4.1.1 Establishment of a Kinetic Model   Existing models disregard many of the significant phenomena that are known to influence SSF. As a result, available models cannot explain the generation of the numerous products that form during any SSF process and the outcome of the process in terms of the characteristics of the final product.

Modeling of growth kinetics. Differentiated and integrated forms of the various empirical growth equations have been applied to SSF systems, including linear, exponential, logistic, and fast-acceleration/slow-deceleration. Mathematical models of SSF bioreactors most

Table 1.2   The Products Produced by SSF

TYPE	PRODUCT	STRAIN	SUBSTRATE	REFERENCE
Enzymes	Cellulase	Fungi *Trichoderma reesei* RUT C30	Lignocellulosic	Sukumaran et al., 2009
	Laccases		Lignocellulosic waste material	Majeau et al., 2010
	α-Amylase	*Bacillus subtilis* DM-03	Potato peel	Mukherjee et al., 2009
	Alphaamylase	*Bacillus amyloliquefaciens*	Wheat bran and groundnut oil cake	Gangadharan et al., 2008
	Xylanase	*Aspergillus terreus* MTCC 8661	Palm industrial waste	Lakshmi et al., 2009
Bioactive products	Single cell oil	*Mortierella isabellina*	Sweet sorghum	Economou et al., 2010
	Lipases and biosurfactants	*Aspergillus* spp	Oil	Colla et al., 2010
	Biosurfactant	*Candida glabrata*	Vegetable fat waste	de Gusmão et al., 2010
	Lipopeptide biosurfactants	*Bacillus subtilis*	Cheap carbon source	Das and Mukherjee, 2007
Organic acids	Citric acid	*Aspergillus niger*	Banana peel	Karthikeyan and Sivakumar, 2010; Singh Dhillon et al., 2011
	Gluconic acid	*Aspergillus niger* ARNU-4	Tea waste	Ramachandran et al., 2008; Sharma et al., 2008
	Lactic acid	*Lactobacillus helveticus*		Bouguettoucha et al., 2009
	Polyunsaturated fatty acids	*Mortierella alpina*	Rice bran	Jang and Yang, 2008
Biopesticides	Biodetoxification	*Penicillium simplicissimum*	Castor bean waste	Godoy et al., 2009; Godoy et al., 2011
	Coffee pulp	*Penicillium commune*		Nava et al., 2006
	Chemical insecticides	*Metarhizium anisopliae*		van Breukelen et al., 2011

continued

Table 1.2 (continued)   The Products Produced by SSF

TYPE	PRODUCT	STRAIN	SUBSTRATE	REFERENCE
Biofuel	Bioethanol	*Saccharomyces cerevisiae*	Water hyacinth biomass	Mohanty et al., 2009; Aswathy et al., 2010; Chen and Qiu 2010; Rodriguez et al., 2010
	Ethanol	*Phanerochaete chrysosporium*	Corn fiber	Shrestha et al., 2008
Aroma compounds	Polyhydroxyalkanoates		Waste material	Castilho et al., 2009
	Feruloyl esterase	*Aspergillus niger*		Zeng and Chen, 2009
	Free gossypol	*Candida tropicalis*		Weng and Sun, 2006
Miscellaneous compounds	Phosphate-solubilizing microorganisms		Agro-industrial wastes	Vassileva et al., 2010
	Conversion of wheat straw into lipid	*Aspergillus oryzae* A-4		Lin et al., 2010
	Compactin	*P. brevicompactum*	Corn husk	Longo and Sanroman, 2006; Shaligram et al., 2008
	Rifamycin B	*Amycolatopsis sp.*	Corn husk	Mahalaxmi et al., 2010
	Cephamycin C	*Nocardia lactamdurans*	The soybean flour	Kagliwal et al., 2009

commonly use the logistic equation, because they give an adequate approximation of the whole growth curve in a single equation, while other kinetic models of the growth curve are separated into various phases with a different equation for each phase. Filamentous fungi, which bind tightly to the substrate, usually make it impossible to determine the dry weight of fungal biomass directly (Rahardjo et al., 2006). Therefore, many of the kinetic studies done to determine the kinetic model are either undertaken in artificial systems that allow microbial biomass measurements, such as membrane culture (Rahardjo et al., 2004) or amberlite resin (Rahardjo et al., 2002), or are undertaken in the real system but using indirect measurements of growth, such as O_2 consumption or glucosamine content. However, both approaches have problems. Efforts are underway to make the growth conditions provided by an artificial system mimic those experienced in the real system.

Modeling of death kinetics. The modeling of death kinetics in SSF systems has received relatively little attention due to the fact that the majority of SSF processes involve filamentous fungi. The mycelial mode of growth of fungi makes the definition of death more problematic than is the case for unicellular organisms. Mitchell et al. (2003) assumed first-order death kinetics and segregated the microbial biomass into living and dead subpopulations. Smits et al. (Smits et al., 1999) brought up an equation that described how the specific respiration activity of the microbial biomass decreased over time, something which would be expected if a part of the biomass were dying.

Modeling of substrate consumption. Some models take this dry weight loss into account to predict the consumption of the solid substrate.

1.4.1.2 Mass Balance and Energy Balance   Heat and mass transfer in bioreactors is one of the continuous research requirements. The SSF fermentation is very complex, and many mathematical models for bioreactors are not yet mature, for they were based on simplified hypotheses. For example, when modeling intraparticle phenomena, we describe the particle as "round" and the space between them as "regular," which is very different from the real situation of the particles. As a result, we can't understand the intraparticle diffusion and its effect on the fermentation.

1.4.2 Control Technology

1.4.2.1 Design of Solid-State Fermentation Reactor Devices   Despite the long history of SSF, the development of SSF bioreactors is far behind LSF, which is reflected in the size of bioreactors. One of the puzzles of the SSF bioreactor is preventing high temperatures to maintain both bed temperature and water content at the optimum values for growth and product formation. Gas flowing is currently the most useful measure, which must combine with stirring to get an optimum result. Another consideration of an SSF bioreactor is how to prevent microbiological contamination during long-time fermentation.

A new generation of small reactors was developed by an INRA-team in France (Singhania et al., 2009). Such reactors have a working volume of about 1 liter, a relative humidity probe, a cooling coil on the air circuit, and a heating cover for the vessel. Each reactor is automatically controlled by a computer (Pandey and Larroche, 2008). A novel bioreactor with two dynamic changes of air (including air pressure pulsation and internal circulation) can increase mass and heat transfers, as well as improve the porosity within the substrate. The reactor can be significant in exploiting numerous socioeconomic advantages of solid-state fermentation (Chen et al., 2002).

1.4.2.2 Strategy of Controlling Parameters   Monitoring SSF processes requires the measurement of environmental parameters (temperature, pH, water content, and activity) and the carbon cycle (biomass, substrate concentration, CO_2). Direct measurements of temperature, pH, and water content are considered employing classical sensors. Indirect measurements of the biomass can be estimated by respirometry or pressure drop (PD). Recently more methods are explored such as aroma sensing, infrared spectrometry, artificial vision, and tomographic techniques (x-rays, magnetic resonance imaging [MRI]; (Bellon-Maurel et al., 2003; Couri et al., 2006).

1.4.3 Application

1.4.3.1 Bioremediation   Bioremediation belongs to biodegradation, which disintegrates or transforms poisonous contamination into a harmless component. Masaphy et al. (Masaphy et al., 1996) inoculated *Pleurotus pulmonarius* in corn and wheat straw hood mixtures in order to weed. Kaštáneka et al. (1999) designed a 15m³ solid-state bioreactor to disintegrate polychlorinated biphenyls and polyvinyl chloride in the polluted water and soil.

1.4.3.2 Biological Pulp   Biological pulp is defined as the pretreatment of wood chips with microorganisms before pulping, which has the advantage of energy conservation and environmentally friendly usage, as well as improving the mechanical pulping technique. *Phanerochaete chrysosporium* and *ceriporiopsis subvermispora* are the most important strains; the former has been industrialized.

1.4.3.3 Biofuel Petroleum fuel contributes to the greenhouse effect while biofuel is friendly to the environment. Ethanol is the major biofuel of SSF with the usage of industrial and agricultural refuse such as cellulose. *Saccharomyces cerevisiae* is the ideal strain in SSF of ethanol. Henk, B. et al. (2010) produces ethanol with sorghum as its substrate while Lopez et al. (2011) use rice hull. Researches show that the step of sterilization has little influence on the ethanol production, which is also one of the SSF.

References

Alexander, M.D., Lara Elize, N.C., and Alex Vinicius, L.M. (2010). A model-based investigation of the potential advantages of multi-layer packed beds in solid-state fermentation. *Biochemical Engineering Journal.* **48**(2), 195–203.

Ali, F.M. and Mahmoodzadeh, V.B. (2009). Modeling of temperature gradients in packed-bed solid-state bioreactors. *Chemical Engineering and Processing.* **48**(1), 446–451.

Ali, H.K.Q. and Zulkali, M.M.D. (2011). Design aspects of bioreactors for solid-state fermentation: A review. *Chemical and Biochemical Engineering Quarterly.* **25**(2), 255–266.

Araya, M.M., Arrieta, J.J., Perez-Correa, J.R., Biegler, L.T., and Jorquera, H. (2007). Fast and reliable calibration of solid substrate fermentation kinetic models using advanced non-linear programming techniques. *Electronic Journal of Biotechnology.* **10**(1), 48–60.

Arturo, F.M., Tristan, E.I., and Gerardo, S.C. (2011). Improvement of heat removal in solid-state fermentation tray bioreactors by forced air convection. *Journal of Chemical Technology and Biotechnology.* **86**(10), 1321–1331.

Bao, Q.A. (1985). Vinegar production technology in ancient China (1). *Food Science (China).* **4**, 12–16.

Barrios-Gonzalez, J. and Mejia, A. (1996). Production of secondary metabolites by solid-state fermentation. *Biotechnology Annual Review.* **2**, 85–121.

Bellon-Maurel, W., Orliac, O., and Christen, P. (2003). Sensors and measurements in solid state fermentation: A review. *Process Biochemistry.* **38**(6), 881–896.

Brijwani, K., Vadlani, P.V., Hohn, K.L., and Maier, D.E. (2011). Experimental and theoretical analysis of a novel deep-bed solid-state bioreactor for cellulolytic enzymes production. *Biochemical Engineering Journal.* **58–59**, 110–123.

Chen, H.Z. (1992). Advances in solid-state fermentation. *Research and Application of Microbiology (China).* **3**, 7–10.

Chen, H.Z. and Xu, J. (2004). *Mechanism and Application of Modern Solid-State Fermentation.* Beijing: Chemical Industry Press.

Chen, H.Z., Xu, F.J., Tian, Z.H., and Li, Z.H. (2002). A novel industrial-level reactor with two dynamic changes of air for solid-state fermentation. *Journal of Bioscience and Bioengineering.* **93**(2), 211–214.

Chisti, Y. (1999). Fermentation (industrial): Basic considerations. In *Encyclopedia of Food Microbiology*, eds. R. Robinson, C. Batt and P. Patel. London: Academic Press, pp. 663–674.

Coradin, J.H., Braun, A., Viccini, G., Luz, L.F.D., Krieger, N., and Mitchell, D.A. (2011). A three-dimensional discrete lattice-based system for modeling the growth of aerial hyphae of filamentous fungi on solid surfaces: A tool for investigating micro-scale phenomena in solid-state fermentation. *Biochemical Engineering Journal.* **54**(3), 164–171.

Couri, S., Merces, E.P., Neves, B.C.V., and Senna, L.F. (2006). Digital image processing as a tool to monitor biomass growth in *Aspergillus niger* 3t5b8 solid-state fermentation: Preliminary results. *Journal of Microscopy-Oxford.* **224**, 290–297.

Dongemi, W. (2002). Detoxification of cottonseeds by em strains. *Chinese Cotton.*

Durand, A. (2003). Bioreactor designs for solid state fermentation. *Biochemical Engineering Journal.* **13**, 113–125.

Gautam, P., Sabu, A., Pandey, A., Szakacs, G., and Soccol, C.R. (2002). Microbial production of extra-cellular phytase using polystyrene as inert solid support. *Bioresource Technology.* **83**(3), 229–233.

Gowthaman, M.K., Ghildyal, N.P., Rao, K.S., and Karanth, N.G. (1993). Interaction of transport resistances with biochemical reaction in packed bed solid state fermenters: The effect of gaseous concentration gradients. *Journal of Chemical Technology and Biotechnology.* **56**(3), 233–239.

Hamidi-Esfahani, Z., Hejszi, P., Shojaosadati, S.A., Hoogschagen, M., Vasheghani-Farahani, E., and Rinzema, A. (2007). A two-phase kinetic model for fungal growth in solid-state cultivation. *Biochemical Engineering Journal.* **36**(2), 100–107.

Han, B., Wang, L., Li, S., Wang, E., Zhang, L., and Li, T. (2010). Ethanol production from sweet sorghum stalks by advanced solid state fermentation (ASSF) technology. *Journal of Biotechnology (Chinese).* **26**(7), 966–973.

Ikasari, L. and Mitchell, D.A. (2000). Two-phase model of the kinetics of growth of *Rhizopus oligosporus* in membrane culture. *Biotechnology and Bioengineering.* **68**(6), 619–627.

Ito, K., Kawase, T., Sammoto, H., Gomi, K., Kariyama, M., and Miyake, T. (2011). Uniform culture in solid-state fermentation with fungi and its efficient enzyme production. *Journal of Bioscience and Bioengineering.* **111**(3), 300–305.

Iyer, P. and Singhal, R.S. (2010). Production of glutaminase (e.C. 3.2.1.5) from *Zygosaccharomyces rouxii* in solid-state fermentation and modeling the growth of z. Rouxii therein. *Journal of Microbiology and Biotechnology.* **20**(4), 737–748.

Jin, J., Shi, S.Y., Liu, G.L., Zhang, Q.H., and Cong, W. (2012). Numerical simulation of flow and particle collision in a rotating-drum bioreactor. *Chemical Engineering & Technology.* **35**(2), 287–293.

Kaštáneka, F., Demnerová, K., Pazlarová, J., Burkhard, J., and Maléterová, Y. (1999). Biodegradation of polychlorinated biphenyls and volatile chlorinated hydrocarbons in contaminated soils and ground water in field condition. *International Biodeterioration and Biodegradation.* **1**(44), 39–47.

Krishna, C. (2005). Solid-state fermentation systems: An overview. *Critical Reviews in Biotechnology.* **25**(1–2), 1–30.

Liao, C.Y. and Zheng, Y.G. (2005). Solid state fermentation bioreactor. *Microbiology (China).* **32**(1), 99–103.

Lopez, Y., Gullon, B., Puls, J., Parajo, J.C., and Martin, C. (2011). Dilute acid pretreatment of starch-containing rice hulls for ethanol production. *Holzforschung.* **65**(4), 467–473.

Lu, M.Y., Maddox, I.S., and Brooks, J.D. (1998). Application of a multi-layer packed-bed reactor to citric acid production in solid-state fermentation using *Aspergillus niger. Process Biochemistry.* **33**(2), 117–123.

Marsh, A.J., Stuart, D.M., Mitchell, D.A., and Howes, T. (2000). Characterizing mixing in a rotating drum bioreactor for solid-state fermentation. *Biotechnology Letters.* **22** (6), 473–477.

Masaphy, S., Levanon, D., and Henis, Y. (1996). Degradation of atrazine by the lignocellulolytic fungus *Pleurotus pulmonarius* during solid-state fermentation. *Bioresource Technology.* **56**, 207–214.

Mazutti, M.A., Zabot, G., Boni, G., Skovronski, A., De Oliveira, D., Di Luccio, M., Rodrigues, M.I., Maugeri, F., and Treichel, H. (2010). Mathematical modeling of *Kluyveromyces marxianus* growth in solid-state fermentation using a packed-bed bioreactor. *Journal of Industrial Microbiology and Biotechnology.* **37**(4), 391–400.

Mitchell, D.A., Von Meien, O.F., and Krieger, N. (2003). Recent developments in modeling of solid-state fermentation: Heat and mass transfer in bioreactors. *Biochemical Engineering Journal.* **13**(2–3), 137–147.

Mitchell, D.A., Von Meien, O.F., Krieger, N., and Dalsenter, F.D.H. (2004). A review of recent developments in modeling of microbial growth kinetics and intraparticle phenomena in solid-state fermentation. *Biochemical Engineering Journal.* **17**(1), 15–26.

Mitchell, D.A., Von Meien, O.F., Luz Junior, L.F.L., and Krieger, N. (2005). Construction of a pilot-scale solid-state fermentation bioreactor with forced aeration and intermittent agitation. *4th Mercosur Congress on Process Systems Engineering.*

Moreira, M.T., Sanroman, A., Feijoo, G., and Lema, J.M. (1996). Control of pellet morphology of filamentous fungi in fluidized bed bioreactors by means of a pulsing flow: Application to *Aspergillus niger* and *Phanerochaete chrysosporium. Enzyme and Microbial Technology.* **19**(4), 261–266.

Nopharatana, M., Howes, T., and Mitchell, D. (1998). Modelling fungal growth on surfaces. *Biotechnology Techniques.* **12**(4), 313–318.

Oostra, J., Le Comte, E.P., Van Den Heuvel, J.C., Tramper, J., and Rinzema, A. (2001). Intra-particle oxygen diffusion limitation in solid-state fermentation. *Biotechnology and Bioengineering.* **75**(1), 13–24.

Oriol, E., Chettingou, E., Niegra-Gonza, S., and Raimbault, M. (1988). Solid-state culture of *Aspergillus niger* on support. *Journal of Fermentation Technology.* **66**, 57–62.

Pandey, A. (1992). Recent process developments in solid-state fermentation. *Process Biochemistry.* **27**, 109–117.

Pandey, A. and Larroche, C. (2008). *Current Developments in Solid-State Fermentation.* Springer-Verlag.

Pandey, A., Soccol, C.R., Rodriguez-Leon, J.A., and Nigam, P. (2001). *Solid-State Fermentation in Biotechnology-Fundamentals and Applications.* New Delhi: Asiatech Publishers.

Rahardjo, Y.S., Weber, F.J., Le Comte, E.P., Tramper, J., and Rinzema, A. (2002). Contribution of aerial hyphae of *Aspergillus oryzae* to respiration in a model solid-state fermentation system. *Biotechnology and Bioengineering.* **78**(5), 539–544.

Rahardjo, Y.S.P., Korona, D., Haemers, S., Weber, F.J., Tramper, J., and Rinzema, A. (2004). Limitations of membrane cultures as a model solid-state fermentation system. *Letters in Applied Microbiology.* **39**(6), 504–508.

Rahardjo, Y.S.P., Tramper, J., and Rinzema, A. (2006). Modeling conversion and transport phenomena in solid-state fermentation: A review and perspectives. *Biotechnology Advances.* **24**(2), 161–179.

Raimbault, M. (1998). General and microbiological aspects of solid substrate fermentation. *Electronic Journal of Biotechnology.* **1**, 1–15.

Rajagopalan, S. and Modak, J.M. (1995). Modeling of heat and mass transfer for solid state fermentation process in tray bioreactor. *Bioprocess Engineering.* **13**(3), 161–169.

Rodríguez-Meza, M.A., Chávez-Gómez, B., Poggi-Varaldo, H.M., Ríos-Leal, E., and Barrera-Cortés, J. (2010). Design of a new rotating drum bioreactor operated at atmospheric. *Bioprocess and Biosystems Engineering.* **33**, 573–582.

Ryoo, D., Murphy, V.G., and Karim, M.N. (1991). Evaporative temperature and moisture control in a rocking reactor for solid substrate fermentation. *Biotechnology Techniques.* **5**(1), 19–24.

Sangsurasak, P. and Mitchell, D.A. (1998). Validation of a model describing two-dimensional heat transfer during solid-state fermentation in packed bed bioreactors. *Biotechnology and Bioengineering.* **60**(6), 739–749.

Selvakumar, P. and Pandey, A. (1999). Solid state fermentation for the synthesis of inulinase from *Staphylococcus* sp. and *Kluyveromyces marxianus. Process Biochemistry.* **34**, 851–855.

Singhania, R.R., Patel, A.K., Soccol, C.R., and Pandey, A. (2009). Recent advances in solid-state fermentation. *Biochemical Engineering Journal.* **44**(1), 13–18.

Smits, J.P., Van Sonsbeek, H.M., Tramper, J., Knol, W., Geelhoed, W., Peeters, M., and Rinzema, A. (1999). Modelling fungal solid-state fermentation: The role of inactivation kinetics. *Bioprocess Engineering.* **20**(5), 391–404.

Spier, M.R., Letti, L.A.J., Woiciechowski, A.L., and Soccol, C.R. (2009). A simplified model for *A. niger* FS3 growth during phytase formation in solid state fermentation. *Brazilian Archives of Biology and Technology.* **52**, 151–158.

Viccini, G., Mitchell, D.A., Boit, S.D., Gern, J.C., Da Rosa, A.S., Costa, R.M., Dalsenter, F.D.H., Von Meien, O.F., and Krieger, N. (2001). Analysis of growth kinetic profiles in solid-state fermentation. *Food Technology and Biotechnology.* **39**(4), 271–294.

Von Meien, O.F. and Mitchell, D.A. (2002). A two-phase model for water and heat transfer within an intermittently-mixed solid-state fermentation bioreactor with forced aeration. *Biotechnology and Bioengineering.* **79**(4), 416–428.

Wang, E.Q., Li, S.Z., Tao, L., Geng, X., and Li, T.C. (2010). Modeling of rotating drum bioreactor for anaerobic solid-state fermentation. *Applied Energy.* **87**(9), 2839–2845.

Xiao, F. (2000). The history and culture of vinegar (1). *China Brewing. (China).* **4**, 31–37.

Xu, Z.Q. (2006). *Technique and Application of Modern Solid-State Fermentation.* Beijing: Chemical Industry Press.

Yan, J. and Xiong, Y. (1997). Shanxi old mature vinegar. *Shanxi Food Industry. (China)* **1**, 33–36.

Zhang, P. (2000). Vinegar and soy sauce: Necessities in life. *China Brewing. (China).* **4**, 33–34.

2

BIOREACTORS OF SOLID-STATE FERMENTATION

GANRONG XU AND BOBO ZHANG

Key Laboratory of Industrial Biotechnology, Ministry of Education, School of Biotechnology, Jiangnan University, Wuxi, China

Contents

2.1 The Fundamental Performance Requirement of a Solid-State Fermentation Bioreactor

The solid-state fermentation bioreactor provides an appropriate environment and space for microbial growth and production of metabolites on the solid substrates (Pandey, 2003; Hölker and Lenz, 2005; Krishna, 2008).

Some fundamental performance requirements should be satisfied when a solid-state fermentation bioreactor is designed (Mitchell et al., 2006). It should

1. Contain the substrates (hermetic or semihermetic).
2. Prevent the contamination from other microorganisms as far as possible.
3. Hold a proper temperature and humidity for the fermentation substrates.
4. Provide enough oxygen for the aerobic microorganisms or exclusion of oxygen for the anaerobic microorganisms.
5. Be convenient for mixing and transportation of the substrates.
6. Be convenient for the isolation process of the fermentation products.
7. Provide a uniform distribution of the substrates as far as possible.

2.2 Types of Solid-State Fermentation Bioreactors

There are many valuable fermentation metabolites produced by solid-state fermentation. Accordingly, different types of solid-state fermentation bioreactors are developed with various constructive materials, shapes, and configurations. The simplest bioreactor could be a hermetic space or container. For example, the traditional solid-state fermentation of *qu* (qu is a product fermented with the growth of molds and contains various hydrolytic enzymes, or the Japanese equivalence of *koji*) in China could be carried out in a simple room (Yue et al., 2007). The function of a traditional solid-state fermentation bioreactor is single, which is only for the process of fermentation. The other functions (i.e., soaking of substrates, sterilization, cooling, and drying) often need the assistance of additional equipment. Modern solid-state fermentation bioreactors consist of complex internal structures and many external supporting facilities, which make their functions more complete.

Based on the manner of mixing and aerating, solid-state fermentation bioreactors can be divided into four types as shown in Figure 2.1 (Mitchell et al., 2006):

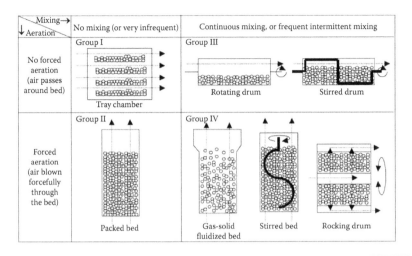

Figure 2.1   Types of solid-state fermentation bioreactors.

1. *Unaerated and unmixed bioreactor:* a bioreactor in which the bed is static, or mixed only very occasionally (i.e., once or twice per day), and air is circulated around the bed but not blown force-fully through it. These are often referred to as "tray bioreactors."

2. *Forcefully-aerated bioreactor without mixing:* a bioreactor of which the shape is usually like a column or a rectangle. The bed is static or mixed only very occasionally (i.e., once per day) and air is blown forcefully though the bed. These are typically referred to as "packed-bed bioreactors."

3. *Unaerated but mixed bioreactor:* a bioreactor in which the bed is continuously mixed or mixed intermittently with a fre-quency of minutes to hours, and air is circulated around the bed, but not blown forcefully through it. Two bioreactors that have this mode of operation, using different mechanisms to achieve the agitation, are *stirred-drum bioreactors* and *rotating drum bioreactors.*

4. *Mixed, forcefully-aerated bioreactor:* a bioreactor in which the bed is agitated and air is blown forcefully through the bed. This type of bioreactor can be typically operated in either of two modes, so it is useful to identify two subgroups: mixed continuously (i.e., gas-solid fluidized beds, rocking drum,

and various stir-aerated bioreactors) and mixed intermittently with intervals of minutes to hours between mixing events (i.e., a disk starter propagation vessel).

2.3 Traditional Solid-State Fermentation Bioreactors in China

Many traditional solid-state fermentation bioreactors are still widely used in China, and some of them are introduced in this section.

2.3.1 Culture Bottles

Glass flasks or plastic bottles, which are commonly used in laboratories, are still used as fermentation containers in the large-scale production of the edible mushroom in China (Ji et al., 2011). The use of culture bottles for solid-state fermentation could guarantee the fermentation process against contamination for a long period, or if the contamination occurs, the loss could be minimized just by discharging the contaminated one from the batch. Because of the fragility of the glass flasks, they could be replaced by thermo-tolerance polyethylene or polypropylene plastic bottles, which are commonly used. The number of culture bottles can be huge and the operation of feeding, discharging, and sterilization all need considerable manpower. To increase the efficiency, special equipment is designed and applied for these fermentation processes (see Figures 2.2 and 2.3).

The opening of the flasks or plastic bottles is covered by a cotton wool tampon, multilayers of gauze, or special plastic caps. In order to control the temperature in the bottle, all the culture bottles are put into a culture room where the temperature and humidity are controlled.

Solid-state fermentation in culture bottles belongs to the natural ventilation static fermentation. The culture rooms should be installed with ventilation devices. Because of the stuffiness at the bottom of the culture bottles, the oxygen essential for the growth of microorganisms and the carbon dioxide produced in the fermentation process only exchange with the surrounding environment through the natural diffusion at the beak of the bottle. Consequently, the diffusion process is inefficient. In order to increase the diffusion efficiency, the culture bottle could be placed horizontally, which is beneficial for the diffusion of air. The mycelia of microorganisms may tangle into blocks in

Figure 2.2   Flask culture of monascus red rice.

Figure 2.3   Plastic bottle culture of edible mushroom.

the fermentation process and hence, shaking the bottles in a timely way is needed. Shaking the bottle means patting the bottles gently to shake the substrates for air exchange.

2.3.2 Rotating Bottles

The rolling bottles are similar to small-scale rotating drum bioreactors (Wang et al., 2010). The fermentation bottles are set between two

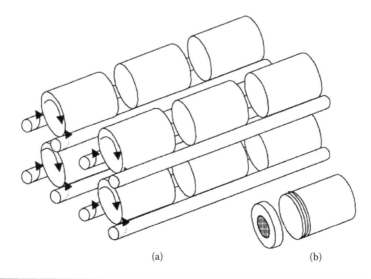

Figure 2.4 The fermentation system of rotating bottles. (a) The fermentation bottles are set between two rods and the drive rods are shown with solid arrows while the slave rods are shown with a dotted arrow. (b) The fermentation bottles with the caps.

rods, one of which is the drive rod that can rotate the bottles and the substrates (see Figure 2.4).

2.3.3 Semipermeable Plastic Bags

Semipermeable plastic bags are the containers made by semipermeable plastic films. Air can freely pass the plastic film but the water vapor cannot. The semipermeable plastic bags are usually used in the culture of edible and medicinal mushrooms. It is similar to the solid-state fermentation in the glass flasks or plastic bottles.

2.3.4 Heap Fermentation

This type of fermentation is usually used in the traditional compost and silage (Camu et al., 2008). In the heap fermentation, just a plain ground is necessary, and then the hay and substrates are heaped on it accordingly. The substrates are covered by mud or plastic cloth tightly and only an air outlet is left for the exhaust of the gas accumulated in the fermentation process. This type of solid-state fermentation is simple, practical, and cheap. However, there are some disadvantages to this method, one of which is the uncontrollable temperature, the main problem.

Figure 2.5 The pits.

2.3.5 Pit Fermentation

Pit fermentation, also called a *wine pit*, has been widely used in the fermentation of a traditional distilled spirit (Figure 2.5) (Yue et al., 2007). There is little on record about the origin of this method.

2.3.5.1 Structure of a Pit A pit is a typical kind of anaerobic fermentation container that provides an anaerobic space for the production of a distilled spirit. The distilled spirit is produced from the distillation of *zaopei*, which is a mixture of fermented grains such as sorghum, corn, wheat, and rice (Shi et al., 2011). The metabolic reactions consist of an interaction of the microorganisms in the qu environment, the pit mud, and numerous enzymes produced by the microbes. Chinese qu is prepared by a natural inoculation of molds, bacteria, and yeasts and their growth on grains. Included are amylases, proteases, lipases, and numerous other enzymes. The decomposition and transformation of various macromolecular substances (mainly starch) are continuous and occur while ethanol and numerous aromatic compounds, such as acids, aldehydes, alcohols, and esters are developing.

The volume of the pit should be appropriate to ensure the specific surface area is as large as possible, which may be beneficial for the interaction between zaopei and the pit mud to produce more aromatic substances (Shi et al., 2011). The general size of the pit is length × width × depth = 2.5 × 3 × 2.5 (m). The sides and bottom of the pit are covered by mud. Distilled zaopei is added to newly steamed grains, mixed with Chinese qu, transferred to the pit, and sealed. The pit may be covered by mud or plastic cloth to avoid the entry of air after the filling of substrates. However, there is some air in the substrates and inevitably, the substrates will perform an aerobic fermentation to some extent at the beginning of the fermentation process. With the progress of the fermentation, the extent of anaerobic reaction is increasing.

The pit uses a layer of soil as a natural temperature control system. Although there is no artificial temperature control system, the fermentation temperature remains stable in the pit and at a lower range. For example, the soil temperature of pits in Yibin is maintained at ~10–20°C all year round, which is a proper temperature for the growth of special microorganisms in the pits.

It is not necessary and also not possible to make the zaopei in a fully homogeneous state in the pit. So it is not needed to rotate the zaopei after it is transferred into the pit.

2.3.5.2 Pit Mud Pit mud, which is an important part of the pit, is a layer of mud in the pit walls and bottom formed by nature or artificial cultivation (Hu et al., 2005). The widely used pit mud is "artificial pit mud," which is blended with specific yellow mud, spring or well water, aged pit mud, and the qu.

Pit mud is the base for liquor-making microorganisms. The microorganisms necessary for the fermentation of distilled alcohol are not only provided by the qu and the brewing of raw materials but also by the pit mud. In a relative anaerobic condition, the pit mud is enriched with caproic acid bacteria as the main functional microbes (Hu et al., 2005). As a result of long-term continuous enrichment, domestication, more and more beneficial microbes accumulated in the aged pit mud could form an excellent functional microbial flora. It is reasonable to consider a pit as ecological equipment for the production of distilled alcohol. The pit is a microbial fermentation vessel.

The reproduction and accumulation of microorganisms in the pit mud make it a liquor-making habitat for the growth of microbes. On one hand, the nutrients and environment for the microbial growth are provided by the pit mud; on the other hand, through the fermentation of the zaopei in the pit, the produced yellow water, aromatic substances, and many kinds of organic substances continue to penetrate into the pit mud and become the nutrition for microbial growth.

The quality of the distilled alcohol is directly dependent on the fermenting pits and the pit mud. The degradation of pit mud would lose its ecological effect and further influence the quality and output of distilled alcohol. The causes of the pit mud degradation mainly include the depletion of required nutrient substances for microbial growth, aged pit mud hardening, the dry climate, poor temperature preservation of the pit, improper control on technical operation, inappropriate building structure of the pit, etc. Therefore, scientific knowledge of how to maintain the pit mud is necessary. Meanwhile, the modern technique of the artificial cultivation of pit mud will be also helpful for the pit maintenance.

2.3.6 Pot Fermentation

Pot fermentation vessels mainly include pottery, concrete pool, and metal tank. The pot fermentation is usually applied for semisolid fermentation. At the beginning of the fermentation, the materials are often in a solid state. However, with the progress of the fermentation, the materials are liquefied and become a semisolid state.

The pot fermentation is the most common ancient brewing method (Haruta et al., 2006), which is often used in traditional fermentation of soy sauce, rice wine, and distilled alcohol (Figure 2.6). Metal fermentation tanks are now commonly used. The pots can be placed either in a room or outdoors (Figure 2.7). The temperature could be

Figure 2.6   The pots for the fermentation of rice wine. (a) Stirring qu. (b) The fermentation of rice wine. (c) Stirring rice wine. (d) The fermentation pots.

Figure 2.7   The outdoor pot fermentation of rice wine.

naturally regulated when the pots are exposed outdoors. Similar to this exposure method, a sun shed is designed for the pot fermentation in the modern. The heat preservation effect in the sun shed is better than that of the natural exposure method.

2.3.7 Thick-Layer Aerated Fermentation Pool

Thick-layer aerated fermentation pools belong to the intermittently-mixed forcefully-aerated bioreactor, which is the most widely used solid fermentation reactor in industry.

Two processes, mixing of materials and static culture, are alternately operated in an intermittently-mixed forcefully-aerated bioreactor. This fermentation operation is more suitable to the microorganisms which are not very sensitive to the shear stresses and not vulnerable to the intermittent mixing (Mitchell et al., 2006). The operation of intermittent mixing should be carried out when the culture substrate begins caking, and meanwhile water could be added into the culture medium if necessary. The pressure drop of the material layer could be reduced with a mixing operation when adopting an intermittently-mixed bioreactor. The material layer is in the static culture process without a mixing operation, which belongs to the packed-bed model. At this point, uniform ventilation should be maintained in the material layer to supply enough oxygen for the microorganisms and exhaust the carbon dioxide.

When the diameter of a reactor is large enough, the radial heat transfer could be ignored and only the axial mass and heat transfer should be mainly considered in this kind of intermittently-mixed operation. During this kind of fermentation process, because the material layer is in the static state most of the time, the temperatures of each material layer are different and fluctuate within a certain range. The temperature rises gradually in the static state and drops quickly to a set point when mixing. After mixing, the temperature rises gradually again, and so on. The microbial growth rate could also be influenced by the mixing operation.

During ventilation, the material moisture content decreases dramatically due to the moisture evaporation; timely water supplementation or the addition of air with high humidity is necessary. To avoid too much fluctuation in the temperature and water activity of the materials, it is critical to control the temperature and humidity of the import and export air, of which the former is more important. The water activity in the material layer could be kept in a small range when the temperature and humidity of the import air is close to that of the export air.

Thick-layer aerated fermentation pools or tanks are widely used in the production of qu used in rice wine, koji for soy sauce, and qu for vinegar fermentation. Equipment body of the bioreactor is the qu pool or tank, of which the general internal dimensions are length × width × height = (7~10) m × 2 m × (0.7 ~ 0.9) m. The length and width of the large-scale qu tank should be increased. The tank body is made of bricks with external cement mortar. In order to facilitate the drainage, uniform ventilation and the direction change of import air, the air duct at the bottom of the tank is made to slope with an appropriate angle of 8°~10°. There are two edges about 10 cm at both sides of the ventilation path in the traditional qu tank. A wooden batten is installed between the edges with a bamboo raft or bamboo screen on it. Stainless steel plates with neat rows of holes are used in the modern version to replace the bamboo raft or bamboo screen and then the fermentation materials are stacked on the steel plates. General thickness for the stacking materials is about 30 cm. The qu tank could be covered or uncovered. The thick-layer aerated fermentation pool or tank is shown in Figure 2.8. In order to facilitate ventilation in each

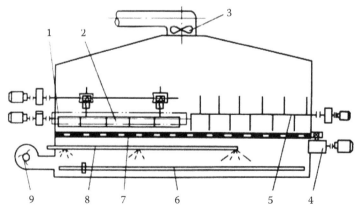

1. Throwing component   2. Scraping component   3. Exhaust fan
4. Rotating component   5. Agitating component   6. Heating component
7. Sieve tray   8. Spraying component   9. Air blower

Figure 2.8 Disc koji-making machine (thick-bed bioreactor).

culture chamber, there are usually two qu tanks, between which are the crosswalks.

The ventilator in the qu pool or tank is generally a medium pressure type. The air volume (m³/h) is usually ~4–5 times to the weight of raw materials (kg). In order to maintain heat and moisture, the qu pool could be equipped with a cover. Meanwhile, some are open. The ventilator channel is set near the end of the ventilator in the qu pool. The path for the drainage water is set at the bottom of the qu pool.

The filling thickness of materials in the thick-layer aerated fermentation pool is ~25–30 cm. The filling amount of materials in each pool should be suitable for the size of the steaming pot. For example, the filling amount is about 1500 kg (dry weight) in a 5 m³ cooking pot. The ventilator in the qu pool or tank is a type of medium pressure, which is about 1 kPa (100 mm H_2O). The air volume is about ~6000–7500 m³/h.

The air temperature and humidity are adjusted in the air conditioning room before entering into the air duct. Temperature is adjusted by the heat sink, while humidity could be adjusted by a fluxing steam or spray. The centrifugal ventilator is the common kind of ventilation machine and the main performance index are air volume and pressure. Mechanical mixing is used in a large-scale thick-layer aerated fermentation pool, while artificial mixing (manually or by a mixing tool) is adopted in a small-scale pool.

A thick-layer aerated fermentation pool or tank still has the main problems of this kind of open fermentation bioreactor—the contamination coming from surrounding bacteria. Air and the indoor environment are the major sources of contamination. The air system is usually without strict sterilization by filtration. Therefore, the number of bacteria in the air is high. When the air passes through the material layer, bacteria are left in the inner room of the fermentation pool. One part of the air is exported outside the system, while the other part of the air is recirculated back to the room of the fermentation pool by the ventilator. Hence, aseptic conditions are difficult to achieve in the fermentation pool (see Figures 2.8 to 2.10).

2.3.8 Solid-State Fermentation Bioreactor with a Double-Dynamic Gas Phase

Compared to the liquid fermentation, the most characteristic of solid-state fermentation is that the substrate is in a solid state and its

Figure 2.9 The bottom structure of a thick-layer aerated fermentation pool.

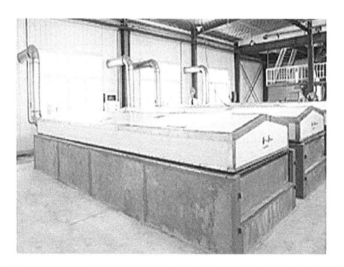

Figure 2.10 Thick-layer aerated fermentation pool with a cover.

mobility is poor. Moreover, when fungi or other filamentous micro-organisms are selected for the fermentation process, the mycelia are relatively vulnerable to the external force. Therefore, the agitation of substrates, which is necessary for the mass and heat transfer, becomes a big problem. The traditional way is carried out by mechanical or manual mixing after the completed growth of mycelia. However, it is difficult to operate in the closed fermenting tank if considering the aseptic conditions.

A novel solid-state fermentation bioreactor with a double-dynamic gas phase (air pressure pulsation solid-state fermentation reactor) was designed by Chen and Li (Chen and Li, 2002) to conquer this problem. This bioreactor includes horizontal solid-state fermentation tanks, air pressure pulsation control system, air circulation system, disk rack system, and mechanical transport system. A double-dynamic gas phase means air pressure pulsation and internal air circulation is static in the solid state.

The solid-state fermentation bioreactor with a double-dynamic gas phase is shown in Figure 2.11. Periodical pulsation of air is regulated by the control system as follows: When the air inlet valve is on and the air outlet valve is off, the compressed and sterilized air enters into the fermentation vessel rapidly until the air pressure reaches the set value; when the inlet valve is off, the air pressure is maintained for a given time; when the set time is over, the outlet valve is on and the air leaves the bioreactor rapidly until the air pressure reaches the set value. At the same time, the outlet valve is off and the air pressure is maintained for a while. When the set-up time is over, the air will start to repeat the above process of air pulsation. The internal air circulation is regulated by the dynamic system: The air in the bioreactor is forced to circulate through the spaces divided by the baffles, and the rate of air circulation can be regulated through a variable speed motor.

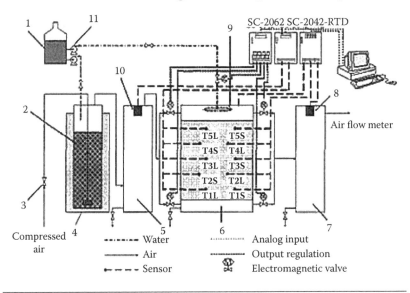

Figure 2.11 The solid-state bioreactor with pressure pulsation.

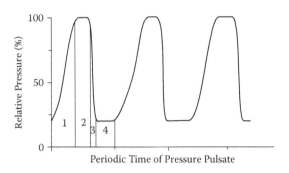

Figure 2.12   The course of pressure pulsation in the bioreactor.

The air pressure pulsation in the bioreactor with the double-dynamic gas phase is realized by charging and releasing the sterile air. The course of pressure pulsation in this bioreactor is shown in Figure 2.12. The pressure in the bioreactor is periodically changed between high and low air pressures. The periodic processes consist of four phases: compressed air phase, stable phase of high air pressure, discharged air phase, and the stable phase of low air pressure. Therefore, the time (t_2) required to maintain the upper limit of air pressure pulsation is comparatively short (generally 2 min) while the time (t_4) required to maintain the lower limit varied according to the stage of microbial metabolism. Discharged air phase forces produced by air pressure pulsation are not shearing but normal. When air pressure in the bioreactor drops suddenly, the gas phase within substrates will swell, and then the swell will loosen the solid substrates. The resultant solid substrate loosen could replace the agitating effect. Hence, the material mixing in the tank could be realized without mechanical agitation equipment. The dynamic changes of air could help to accelerate heat and mass transfers and minimize the temperature gradients, as well as avoid damage of the selected microorganism and hence shorten the fermentation cycle.

2.3.9 Fermentation Room for Qu

Chinese qu essentially is a kind of material containing crude enzymes that are saccharifying agents in wine brewing. Qu is divided into five categories: *da qu, mai qu, xiao qu, fu qu,* and *monascus*. According to the shape of qu, it can be divided into powder or block qu. Block qu is made into a certain shape. The da qu for making a distilled spirit and block qu for making rice wine have their own shape, size, and standard. The general sizes are length × width × height = (~15–40) cm × (~10–30) cm × (~5–15) cm. For example, the size for traditional maotai qu is length × width × height = 36 cm × 23 cm × 7 cm. The shape of block qu is very different. For example, the shape of xiao qu is quite small, usually smaller than an egg. It could also be made in a ball or a cube shape. The weights are also quite different among different types; a small one can be only about 5 g while a big one is about 50 g. The production of qu usually uses cereals (wheat, rice, etc.) as raw material with various molds (mainly *Aspergillus* spp.) growing on them.

In order to facilitate maintaining heat and moisture, most of the traditional solid fermentation of qu is carried out in the qu room. Hence, the qu room could be regarded as a special solid-state fermentation bioreactor. Generally, the size of qu room may be 20 m², whereas larger than 100 m² is not suitable. The roof of the qu room is covered with straw or grass in order to maintain the heat and moisture. Windows and doors are required, some of which still allow a skylight. The indoor temperature and humidity could be adjusted by opening or closing the windows and doors, by which the supply of fresh air and exhaust of carbon dioxide can be achieved. Traditional methods for maintaining temperature is by covering with straw or grass, but now it could also be done by a heater or steam pipe setting in the room. The adjustment of moisture is generally added manually.

The mold for qu blocks is usually made by wood and with a size of 37 × 23 × 6.5 cm (Figure 2.13). The production materials of block qu mainly include barley, wheat, rice, and peas in a certain proportion. Raw materials generally require crushing, adding water and mixing

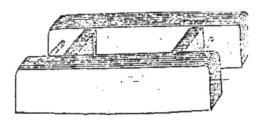

Figure 2.13 The module for making qu.

Figure 2.14   The block of da qu.

Figure 2.15   Qu cultivated on the ground in the cave.

well, and then finally finishing the raw block qu. After that, the raw blocks are sent into the qu room, culturing at a certain temperature.

The traditional cultures of block qu are usually carried out in the qu room. Blocks of qu could be cultured by suspension, and also could be stacked in the qu room according to a certain style (see Figure 2.14).

The cultivation of qu in the qu room is usually carried out

—In the culture room or pits, placed on the ground (Figure 2.15).
—Placed on the bamboo plaque (Figure 2.16).
—Placed on the shelf (Figure 2.17).

Figure 2.16   Qu in the culture room.

Figure 2.17   Small block of qu on the shelf.

—Hung from the shelf (Figure 2.18). Block qu is a traditional wheat qu in the South Yangtze area of China. Qu is put into a wooden box, trodden tightly by foot, and cut into a long block shape with a knife. In the qu room, several shelves are set up by a wooden (or iron piped) frame, with bamboo or wooden beams on them. And then the qu is bundled with straw, hung

Figure 2.18 Qu hung on the shelf.

on the beam, and then cultured on the frame. The style of "this block qu" facilitates the ventilation and improves the space utilization. The indoor temperature and humidity could be adjusted by opening or closing the windows and doors. It also could be equipped with a temperature control device.

—Stacked and stored (Figures 2.19, 2.20, and 2.21); Blocks of da qu could be stacked according to a certain style in the qu room. The stacking can be single- or multilayer. Each layer

Figure 2.19 Blocks of qu placed on the ground.

Figure 2.20   Blocks of qu stacked on the ground.

Figure 2.21   Blocks of qu placed on the shelf.

of the qu block should keep a certain distance to facilitate ventilation. Some manufacturers usually stack the blocks of qu with 4 to 6 layers. In later periods of fermentation, the distance could be reduced and made the block stacking more closely for thermal insulation. Moreover, the blocks of qu also could be placed on the wooden shelf.

—Placed in a specially designed box (Figure 2.22).

Figure 2.22   Blocks of qu stacked storage.

2.3.10 The Modern Computer-Controlled Shelf for "Hang Qu" Technology

There are many disadvantages in making qu by manual operation or other traditional fermentation processes, such as the labor intensity, poor hygiene conditions, low production efficiency, and large area required for the qu room. In the early 1990s, some efforts were made to overcome these disadvantages by applying a computer monitoring system. However, the quality of qu was worse than that of manual operation. Thereafter. producers continued efforts to improve the qu-making technology by computerization.

Recently, it has been reported that the computer-controlled shelf *hang qu* technology could solve the quality problem quite well (Zhong, 2002).

Microbiological technology, electronic technology, computer technology, automatic control technology, and simulation of traditional qu-making technology are applied in the computer-controlled shelf with hang qu technology. All the systems in this technology are working to provide a suitable and consistent ecological environment for microbial growth and production. Wheat is used as the raw material for making da qu. The powder of da qu is first made into a certain shape and then placed on a special multilayered rack for the fermentation process. The optimum temperature and humidity for culturing the beneficial microorganisms in da qu is set by a computer.

The relevant parameters are monitored through multiple sensors and adjusted (automatically increase or decrease the humidity, automatic heating or cooling) by the computer at real time to keep the best conditions for the cultivation of da qu.

The qu room with computer-controlled shelf technology is different from the traditional culture room. In order to simulate the natural environment, the computer-controlled room is also covered by straw to strengthen the control of temperature, humidity, and growth of microorganisms. Other measures are also applied to increase the productivity of the qu room. Through contrasting experiments, it is demonstrated that the improved computer-controlled shelf hang qu technology can greatly enhance the quality of da qu and the efficient use of the qu room. The quality of each batch is stable. It is not necessary to mixing qu manually. As a result, this modern technology could greatly reduce the labor needs and increase productivity when compared to traditional methods.

References

Camu, N., González,, Winter, T.D., Schoor, V.A., Bruyne, K.D., Vandamme, P., Takrama, J.S., Addo, S.K., and Vuyst, L.D. (2008). Influence of turning and environmental contamination on the dynamics of populations of lactic acid and acetic acid bacteria involved in spontaneous cocoa bean heap fermentation in Ghana. *Applied and Environmental Microbiology.* **74**, 86–98.

Chen, H.Z. and Li, Z.H. (2002). Study on solid-state fermentation and fermenter. *Chemical Industry and Engineering Progress.* **21**, 37–42.

Hölker, U. and Lenz, J. (2005). Solid-state fermentation: Are there any biotechnological advantages? *Current Opinion in Microbiology.* **8**, 301–306.

Haruta, S., Ueno, S., Egawa, I., Hashiguchi, K., Fujii, A., Nagano, M., Ishii, M., and Igarashi, Y. (2006). Succession of bacterial and fungal communities during a traditional pot fermentation of rice vinegar assessed by pcr-mediated denaturing gradient gel electrophoresis. *International Journal of Food Microbiology.* **109**, 79–87.

Hu, C., Ying, H., Xu, D.F., and Hu, Y.S. (2005). Research on microflora in pit mud and its application. *Liquor-Making Science and Technology.* **3**, 34–38.

Ji, K.P., Cao, Y., Zhang, C.X., He, M.X., Liu, J., Wang, W.B., and Wang, Y. (2011). Cultivation of *Phlebopus portentosus* in southern China. *Mycological Progress.* **10**, 293–300.

Krishna, C. (2008). Solid-state fermentation systems: An overview. *Critical Reviews in Biotechnology.* **25**, 1–30.

Mitchell, D.A., Krieger, N., and Berovič, M. (2006). *Solid-state Fermentation Bioreactors: Fundamentals of Design and Operation*. Berlin, Germany: Springer.

Pandey, A. (2003). Solid-state fermentation. *Biochemical Engineering Journal.* **13**, 81–84.

Shi, S., Zhang, L., Wu, Z.Y., Zhang, W.X., Deng, Y., Zhong, F.D., and Li, J.M. (2011). Analysis of the fungi community in multiple- and single-grains zaopei from a luzhou-flavor liquor distillery in western China. *World Journal of Microbiology and Biotechnology.* **27**, 1869–1874.

Wang, E.Q., Li, S.Z., Tao, L., Geng, X., and Li, T.C. (2010). Modeling of rotating drum bioreactor for anaerobic solid-state fermentation. *Applied Energy.* **87**, 2839–2845.

Yue, Y.Y., Zhang, W.X., Yang, R., Zhang, Q.S., and Liu, Z.H. (2007). Design and operation of an artificial pit for the fermentation of Chinese liquor. *Journal of the Institute of Brewing.* **113**, 374–380.

Zhong, F.D. (2002). Soundness and development of computer-controlled shelf starter-making. *Liquor-Making Science and Technology.* **4**, 31–33.

3

PROCESS ENGINEERING OF SOLID-STATE FERMENTATION

LONG LIU

State Key Laboratory of Food Science and Technology, School of Food Science and Technology, Jiangnan University, Wuxi, Jiangsu Province, China

Contents

3.1 Introduction

Solid-state fermentation (SSF) is a microbial process where a solid material is used as the substrate or the inert support over which microorganisms are immobilized.

The SSF system has gas, liquid, and solid phases, in which microbial growth and heat/oxygen transfer result in temperature gradients, where transfer of heat and mass are obstacles. Therefore, the control of process parameters of the SSF is critical. These parameters include temperature, pH value, aeration, water activity, and moisture. The physical–chemical properties of a medium are influenced by water activity and moisture, and provide a humid environment for the growth of microorganisms. Water activity and moisture will also pose an effect on the utilization of oxygen and heat transfer.

3.1.1 Water Activity and Moisture Content

A substrate must possess enough moisture to support growth and metabolism of the microorganisms. The growth of microorganisms depends on the water activity of the substrate.

Water activity a_w is defined as

$$a_w = \frac{p}{p_w},$$

where p is the saturation vapor pressure of humid substrates, and p_w is the saturation vapor pressure of water under the same temperature.

In general, the value of a_w allowing bacterial growth is ~0.90–0.99, but most yeasts tolerate a lower a_w of ~0.80–0.90; most fungi and some yeasts can grow at an a_w of ~0.60–0.70, thus fungi are suited to culturing in SSF because of their tolerance for low a_w. On the other hand, the environment of low a_w can prevent bacterial infection. High a_w can reduce the porosity of a substrate, prevent the diffusion of oxygen and carbon dioxide, multiply the chance of infection, and in turn affect the growth of fungi. In the fermentation process, the value of a_w will reduce because of evaporation and temperature rise. Adding sterile water into the substrate, moisturizing the air, and equipping water sprays can increase a_w. The value of a_w also can affect water molecular and solute to cross membranes.

The establishment of the relationships between the physiology of the microorganisms and the physical chemical factors is the aim for the development of proper models. These factors include temperature, pH, aeration, water activity and moisture, bed properties, nature of solid substrate employed, etc. Among several critical factors, moisture and nature of solid substrate employed are the most important factors affecting SSF processes. Selection of moisture depends on the microorganism employed and also the nature of a substrate. Fungi can grow at lower moisture, 40–60% moisture could be sufficient but selection of a substrate depends upon several factors mainly related to cost and availability and thus may involve the screening of several agro-industrial residues. High moisture percentage results in low substrate porosity which in turn prevents oxygen penetration, whereas low moisture content may lead to poor accessibility of nutrients resulting in hampered microbial growth. The consequences of these difficulties for bioreactor performance have been well explored.

Evaporative cooling has been incorporated for the large-scale operation of a rotating drum bioreactor (RDB). The mass transfer coefficient for evaporation from the bran bed to the head space has been determined by measuring outlet water vapor concentrations for a RDB containing wet wheat bran. Sangsurasak and Mitchell (1995) suggested that evaporation can remove as much as 78% of the heat from the bed during the time of peak heat generation, even when the bed is aerated with saturated air. This is due to the increase of water vapor pressure at increasing temperature. It results in large moisture losses and drying of the solid. A new generation of small reactors was developed by an INRA-team in France. Such reactors have a working volume of about 1 L, a relative humidity probe, a cooling coil on the air circuit, and a heating cover for the vessel. Each reactor is automatically controlled by a computer. A novel bioreactor with two dynamic changes of air (including air pressure pulsation and internal circulation) can increase mass and heat transfers, as well as improve the porosity within the substrate. The reactor can be significant in exploiting numerous socioeconomic advantages of solid-state fermentation.

Xylanase and pectinase activities were lower when the moisture content was higher than 65%. The low enzyme activity at high substrate moisture levels could be attributed to the decreased porosity, alteration in particle structure, gummy texture, lower oxygen transfer, or increased formation of aerial hyphae. Likewise, lower moisture levels lead to reduced diffusion of the nutrients in the solid substrate, lower degree of swelling, and higher water tension. The optimum moisture contents for xylanase production by *Trichoderma longibrachiatum* and *Aspergillus tereus* were 55% and 75%, respectively.

3.1.2 *Temperature and Heat Transfer*

Temperature exerts a tremendous influence on the SSF process. High temperature affects germination, growth of spores, product formation, and spore forming, while low temperature is unfavorable for the growth of microorganisms and other biochemical reactions. In the initial stage of fermentation, the temperature in all parts in the solid substrate is the same. While the fermentation process goes on, a temperature gradient is formed due to the poor thermal conductivity of the substrate. At the same time, the contraction of a substrate will occur during the fermentation process, which adds to the difficulties of heat transfer. So temperature gradients are developed, as the solid bed is aerated from one end, significant axial temperature gradients develop, which can limit the performance of a packed-bed bioreactor. The axial temperature gradients existing in the bioreactor pose a potential threat to the growth of microorganisms and hence metabolite production, since evaporation of water from the solid bed is enhanced, leading to eventual drying out of the bed. So the heat transfer is a difficult problem to deal with, especially for large-scale SSF. The metabolic heat production rate of SSF is related to the metabolic intensity of microorganisms. There is a huge difference among the various fermentation systems. One of the most noticeable differences between the submerged fermentation and SSF is that the whole SSF system cannot attain the optimal temperature because of the heat transfer barrier; especially during the logarithmic growth phase, the temperature gradients between the external and internal of solid material can reach 3°C/cm, which is unfavorable to microbial growth and enzyme production. Generally, the heat transfer rate is positively correlated to the forced aeration rate. Without forced aeration, the main pattern of heat transfer from the solid material outside to the vapor phase is heat conduction. The conventional heat transfer pattern of conduction and convection cannot meet the requirements of SSF. Evaporative cooling is a major method for temperature control, which can dissipate 80% of the fermentation heat. Forced aeration not only can provide oxygen and send out carbon dioxide, but also can enhance heat transfer. The effects of forced aeration include two aspects: the first one is transferring heat directly; the second is transferring heat by means of influencing the amount of evaporation. The rate of evaporation can be adjusted by changing the rate of forced aeration and the moisture content of the medium. In a general way, the temperature gradients can be alleviated by forced aeration; more heat can be taken away by dry sterile air than water saturated sterile air. In summary, the coupling of forced aeration, temperature, and moisture content by the measures generally available today is the major approach to temperature control of large-scale SSF.

3.1.3 pH

The pH value of the solid-state fermentation system is difficult to measure by using a pH probe directly. A method for measuring SSF pH values is as follows: fetch some wet substrate, insert the pH probe, squeezing the substrate gently, and then read the pH value of water squeezed from the substrate. Another method is to add some distilled water to the substrate, mix thoroughly, and read the pH value by direct measurement. The measured value by this method is ~0.1–0.2 higher than the value measured by the previous method.

The acid or alkaline substances in the substrate of a stationary SSF system have a negative impact on fermentation, but in order to avoid the fluctuation of pH, solutions containing strong alkali or acid are used to maintain the pH value. Other pH control measures are by adding buffer into the substrate and by using urea instead of ammonium salt as the main nitrogen source. In most cases, the control of pH is achieved by its own buffering capacity of the substrate.

3.1.4 Aeration

In the SSF fermentation process with aerobic microorganisms, microbial growth and biomass are often limited by oxygen transfer. Oxygen transfer both in the solid medium and among the particles of mass are the main transfer processes.

In SSF fermentation, the most important oxygen transfer process among the mass particles is the oxygen transfer from the void volume of the medium to microorganisms. The size of void volume is determined by the property, size, and the water ratio of particles. Chaff is added to the medium to increase the void volume. The water ratio is related to the transfer of oxygen to the void, because the water in the medium prevents air from flowing. To a certain size of void, oxygen transfer in the particles is realized by agitation and aeration. Timed agitation and intermittent, continuous aeration are common methods often used to increase the dissolution oxygen.

With respect to aeration, two strategies are available: The air can either be circulated around the substrate bed or blown forcefully through it. Unfortunately, however, in many cases, mixing causes shear damage to the mycelium and loss of productivity, so this has to be balanced against the loss of productivity resulting from improper temperature control or bed drying. Constant aeration also makes it more difficult to maintain a stable environment inside the bioreactor with respect to oxygen and carbon dioxide. This is because the air flow rates are linked to heat removal rates and are not therefore easily controlled in an independent manner to provide the desirable oxygen to carbon dioxide balance that may be optimal for high productivity.

The production of citric acid depends strongly on an appropriate strain and on operational conditions. Oxygen level is an important parameter for citric acid fermentation. Several researchers have studied the influence of forced aeration on citric acid production and the metabolic activity of *A. niger* in SSF by respirometric analysis. They showed that citric acid production was favored by a limited biomass production, which occurred with low aeration rates. Both works showed the feasibility of using the strain *A. niger* for citric acid production by SSF.

Generally, some techniques can be used to improve the mass transfer status: grainy and fibriform materials are utilized as a substrate;

decrease the thickness of a substrate, increase the void volume of a substrate, using a porous shallow pan or drum bioreactor, agitating the substrate. Nowadays, the common methods are the coupling of forced aeration and agitation, but it is necessary to time aeration to prevent the occurrence of channel flow.

3.2 Mass and Heat Transfer, and Energy Equation Calculation of Solid-State Fermentation

Solid-state fermentation (SSF) involves the growth of microorganisms on moist solid substrates in the absence of free-flowing water. The necessary moisture in SSF exists in an absorbed or complex form within the solid matrix, which is likely to be more advantageous for growth with a more efficient oxygen transfer process. In SSF, the water content is quite low and the microorganism is almost in contact with gaseous oxygen in the air, unlike in the case of submerged fermentation (SmF). The water activity in the substrate is also important. The principles and engineering aspects of SSF have been earlier reviewed by Moo-Young and Blanch (1981) and Raghava (1993) (Moo-Young and Blanch, 1981; Raghava, 1993).

In recent years, research interest in SSF has addressed several problems on the production of protein-enriched feed from starchy materials, single cell protein (SCP) production from a variety of wastes, ethanol from cassava roots and sugar beets, enzymes, organic acids, biogas, antibiotics, surfactants, bioremediation agents, mushrooms, microbial polysaccharides, biocides, and mycotoxins from corn, etc. The use of a mixed culture on cellulosic wastes to improve the flavor and protein content of such food products is a new interesting approach. SSF is often simpler and requires less processing energy than SmF. The low volume of water present in the media per unit mass of substrate can substantially reduce the space occupied by the fermentor without severely sacrificing the yield of the product. Aeration and mixing requirements may also be easily met. On the negative side, SSF processes are slower than the liquid fermentations due to the additional barrier from the bulk solid. They also present heat dissipation problems, which can be limited by inter- as well as intraparticle resistances and are more difficult to control due to the lack of adequate sensors and efficient solid-handling techniques especially for continuous operations.

Many microorganisms are capable of growing on solid substrates, but only filamentous fungi can grow to a significant extent in the absence of free water. Bacteria and yeasts grow on solid substrates at a 40–70% moisture level, such as in composting, and anaerobic and aerobic ensiling, but the growth and propagation of single cell organisms always require free water.

While there have been reviews on various aspects of SSF by Raghava (1993), Mitchell et al. (2003), and Pandey (2003), the following presents an overview of the recent developments on the mass and heat transfer in SSF (Raghava, 1993; Mitchell et al., 2003; Pandey, 2003). An attempt has also been made to enumerate the applications of SSF in various phases of development.

3.2.1 Mass Transfer in SS

The mass transfer processes involved in SSF can be divided into microscale phenomena and macroscale phenomena. At the microscale, the mass transfer processes depend on the nature of the microorganisms, that is, whether growth occurs as mycelium or a biofilm of unicellular organisms. At the macroscale, mass transfer processes occurring include:

1. The bulk flow of air into and out of the bioreactor and, as a consequence, changes in the sensible energy and concentrations of O_2, CO_2, and water.
2. Natural convection, diffusion, and conduction taking place in a direction normal to the flow of air during unforced aeration.
3. Conduction across the bioreactor wall and convective cooling to the surroundings.
4. Shear effects caused by mixing within the bioreactor, including damage to either the microorganism or the integrity of the substrate particles.

3.2.1.1 Diffusion of Oxygen and Nutrients   The transfer of oxygen from the void fraction within the solid phase to the growing microorganism is an interparticle mass transfer (Moo-Young and Blanch, 1981). The volume occupied by the air within the substrate gives the void fraction, which itself is dependent on the substrate characteristics and the moisture content. The moisture content should be optimal. If it is too high, the void space is filled with water and the air driven out, which creates anaerobiosis. At the other extreme, if the moisture content is too low, the growth of microorganism will be hindered. Mixing and aeration provide a means of achieving interparticle oxygen transfer, the efficiency of which is influenced by the void fraction and moisture content of solid substrate. However, at high values of the void fraction, mixing and aeration may not be critical as the voids contain enough oxygen to sustain the growth of cells.

Intraparticle mass transfer refers to the transfer of nutrients and enzymes within the substrate solid mass (Moo-Young and Blanch, 1981). The main aspects that need to be considered here are the diffusion of oxygen into the substrate containing the biomass and the degradation of a solid substrate by enzymes secreted by the growing microorganisms. In dealing with the intraparticle mass transfer, the effectiveness factor (Ef) is a very important concept. It is defined as a ratio of the observed reaction rate (r_{obs}) to the rate in the absence of any substrate concentration gradients. An important parameter required for the evaluation of Ef is the Theile modulus (φ), which is a measure of the ratio of biochemical reaction rate to the rate of differential mass transfer within the solid. By making use of this concept, a criterion is developed to evaluate intraparticle mass transfer limitation for the case of first-order reaction rate kinetics (Weisz and Prater, 1954).

Mitchell et al. (2004) studied the diffusional limitation of glucoamylase in a gel substrate in which starch was embedded (Mitchell et al., 2004). The results showed that due to the diffusional limitation, the enzyme concentration tended to be higher at the surface, resulting in a rapid utilization of starch at the surface. This resulted in a drastic reduction in the rate of glucose production, reaching a value as low as 20% of the activity that would be expected if all the enzymes were in contact with the starch. In their work, the biomass had no structure and the diffusion of glucose within the biomass layer was not considered. Furthermore, the role of oxygen diffusion and consumption at the intraparticle level was not considered. Rajagopalan and Modak (1995) developed a model for the growth of a unicellular organism in a biofilm of constant density (Rajagopalan and Modak, 1995). They concluded that oxygen is more likely to limit the growth within the film than lack of glucose, even under conditions where the outer edge of biofilm is high in oxygen concentration.

Bischoff (1967) developed an extended hypothesis to assess the mass transfer limitations by defining a generalized form of the Theile modulus valid for any reaction order and particle shape (Bischoff, 1967). This could be very useful since in most of the cases the rates of the biochemical reactions are highly nonlinear. Experimentation to check the probable application of these criteria to SSF as well as the development of more appropriate criteria suitable to SSF would be of immense use.

Oxygen diffusion into mold pellets has been extensively studied in SmF. While the mass transfer phenomenon in SSF is different, involving the growth of microorganisms on and into the solid substrate particles, the analysis of oxygen diffusion in mold pellets is useful in understanding the situation in SSF. This is because of the fact that intraparticle concentration gradients, coupled with mass transfer limitations, affect the performance of a process. Oxygen is an important substrate for fungal growth. In SSF, as the fungal mycelium develops on a solid surface, the void spaces between the hyphae can either be fully or partially filled with water. In the former case, the situation is similar to SmF, resulting in severe oxygen limitation and anaerobic conditions (Oostra et al., 2001). Moo-Young et al. (1983) and Mitchell et al. (1991) studied the growth of *Rhizopus oligosporus* on model substrates in SSF, and the results showed that the diffusive process limited the rate of growth, especially within the substrate (Moo-Young and Blanch, 1981; Mitchell et al., 1991). This is because, to reach the interior, oxygen must pass through the actively respiring biomass at the substrate particle surface and then diffuse through the aqueous phase within the substratum. Literature information on oxygen transfer capabilities of fermentors involving complex heterogeneous three-phase systems such as SSF is sparse (Charles, 1978; Metz et al., 1979). This is partly due to the inadequacy of the existing techniques for measuring the oxygen transfer coefficient (kL a) in these complex situations. Andre et al. (1981) suggested an improved method for the dynamic measurement of kL a for SSF systems (Andre et al., 1981). Durand et al. used sulfite oxidation rates to estimate the mass transfer coefficient in a packed-bed bioreactor (Durand et al.,

1988). Gowthaman et al. (1995) reported a method for the estimation of overall kL a in SSF on the basis of the inlet and outlet oxygen concentrations, assuming that the decrease in oxygen concentration within the liquid film was always 10% of the saturation concentration (Gowthaman et al., 1995). For mycelial growth in SSF, the fungal hyphae exposed directly to the air can probably take up oxygen directly from the air. It is well reported that the productivity in SSF is much higher than in SmF. One of the main reasons is the high interfacial area per unit volume. But this alone cannot explain completely the observed effect. There is no free-flowing water in SSF, and the water content is just sufficient to moisten the substrate. It is practically impossible to have water film uniformly around all the substrate particles or the substrate clusters or lumps. The microorganism takes oxygen directly in the places of substrate where the water film is not present and hence there is much less mass transfer resistance and in turn higher productivity.

In systems with forced aeration, oxygen transfer to these aerial hyphae is less likely to be rate-limiting (Nopharatana et al., 1998). In this case, nutrient movement within the aerial hyphae layer might be the factor controlling growth. Diffusion is the most probable mechanism for translocation of at least some nutrients, such as glucose and orthophosphate, within the hyphae of some fungi-like *Rhizopus nigricans* (Olsson and Jennings, 1991). Oostra et al. (2001) studied the effect of intraparticle oxygen limitation on the growth of *R. oligosporus* on different media. The results indicated that the optimal oxygen transfer and hence aerobic growth in SSF depended on the gas–liquid interfacial area and the thickness of the wet fungal layer. The interfacial area could be increased by either reducing the particle size of the substrate or by some pretreatment techniques such as steaming, puffing, extrusion, etc., that help to increase the pore size of the particles. This not only aids easy oxygen transfer but also helps in making the substrate more accessible for the action of enzymes of the fungi. However, the optimal particle size needed for such an effect needs to be experimentally investigated. The thickness of the fungal layer depends on the moisture content of the substrate. Hence, the optimum intraparticle oxygen transfer could be achieved by controlling the moisture content of the particle, which in turn is dependent on the flow rate and relative humidity of the incoming air.

3.2.1.2 Diffusion of Enzymes   Diffusion of enzymes and substrate fragments is another important aspect of intraparticle mass transfer in SSF. For the most part, the substrate is water insoluble, whereas the organism can utilize only a water-soluble substrate for growth (Huang, 1975; Suga et al., 1975; Mandels et al., 2004). For this reason, the action of extracellular enzymes in degrading the solid substrate into soluble fragments is a very important step in SSF. If the mass transfer resistance is very high, this could even be the rate-controlling step.

The diffusion of enzymes is facilitated by the open pore structure of the substrate, and the degradation can happen inside the substrate. In this case, the water-soluble fragments of the substrate will have to diffuse out of the solid matrix into the bulk region, where further enzymatic action will take place and metabolizable compounds are formed. However, when the porosity of the substrate is low, a major portion of the degradation will occur at the outer surface of the substrate (Humphrey et al., 1977; Knapp and Howell, 1980). In either mode of enzymatic action, the solid and polymeric materials are modified so that they can enter the cell and serve as carbon or energy sources. Thus, the utilization of solid substrates by the microorganisms is affected by factors that are relatively unimportant for the growth of microorganisms in SmF where the substrate is soluble and can penetrate the cell membrane. As the substrate cannot reach the microorganism by circulation currents caused by mixing, as in the case of SmF, the substrate bed needs to be mixed intermittently, but very gently so that the hyphae are not broken to a very great extent.

These theoretical studies on mass transfer indicate the following directions for the design and operation of the bioreactor:

1. Since the O_2 transfer is mainly achieved by diffusion (which is a slow process), superimposition of convective flow enhances the mass transfer (by forced convection of air through the SSF bed). This can enhance the O_2 gradients very effectively.
2. Similarly, CO_2 dissipation could be easily achieved by forced circulation.
3. The SSF bed needs to be of a smaller thickness, instead of one very thick bed, in order to overcome the problems of reduction in porosity with progress of fermentation.

3.2.2 Heat Transfer in SSF

In SSF, a considerable amount of heat as a function of the metabolic activities of microorganisms is generated (Chahal, 1983). In the initial stages of fermentation, the temperature and oxygen concentrations are the same at all the locations of the SSF bed. As the fermentation progresses, oxygen diffuses and undergoes bioreactions, liberating heat, which is not easily dissipated due to the poor thermal conductivity of the substrate. With the progress of the fermentation, respiration during the growth, which is dependent on the oxygen consumption and CO_2 formation, is highly exothermic, and heat generation is directly related to the level of metabolic activities of the microorganisms. The shrinkage of the substrate bed occurs and the porosity also decreases, further hampering the heat transfer. Under these circumstances, temperature gradients develop in the SSF bed, which can sometimes be steep, giving rise to high temperatures. For example, in the case of composting in heaps, temperature can rise to as high as 70°C. Thus, the removal of metabolic heat during large-scale SSF is one of the most critical issues, because microbial growth is particularly sensitive to the rise of temperature, given that the heat generation produces thermal gradients.

The transfer of heat into or out of the SSF system is closely associated with the metabolic activity of the microorganism, as well as the aeration of the fermentation system. The temperature of the substrate is very critical in SSF. High temperatures affect spore germination, growth, product formation, and sporulation, whereas low temperature is not favorable for the growth of microorganisms and the other bioreactions (Humphrey et al., 1977). The low moisture content and poor conductivity of the substrate make it difficult to achieve a good heat transfer in SSF. Significant temperature gradients exist even when small depths of the substrates are employed, thus it is very difficult to control the temperature of the fermentors on a large scale (Chahal, 1983). In fact, the limited heat dissipation is one of the major drawbacks of SSF in comparison with conventional SmF, where complete mixing is favorable for efficient dispersal of sparged oxygen and also serves to give better temperature control. Mixing not only aids homogeneity of the bed but also ensures an effective heat and mass transfer.

Thus, water addition coupled with continuous mixing is advantageous for simultaneous control of temperature and moisture control in a large-scale SSF.

Many approaches are available to enhance the heat transfer and control the microbial growth, such as mixing, forced aeration, evaporative cooling, and utilization of cooling jackets and additional cooling surfaces, which are the commonly used systems to achieve the best thermal conditions.

The conventional techniques used for temperature control in SmF are not applicable to SSF. Conventionally, temperature control in SSF is mainly accomplished by adjusting the aeration rate. If the temperature is too low, decrease the aeration rate to enable the temperature rising due to the respiration of microorganisms. However, enough care has to be taken in order to prevent the oxygen concentration from falling below the critical level that would adversely affect metabolic activity of the cells. On the other hand, if the temperature of the substrate is high, increase the aeration rate to promote the cooling of the substrate. To compensate for this, air of partially saturated is used for aeration. This method, namely, evaporative cooling of the substrate or the biomass, is effective if uniform aeration exists. Airflow causes variation in the water content profile and hence affects the biomass production and substrate consumption. Sargantanis et al. (1993) studied the growth of *R. oligosporus* on corn grits in a rocking drum bioreactor, and the results indicated that the temperature of the substrate bed could be effectively controlled by evaporative cooling for an improved biomass production (Sargantanis et al., 1993).

The importance of evaporative cooling and moisture content of the substrate on the performance of SSF bioreactor has also been highlighted by Nagel et al. (Nagel et al., 2000a,b). Experiments were conducted on *Aspergillus oryzae* grown on wheat grains as another experimental membrane-based model system. During scale-up, removal of the metabolic heat of the actively growing microorganisms by conduction in static beds is hampered due to the poor thermal conductivity of the substrate and also due to the lack of heat exchange surfaces. Evaporative cooling has not only been found to affect the removal of heat but has also been found to cause drying of the substrate/biomass, leading to a fall in the bioreactor performance. Water activity could also drop due to accumulation of solutes such as glucose, amino acids, etc. This could be prevented by spraying of water on to the solid substrate, coupled with the mixing. Water addition also aids the growth of new fungal cells. Studies show that the growth of fungi is hampered due to limited water availability, even if the total water content of the fermenting mass increases with time. This is because water is rapidly taken by the growing spores and as a result the residual water outside the cells, or in other words the water activity of the substrate, becomes limiting.

3.2.3 Mathematical Models of Mass and Heat Transfer

Mathematical modeling is an essential tool for optimizing bioprocesses. Not only can models guide the design and operation of bioreactors but also they can provide insights into how the various phenomena within the fermentation system combine to control the overall process performance. However, it is only over the last decade that concerted efforts have been made to develop mathematical models of SSF processes.

Mathematical modeling in SSF is reviewed in two parts. The current part reviews the balance/transport submodels of bioreactors. Part II reviews both the kinetic submodels of bioreactor models and microscale models of those processes associated with growth that occur at the level of individual particles (Mitchell et al., 2003). Both of these reviews aim to give an insight into how the various phenomena that occur within the system can be described mathematically. To this end, key model equations are reproduced and discussed. However, the sets of equations constituting the various models that are discussed are not reproduced entirely. Readers wishing to understand individual models in detail must refer to the original works. Also discussed are the insights that have been achieved through the modeling work, and improvements to models that will be necessary in the future. The current part of the review is organized on the basis of the various bioreactor types that are used in SSF. In presenting the balance equations, the terms linked with growth—for example, heat production and O_2 consumption—are replaced with a simple rate term, since the mathematical form of such terms is the subject of Part II (Mitchell et al., 2003).

3.2.3.1 Oxygen (Mass) Balances for SSF (Tray Bioreactors) The balance equations for O_2 within the bed have terms for O_2 diffusion within

the pores and uptake by the microorganism. However, due to the different ways in which the system was visualized, the equations appear differently. The equation used by Smits et al. (1999) to describe the O_2 balance within the bed is relatively simple:

$$\frac{\partial C^b_{O_2}}{\partial t} = D^b_{O_2} \frac{\partial^2 C^b_{O_2}}{\partial z^2} - ro_2, \tag{3.1}$$

where t is the time, $C^b_{O_2}$ the concentration of O_2 per unit volume of the bed, z the vertical coordinate, and ro_2 is the rate of O_2 uptake by the microorganism (Smits et al., 1999). The first term on the right-hand side describes the diffusion of O_2 within the pores, the diffusivity $D^b_{O_2}$ being an effective diffusivity that accounts for the fact that the pores represent only part of the overall volume of the bed and provide a tortuous path for diffusion.

The model of Rajagopalan and Modak (1995) recognizes the existence of a biofilm of biomass covering the particles within the bed. In this case, the equation for O_2 transfer within the pores of the bed is

$$\frac{\partial C_{O_2}\varepsilon}{\partial t} = D^b_{O_2} \frac{\partial^2 C_{O_2}}{\partial z^2} - K_a a_x \left(C_{O_2} - H C^f_{O_2} \right), \tag{3.2}$$

where C_{O_2} is the concentration of O_2 within the pores, ε the porosity of the bed, K_a the mass transfer coefficient for O_2 at the air/biofilm interface, a_x the area of the air/biofilm of the bed, and H the height of the headspace. The first term on the right-hand side describes the convection in the headspace across the surface of the bed, and the second term describes the transfer from the headspace to the bed.

In their later version of this model (Rajagopalan and Modak, 1995), the porosity is not a constant but a function of growth. The unsteady state balance equation was shown, and such that the O_2 concentration at the surface of the biofilm $(C^f_{O_2}|R(t))$, appears explicitly in the term that describes the transfer of O_2 across the gas/biofilm interface:

$$\frac{\partial C_{O_2}\varepsilon(t)}{\partial t} = D^b_{O_2} \frac{\partial^2 C_{O_2}}{\partial z^2} - k_a a_x \left(C_{O_2} - H C^f_{O_2} \middle| R(t) \right), \tag{3.3}$$

and another partial differential equation is used to describe the O_2 diffusion and reaction within the biofilm itself:

$$\frac{\partial C_{O_2}^f}{\partial t} = \frac{D_{O_2}^f}{r^2} \frac{\partial}{\partial r}\left(r^2 \frac{\partial C_{O_2}^f}{\partial r}\right) - r_{O_2}, \tag{3.4}$$

where r is the radial coordinate within the substrate particle. The first term on the right-hand side describes the diffusion of O_2 within a sphere, the radius of which increases during growth due to expansion of the biofilm.

Which of these approaches is better for the modeling of O_2 transfer within bioreactors depends on the aim of the modeling and the experimental information available. The necessity to write an O_2 balance for the headspace as was done by Rajagopalan and Modak (1994) is unclear since experimental attention has not been given to headspace O_2 concentrations (Rajagopalan and Modak, 1994). The later model of Rajagopalan and Modak (1995) explicitly describes the various steps in O_2 transfer to and within a biofilm growing at the particle surface and therefore is more mechanistic, providing the basis for an investigation into whether diffusion of O_2 within the pores, transfer from the pores to the biofilm, or diffusion within the biofilm is a limiting step in O_2 transfer. However, if there is good empirical data on which to express O_2 consumption rates, then the approach of Smits et al. might be preferable due to its simplicity (Smits et al., 1999).

3.2.3.2 Energy (Heat) Balances for SSF (Tray Bioreactors)   The energy balance in the model of Rajagopalan and Modak (1995) takes into account conduction (the first term on the right-hand side) and metabolic heat production (the second term on the right-hand side):

$$\rho_S C_{PS} \frac{\partial T}{\partial t} = k_b \frac{\partial^2 T}{\partial z^2} + r_Q, \tag{3.5}$$

where ρ_S is the density of the bed, C_{PS} the heat capacity of the bed, T the bed temperature, k_b the thermal conductivity of the bed, and r_Q the rate of metabolic heat production by the microorganism. The balance of Smits et al. (1999) has an additional term to describe evaporative heat removal (the last term on the right-hand side):

$$\frac{\partial H}{\partial t} = K_b \frac{\partial^2 T}{\partial z^2} + r_Q + \lambda D_{VAP}^* \frac{\partial^2 C_{VAP}}{\partial z^2}, \tag{3.6}$$

where H is the enthalpy of the bed, which is a function of temperature and λ the enthalpy of vaporization of water. The second term on the right-hand side describes the conduction within the bed, whereas the third term on the right-hand side describes the removal of heat by the evaporation and diffusion of water, with the airspace in the pores of the bed assumed to be in moisture and thermal equilibrium with the solid. However, in the typical use of tray bioreactors, the evaporation term is unlikely to be necessary, since the simulations of Smits et al. (1999) shows that the contribution of evaporation to cooling is negligible if the tray is incubated in a 98% relative humidity environment.

Over the past decade or so, there have been significant advances in the modeling of heat and mass transfer phenomena in SSF bioreactors. Further, although initial work tended to focus on the energy balance, with the aim of preventing high temperatures, current models consider both the energy balance and the water balance, allowing their use in bioreactor control schemes designed to maintain both bed temperature and water content at the optimum values for growth and product formation. Various improvements that are desirable for the current models of specific bioreactor types have already been identified above. Some of the future advances in modeling of bioreactors may come from describing the kinetics of growth and the phenomena at the particle level in a less empirical manner than is done presently.

Currently, there are no reports in the literature describing the use of mathematical models in the construction of large-scale bioreactors. However, mathematical models have now reached a level of sophistication that makes this not only possible, but also necessary; it is only through the use of mathematical models as tools during the design process and in the optimization of operation that SSF bioreactors will perform at their full potential and thereby maximize the economic performance of SSF processes.

3.3 Dynamic Model of Solid-State Fermentation

In recent years, several excellent reports have appeared providing a great deal of knowledge and understanding of the fundamental aspects of SSF, and today we have much better information about the heat and mass transfer effects in the SSF processes, which have been considered as the main difficulties in handling SSF systems. However, there still remains much to be done in this regard. During SSF, a large amount of heat is generated, which is directly proportional to the metabolic activities of the microorganism. The solid materials matrices used for SSF have low thermal conductivities, hence heat removal from the process could be very slow. Sometimes accumulation of heat is high, which denatures the product formed and accumulated in the bed. Temperature in some locations of the bed could be 20°C higher than the incubation temperature. In the early phases of SSF, temperature and concentration of oxygen remain uniform throughout the substrate but as the fermentation progresses, oxygen transfer takes place, resulting in the generation of heat. The transfer of heat into or out of the SSF system is closely related with the aeration of fermentation system. The temperature of the substrate is also very critical in SSF as it ultimately affects the growth of the microorganism, spore formation, and germination and product formation. High moisture results in decreased substrate porosity, which in turn prevents oxygen penetration. This may help bacterial contamination. On the other hand, low moisture content may lead to poor accessibility of nutrients, resulting in poor microbial growth. Water relations in SSF must be critically evaluated. Water activity (aw) of the substrate has a determinant influence on microbial activity. In general, the type of microorganism that can grow in SSF systems are determined by aw. The importance of aw has widely been studied by various authors. The aw of the medium has been attributed as a fundamental parameter for mass transfer of the water and solutes across the microbial cells. The control of this parameter could be used to modify the metabolic production or excretion of a microorganism.

Modeling in the SSF system is another important aspect, which needs to be studied in detail. Not enough information is available on the kinetics of reactions in SSF systems. This is mainly because of difficulties involved in the measurements of growth parameters, analysis of cellular growth and determination of substrate consumption, etc., which is caused due to the heterogeneous nature of the substrate, which is structurally and nutritionally complex. Among the several approaches to tackle this problem, an important one has been to use a synthetic model substrate. It is well known that the fermentation kinetics are sensitive to the variation in ambient and internal gas compositions. The cellular growth of the microorganisms can be determined by measuring the change in gaseous compositions inside the bioreactor. This can also be determined by substrate digestion, heating and centrifuging substrate, using light reflectance, DNA measurement by glucosamine level, protein content, oxygen uptake rate, and carbon dioxide evolution rate.

In the solid phase, little water and air were included during solid fermentation, and evaporation of water due to heat quantity caused by microorganisms resulted in heterogeneous three-phases (gas phase, liquid phase, and solid phase) in the fermentation system. Water activity, pH, and the most suitable reaction temperature were difficult to control, leading to a decrease in production. On the other hand, measured values of several factors could not reflect the truth of the reactor. The mathematical models were necessary tools of the optimizing bioprocess (Mitchell et al., 2003). The mathematical models could offer opinions for how to hold together several phenomena to control the whole process in the fermentation system. At the same time, the mathematical models could also guide design and operation of the bioreactor to obtain ideal productivity. The study and development of mathematical models largely determines the advance or retreat of solid-state fermentation.

The mathematical models of solid fermentation includes two kinds: macroscopic models and microscopic models. Macroscopic models refer to operations of the bioreactor and describe the mass transfer and heat transfer of a substrate. Microscopic models refer to several phenomena happening on the surface and inside of the grains, and do not describe the operation of the bioreactor as a whole. Owing to their differences, the two mathematical models are very important for the bioreactor. The models of the solid fermentation bioreactor were used to describe different operational variables as they influenced the properties of the reactor. The model of the bioreactor was made of two submodels: kinetics submodels and balance submodels. Balance submodels described mass transfer and heat transfer in different internal phases or interphases of the bioreactor. Kinetics submodels describe the growth rate of microorganisms' dependence on key environmental parameters. The solid fermentation models describe the balance of several factors of macroscopic and microscopic phenomenons happening in solid fermentation (Pandey, 2003).

3.3.1 Balance of Oxygen

Smits et al. proposed the model describing the oxygen balance relationship of the substrate bed in the tray reactor (Smits et al., 1999):

$$\frac{aC_{O_2}^b}{at} = D_{O_2}^b \frac{aC_{O_2}^b}{az^2} - r_{O_2}, \tag{3.7}$$

where means time, $C_{O_2}^b$ means oxygen concentration of unit volume bed (vertical coordinate), $D_{O_2}^b$ means diffusion rate, and r_{O_2} means ingesting the oxygen rate of the microorganism.

Rajagopalan and Modak (1995) considered the effect of biomembrane covering the bed face in the models. Oxygen transmission through the bed airgap was described in the equation:

$$\frac{\partial C_{O_2} \varepsilon}{\partial r} = D_{O_2}^b \frac{\partial C_{O_2}}{\partial z^2} - K_z a_x \left(C_{O_2} - HC_{O_2}^f \right), \tag{3.8}$$

where CO_2 means oxygen concentration of the bed, ε means porosity, Ka means oxygen transmission coefficient of air/biomembrane interphase, a_x means area of unit volume air/biomembrane interphase in the reactor, H means inductance constant, and $C_{O_2}^f$ means oxygen concentration in the biomembrane.

A single-layer membrane formed on the face of the solid substrate after some time due to the growth metabolism of the microorganism, which could influence oxygen transmission, heat radiation, etc. Oostra et al. propounded an oxygen concentration equation of the solid substrate interface biomembrane, which was verified during the experiments of Rahardjo et al. (Oostra et al., 2001; Rahardjo et al., 2002).

3.3.2 Balance of Water

Smits et al. considered that balance of water in a bed substrate was described for the equation (Smits et al., 1999):

$$\frac{\partial C_w}{\partial r} = r_{H_2O} - \left[\frac{\partial C_{VAP}}{\partial r} - D_{VAP}^* \frac{\partial^2 C_{VAP}}{\partial Z^2} \right], \tag{3.9}$$

where C_w means concentration of unit volume liquid water in the bed, D_{VAP}^* means effective diffusion coefficient of water vapor in the bed, and r_{H_2O} means metabolism producing water rate.

Stuart's revolving drum bioreactor model described the relationship quality balance of water in substrate bed (Stuart et al., 1999):

$$\frac{dMW}{dt} = -kA_{sa}(C_1 - C_B) + r_{H_2O}, \tag{3.10}$$

where M means dry weight of substrate, W means water concentration of the bed, k means transfer coefficient of quality, A_{sa} means air contact area of wall and headspace, C_1 means water vapor concentration used for balance around the substrate, C_B means water vapor concentration of headspace, and r_{H_2O} means metabolism rate of water.

Von Meien and Mitchell (2002) established an intermittent agitation reactor model that included water and energy balance of every solid phase and gas phase axial strain, also describing the influence of temperature and water activity in the solid phase for growth kinetics. The model could show characteristics of temperature, humidity, and biomass when the static bed was a forced draft. The temperature characteristics of time and space by model prediction were similar to them of size of experiment solid fermentation reactor by observing during growth kinetics experiment. The temperature of the bed obviously fell during mixing, and when recovering the static operator, the temperature obviously was raised. The model could not include all phenomena that happened in the solid fermentation reactor. For example, growth resulted in a void ratio of the solid bed, the isotherm in a solid substrate, and a change of mass transfer and heat transfer coefficients, etc. But it captured many relative characteristics in reaction to provide important views for control of large-scale reactors. Garcia (1983) first adopted the distributed parameter model, and adopted the dynamic matrix operation method instead of the classical integral differential quotient method to control temperature and water concentration of batch mixing and forced draft solid fermentation reactor. Peña Y Lillo et al. (2001) established an advanced model that could make use of an indirect online measuring method to assess water concentration, and water concentration was looked upon controlled parameter, although water concentration could not be measured online. Productivity of CO_2 and inside and outside air condition were measured through experiments, and utilized a model to predict temperature and water concentration. Experiments provided that this model was a very useful software sensor for measuring water concentration in a reactor. Nagel et al. (2000a,b) considered that water concentration in the biomass was different from that of the residual substrate, and provided a mathematic model assessing water concentration outside of the cell. Water concentration outside of the cell was used as a parameter of the control process.

3.3.3 Balance of Energy

Rajagopalan and Modak (1995) established an energy balance model of the pallet reactor, and considered heat conduction and metabolism producing heat.

$$\rho_s C_{ps} \frac{\partial T}{\partial r} = k_b \frac{\partial^2 T}{\partial z^2} + r_Q, \tag{3.11}$$

where ρ_s means density of the bed, C_{ps} means heat capacity, T means temperature of the bed, k_b means thermal conductivity in the bed, and r_Q means microorganism metabolism producing heat rate.

Smits et al. (1999) added the description that vapor impacted heat on the base of that model:

$$\frac{\partial H}{\partial r} = k_b \frac{\partial^2 T}{\partial z^2} + r_Q + \lambda D^*_{VAP} \frac{\partial^2 C_{VAP}}{\partial z^2}, \tag{3.12}$$

where λ means enthalpy of the bed, and means enthalpy change of vaporization of water. The right second part of the equation describes heat conduction of the bed, and the third part of it describes heat taken away by evaporation and diffusion of water. However, if the microorganism was cultured in the environment, relative humidity was 98%. Heat reduced by evaporation could be ignored.

Sangsurasak and Mitchell (1995) established an energy balance equation for a packed solid fermentation reactor:

$$\rho_b C_{pb} \left(\frac{\partial T}{\partial r} \right) + \rho_a \left(C_{pa} + f\lambda \right) V_z \left(\frac{\partial T}{\partial r} \right)$$

$$= \left[\frac{k_b}{r} \left(\frac{\partial T}{\partial r} \right) + k_b \left(\frac{\partial^2 T}{\partial r^2} \right) \right] + k_b \left(\frac{\partial^2 T}{\partial r^2} \right) + r_Q, \tag{3.13}$$

where C_{pb} means thermal capacity of material layer, ρ_b means density of material layer, C_{pa} means thermal capacity of moist air, ρ_a means density of air, and V_z means rate surface of air. If the material bed was wide enough and the horizontal heat conduction could be ignored, the equation could be simplified as:

$$\rho C_{pb}\left(\frac{\partial T}{\partial r}\right)+\rho_a\left(C_{pa}+f\lambda\right)V_z\left(\frac{\partial T}{\partial z}\right)=k_b\left(\frac{\partial^2 T}{\partial r^2}\right)+r_Q \quad (3.14)$$

Weber et al. established models such as shown in the following equation, and took energy and water into account (Weber et al., 1999):

$$0=r_Q+F_{air}\frac{d\left(C_{P_g}\left(T-T_{ref}\right)y_{VAP}\left(C_{pVAP}\left(T-T_{ref}\right)+\lambda\right)\right)}{dz} \quad (3.15)$$

$$(1-\varepsilon)C_s\frac{dX_{WS}}{dt}=r_{H_2Oext}-(1-\varepsilon)X_{WS}\frac{dC_s}{dt}-F_{air}\frac{y_{out}-y_{in}}{H} \quad (3.16)$$

Oscar et al. established the gas energy balance equation and solid phase energy balance equation, shown, respectively, as follows (Garcia, 1983):

$$S\left(C_{P_s}+\phi_s C_{P_w}\right)\frac{\partial T_s}{\partial t}=ha\left(T_g-T_s\right)-\lambda K_a\left(\phi_s-\phi_s^*\right)+Y_Q\left(S\frac{\partial b}{\partial t}+b\frac{\partial S}{\partial t}\right)$$

$$(3.17)$$

$$\varepsilon\rho_g\left(C_{P_g}+\phi_g C_{P_v}\right)+\left(C_{P_g}+\phi_g C_{P_v}\right)G\frac{\partial T_g}{\partial z}=ha\left(T_g-T_s\right) \quad (3.18)$$

F_{air} means air velocity, T_{ref} means enthalpy fiducial temperature, y_{VAP} means humidity of gas, C_{P_g} means heat capacity of dry gas, $C_{pVAP}(C_{P_v})$ means heat capacity of water vapor, C_s means quality of dry substrate in the reactor, X_{WS} means water concentration of dry substrate, r_{H_2Oext} means production ratio of outside of cell during growth of microorganism, H means the height of the bed layer, S means particle concentration of dry substrate, C_{P_s} means heat capacity of dry substrate, ϕ_s means water concentration of solid phase, C_{P_w} means heat capacity of liquid water, T_s means temperature of the solid phase, ha means heat transfer coefficient of gas and solid phases, T_g means temperature of the gas phase, λ means enthalpy of water vapor, K_a means water transfer coefficient of gas and solid phase, ϕ_s^* means water concentration of solid phase when the gas phase and solid phase were balance in the temperature of the gas phase, Y_Q means production of heat due to growth of the microorganism, b means concentration of the biosubstrate, ε means the gap of grain, ρ_g means density of gas, ϕ_g water concentration of gas phase, G means airflow rate of entrance, and z means the position of axial direction.

Applicable models of different reactors were different, and even models of the same reactor differed because marcato-considered factors were different. Stuart established an energy and quality balanced model of a revolving drum solid fermentation reactor (Stuart et al., 1999). Sargantanis et al. (1993) established mixing solid fermentation reactor model (Sargantanis et al., 1993).

3.3.4 Other Models

The power source of traditional bioreactors is the tangential friction force of fluid dynamics, which is effective for lifeless chemical reaction grains. Metabolic fluxes produced by the growth of a microorganism affects temperature and heat transfer in the substrate. There are many studies on mass transfer and heat transfer for established models, but those are not about microorganism growth metabolism. Microorganism growth metabolism kinetics are expressed by the process of metabolic production of fermentation. One of these forms of kinetics, based on the releasing and change of CO_2, was used to express the growth rate of thalli. Other kinetics are based on enzyme activity. Enzyme activity, production, and growth of a microorganism are relative. Mitchell et al. (2004) found that distribution and activity of glucose amylase could affect the growth of microorganisms, and affect the production of glucose (Mitchell et al., 2004). Zhang Siliang et al. used the bioreaction engineering position focusing on the analysis and control of cell metabolic flux, and propounded an optimized technology of multiparameter in a study based on parameter correlation and enlarged technology of the multiparameter adjustment in fermentation process (Zhang et al., 2006). This novel bioreactor based on a test of material flow and optimization and enlargement of process was used in dense and highly-efficient synthesis, and the fermentation level was improved by a large margin. Van de Lagemaat and Pyle (2001) established a model on different phases of solid cultivation of tannase, and described biomass, tannase, and the production of spores; they found that their model was very conconent with experiment results. Morteza Khanahmadia et al. (2006) established a model based on types of substrate flow in continuous solid fermentation, and to some extent, verified the rationality of this mode.

There were several studies referring to shrinkage of a substrate, consumption of a substrate, distribution and diffusion enzyme and production in a solid substrate, growth of thalli, and change of porosity.

3.4 Control of Solid-State Fermentation

There are various important factors that produce a considerable impact on the success of a particular technology, hence they need to be considered for the development of any bioprocesses, as well as the SSF. These factors include selection of microorganism and substrate, optimum process parameters, and also purification of the end product, which has been a challenge for this technology. Fungi and yeast were termed as suitable microorganisms for SSF according to the theoretical concept of water activity, whereas bacteria have been considered unsuitable. Still, several research articles prove that bacterial cultures can also be well manipulated and managed for SSF process, even for scarcely produced tannase enzyme. *Bacillus thuringenesis* production was standardized by SSF on wheat bran to obtain maximum toxins and was found to be cost effective.

The establishment of the relationships between the physiology of the microorganisms and the physical–chemical factors is the aim in developing proper models. The factors involved include temperature, pH, aeration, water activity and moisture, bed properties, nature of solid substrate employed, etc. Among several critical factors, moisture and the nature of the solid substrate employed are the most important affecting the SSF processes. Selection of moisture depends on the microorganism employed and also the nature of the substrate. Fungi needs lower moisture; 40–60% moisture could be sufficient but the selection of substrate depends upon several factors mainly those related to cost and availability, and thus may involve the screening of several agro–industrial residues.

Today's environment is rapidly changing. We can expect constant technological advancement backed by innovation as a major catalyst.

SSF appears to possess several biotechnological advantages, though at present mostly on a laboratory scale, such as higher fermentation productivity, higher end-concentration of products, higher product stability, lower catabolic repression, cultivation of microorganisms specialized for the water-insoluble substrates or mixed cultivation of various fungi, and last but not least, a lower demand for sterility due to the low water activity used in SSF. Viniegra-González et al. (2003) have attempted to develop a general approach for the comparison of productivity of enzymes employing SSF and SmF (submerged fermentation), and have tried to explain the reason for higher production in SSF. Higher biomass, high enzyme production, and lower protein breakdown contribute to better production in SSF.

Scale-up, purification of end products and biomass estimation are the major challenges that lead researchers to strive hard to find the solutions. Scale-up in SSF has been a limiting factor for some time, but recently, with the advent of biochemical engineering, a number of bioreactors have been designed that could overcome the problems of scale-up and to an extent also the online monitoring of several parameters, as well as heat and mass transfer.

Separation of biomass is a major challenge in SSF, essential for kinetic studies. However certain indirect methods are there, such as glucosamine estimation, ergosterol estimation, protein estimation, DNA estimation, dry weight changes and CO_2 evolution. All of them have their own weaknesses. Recently, digital image processing has been developed as a tool for measuring biomass in SSF. The images were acquired by stereomicroscope and a digital camera, and processed using KS400 software. In recent times, estimation of oxygen intake and carbon dioxide evolution rate are considered to be most accurate for the determination of growth of the microorganism. Although product recovery and purification processes are more expensive, employing natural supports, their utilization supposes a reduction in production costs and usually much higher activities are obtained. Hence, an economical evaluation of the overall process should be done in order to determine its feasibility for a specific purpose. This system is especially suitable for the production of high-value products like enzymes. There are particular applications where concentrated end products with high titers are required rather than degree of purity,. For example, bioconversion of biomass requires concentrated crude cellulose; in the leather industries, crude protease are enough to remove the hairs from the leather.

3.4.1 Control of Fermentation Conditions

3.4.1.1 Heat Transfer As the solid bed is aerated from one of the ends, significant axial temperature gradients are developed, and this may limit the performance of a packed-bed bioreactor. The axial temperature gradients existing in the bioreactor pose a potential threat to the growth of microorganisms and hence metabolite production, for the enhanced evaporation of water from the solid bed can lead to the eventual drying out of the bed. To address the problem of heat removal, novel reactor designs for a packed bed like Zymotis have been developed. Fixed-bed reactor models have been primarily classified as pseudohomogenous and heterogeneous. The basic pseudohomogenous models for fixed-bed bioreactors primarily assume that the temperature and concentration gradients only occur in the axial direction. In the light of analysis of a packed-bed bioreactor as a pseudohomogenous fixed-bed reactor model, air and the solid bed are not explicitly taken into account as two distinct phases. Thermal and moisture equilibrium is assumed to exist between the air and solid bed. This simplified assumption can be justified if saturated air is used to aerate the bioreactor. The mathematical models for a pseudohomogenous single-phase system primarily involve the solution of an energy balance equation. Saucedo-Castañeda et al. (1990) considered the radial temperature gradients to be dominant in their one-dimensional mathematical model and established that the conduction through the packed bed made a significant contribution to heat transfer resistance. In the two-dimensional model developed by Sangsurasak and Mitchell (1995), the effect of death of microorganisms was also considered. They emphasized the existence of significant axial temperature gradients and the importance of heat transfer through convection in packed-bed solid-state fermentation bioreactors. They extended their two-dimensional mathematical model by the inclusion of a term that described the heat transfer effects occurring as a consequence of evaporation of water. Hasan, A. and Kabir, C. (1994) had developed a model that described heat transfer in a packed-bed solid-state fermentation bioreactor in rice bran. Mathematical models based on the pseudohomogenous single-phase system have been used for studies on scale-up, determination of mechanisms for preventing overheating in bioreactors, and for the analysis of Zymotis packed-bed bioreactors. The mathematical models proposed till date for pseudohomogenous single-phase systems involve the solution of partial differential equations that are solved by discretization using orthogonal collocation, and a fully implicit finite difference method. The focus of studies on packed-bed solid-state fermentation bioreactors has been primarily on the mitigation of axial temperature gradients. Saucedo-Castañeda et al. (1990) used a 6 cm diameter and 35 cm long jacketed packed-bed bioreactor, operated at low superficial velocities to promote radial heat transfer. Axial heat transfer can hence be appropriately neglected only in thin columns operated at low aeration rates; such columns would be impractical for large-scale processes as they would have to be very tall to hold a large quantity of substrate.

In the category of two-phase models, such as the one proposed by Von Meien and Mitchell (2002), the model equations have been solved by the approximation of spatial derivatives by finite differences.

The same model equations were also employed for the analysis of control strategies. In another study concerning two-phase systems, the mass and energy balances developed for packed-bed bioreactors were highly simplified, in which the temporal variations in temperature and enthalpy accumulation terms were rendered negligible.

In the aforesaid discussion, it is clearly evident that for mathematical modeling of packed-bed solid-state fermentation bioreactors, either a rigorous mathematical procedure has to be employed or the ensuing mass and energy balances have to be simplified. In the present study, we have attempted to ease the complexities encountered in computation and simultaneously attempted to avoid oversimplification with the help of concepts envisaged in chemical reaction engineering. In the method of orthogonal collocation employing Jacobi polynomials, Sangsurasak and Mitchell (1995) solved their two-dimensional model by discretizing the spatial coordinates as an axisymmetric problem. In their case, the number and position of internal collocation points were varied by changing the values of exponents of the Jacobi polynomials. The collocation points were manipulated to coincide with positions where experimental data were available.

Recently, a novel double dynamic system, including air pressure pulsation and internal circulation of air, for SSF of thermophilic microorganisms was designed by Chen. Changes of air mean that the gas phase is double dynamic (including air pressure pulsation and internal air circulation) while the solid phase is static. Forces produced by changes of air are not shearing but normal. When air pressure in the fermenter drops suddenly, the gas phase within substrates will swell, and then the swell will lose the solid substrates. The strategy is very beneficial to mycelia, because the method can not only provide sufficient O_2 but also expand more room for fungal propagation and improve heat transfer within the substrate. Because heat removal was more difficult than O_2 supply and higher temperature could strongly restrict fungal growth, making enzyme production unbeneficial, temperature monitor was regarded as the principal reference tool for the optimization of the culture conditions.

This novel SSF bioreactor has been reported in successfully producing biopesticide, using *Bacillus thuringiensis* CM-1, and cellulose, using *Penicillum decumbens* JUA10. The result in cellulase production indicated that the average enzyme activity in the dynamic culture could reach the maximum of 20.4 IU/g, nearly two times that of the static culture (10.8 IU/g).

3.4.1.2 Moisture Content   Substrate must possess enough moisture to support growth and metabolism of the microorganism. The establishment of the relationships between the physiology of the microorganisms and the physical chemical factors is the aim of the development of proper models. These factors include temperature, pH, aeration, water activity and moisture, bed properties, nature of solid substrate employed, etc. Among several critical factors, moisture and nature of solid substrate employed are the most important factors affecting SSF processes. Selection of moisture depends on microorganism employed and also the nature of the substrate. Fungi needs lower moisture, 40–60% moisture could be sufficient but selection of a substrate depends upon several factors mainly related with cost and availability, and thus may involve the screening of several agro-industrial residues. High moisture percentage results in low substrate porosity which in turn prevents oxygen penetration, whereas low moisture content may lead to poor accessibility of nutrients resulting in hampered microbial growth. The consequences of these difficulties for bioreactor performance have been well explored.

Evaporative cooling has been incorporated for large-scale operation of rotating drum bioreactor (RDB). The mass transfer coefficient for evaporation from the bran bed to the head space has been determined by measuring outlet water vapor concentrations for a RDB containing wet wheat bran. Sangsurasak and Mitchell (1995) suggested that evaporation can remove as much as 78% of the heat from the bed during the time of peak heat generation, even when the bed is aerated with saturated air. This is due to the increase of water vapor pressure at increasing temperature. It results in large moisture losses and drying of the solid. A new generation of small reactors was developed by an INRA-team in France. Such reactors have a working volume of about 1 L, a relative humidity probe, a cooling coil on the air circuit, and a heating cover for the vessel. Each reactor is automatically controlled by a computer. A novel bioreactor with two dynamic changes of air (including air pressure pulsation and internal circulation) can increase mass and heat transfers, as well as improve the porosity within the substrate. The reactor can be significant in exploiting numerous socio-economic advantages of solid-state fermentation.

Xylanase and pectinase activities were less low when the moisture content was higher or lower than 65%. The low enzyme activity at high substrate moisture levels could be attributed to the decreased porosity, alteration in particle structure, gummy texture, lower oxygen transfer, or increased formation of aerial hyphae. Likewise, lower moisture levels lead to reduced diffusion of the nutrients in the solid substrate, lower degree of swelling, and higher water tension. The optimum moisture contents for xylanase production by *Trichoderma longibrachiatum* and *Aspergillus tereus* were 55% and 75%, respectively.

3.4.1.3 Fermentable Substrates   In the recent past, many agro–industrial byproducts such as wheat bran, rice bran, molasses bran, barley bran, maize meal, soybean meal, potato peel, coconut oil cake, etc., have been screened as low-cost solid substrates for microbial production of amylase in SSF. A perusal of literature shows that some of the solid substrates possess higher potential as compared to the other solid substrates for supporting/inducing the amylase (or even other enzymes) production by microbes. For example, wheat bran and potato peel are reported to be superior as compared to the other tested solid substrates for optimum amylase production by many microorganisms under the identical SSF condition. However, the exact reason(s) for such a higher enzyme production by microbes while growing on some specific substrates has never been explored and till date no logical explanation has been provided for the above observed effect.

3.4.1.4 Effect of Aeration   With respect to aeration, two strategies are available: The air can either be circulated around the substrate bed or blown forcefully through it. Categorizing SSF bioreactors in this way represents a division on the basis of operation rather than on physical design features of the bioreactor, and the result is that the models describing a bioreactor operation have similar features within each group.

Unfortunately, however, in many cases, mixing causes shear damage to the mycelium and loss of productivity, so this has to be balanced against the productivity loss resulting from improper temperature control or bed drying. Constant aeration also makes it more difficult to maintain a stable environment inside the bioreactor with respect to oxygen and carbon dioxide. This is because the air flow rates are linked to heat removal rates and are not therefore easily controlled in an independent manner to provide the desired oxygen to the carbon dioxide balance that may be optimal for high productivity.

The production of citric acid depends strongly on an appropriate strain and on operational conditions. Oxygen level is an important parameter for citric acid fermentation. Several researchers have studied the influence of forced aeration on citric acid production and the metabolic activity of *A. niger* in SSF by respirometric analysis. They showed that citric acid production was favored by a limited biomass production that occurred with low aeration rates. Both works showed the feasibility of using the strain *A. niger* for citric acid production by SSF.

There are four types of reactors to perform SSF processes, and each with their own design tries to make conditions more favorable for fermentation under solid-state conditions. The bioreactors commonly used, which can be distinguished by the type of aeration or the mixed system employed, include the following:

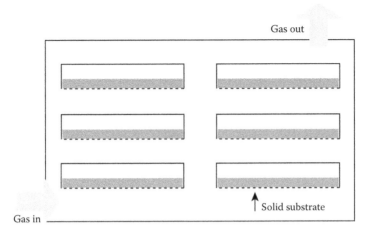

Figure 3.1 Scheme of a tray bioreactor (passive aeration, static).

Tray: Consists of flat trays. The substrate is spread onto each tray, forming a thin layer only a few centimeters deep. The reactor is kept in a chamber at a constant temperature through which humidified air is circulated. The main disadvantage of this configuration is that numerous trays and large volume are required, making it an unattractive design for large-scale production (Figure 3.1).

Packed bed: Usually composed of a column of glass or plastic with the solid substrate retained on a perforated base. Through the bed of substrate, humidified air is continuously forced. It may be fitted with a jacket for circulation of water to control the temperature during fermentation. This is the configuration usually employed in commercial koji production. The main drawbacks associated with this configuration are difficulties in obtaining the product, nonuniform growth, poor heat removal, and scale-up problems (Figure 3.2).

Horizontal drum: This design allows adequate aeration and mixing of the substrate, while limiting the damage to the inoculum or product. Mixing is performed by rotating the entire vessel or by various agitation devices such as paddles and baffles. Its main disadvantage is that the drum is filled to only 30% capacity, otherwise mixing is inefficient (Figure 3.3).

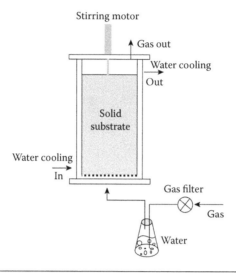

Figure 3.2   Scheme of a packed-bed bioreactor (humidified air; static).

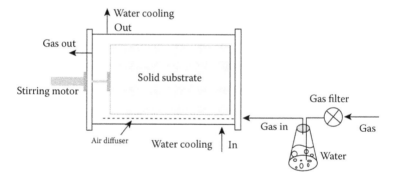

Figure 3.3   Scheme of a horizontal drum bioreactor (humidified air; mechanical agitation).

Fluidized bed: In order to avoid the adhesion and aggregation of substrate particles, this design supplies a continue agitation with forced air. Although the mass heat transfer, aeration, and mixing of the substrate is increased, damage to inoculum and heat build-up through sheer forces may affect the final product yield (Figure 3.4).

3.4.1.5 Biomass   Several approaches have been proposed as indirect measurements of biomass quantification, such as the production of primary metabolites or carbon dioxide, the variation in the electrical conductivity between the biomass and the solid substrate, the changes in the color of the fermentation medium determined using reflected light, the

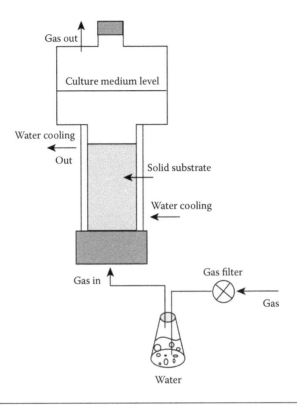

Figure 3.4 Scheme of a fluidized-bed bioreactor (humidified air; pneumatic agitation).

quantification of intracellular protein, the quantification of other compounds such as ergosterol, and the quantification of antibody reactivity in the mycelium cell wall via enzyme-linked immune sorbent assays. There is a need for the development of an efficient and reliable method for directly estimating biomass in the presence of a solid substrate.

Auria et al. (1993), emphasizing that biomass growth has been estimated only indirectly, studied the influence of mold growth on the pressure drop in aerated SSF using Amberlite as support, and concluded that when agricultural products are used as both substrate and support, the pressure drop cannot be directly used to obtain an estimate of the biomass produced. Borzani has presented a general equation correlating the rate of variation in the weight of the fermenting medium (dry matter) and the cell biomass growth rate in solid-state fermentation tests with the main purpose of identifying the microbial growth phases using *Rhizopus oryzae* ATCC 10404 cultivated on a mixture of wheat bran and ground corn. Because there are serious difficulties associated with the direct measurement of cell biomass in SSF systems, a weighing method could provide improvement in kinetic studies of such processes. In fact, bioprocess modeling including cells and cultures can be valuable in optimizing and controlling actual production operations in addition to managing a system description. Also, several challenges exist at each stage in the development of models of enzyme production kinetics.

3.4.2 Prospect

Over the past couple of years, SSF involving growth of microbes on moist solid substrate(s) in the absence of free flowing water has gained a tremendous momentum owing to certain advantages over the conventional SmF, such as low production cost, saving of water and energy, less waste effluent problem, and stability of the product due to less dilution in the medium. Additionally, SSF technology has a tremendous scope for employment generation particularly in developing countries. Furthermore, it has been reported that wild-type strains of microorganisms (fungi or bacteria) have a tendency to perform better in SSF systems compared to genetically altered microorganisms, thus reducing the energy and cost requirement even further.

References

Andre, G., Moo-Young, M., and Robinson, C. (1981). Improved method for the dynamic measurement of mass transfer coefficient for application to solid–substrate fermentation. *Biotechnology and Bioengineering.* **23**(7), 1611–1622.

Auria, R., Morales, M., Villegas, E., and Revah, S. (1993). Influence of mold growth on the pressure drop in aerated solid state fermentors. *Biotechnology and Bioengineering.* **41**(11), 1007–1013.

Bischoff, K. (1967). An extension of the general criterion for importance of pore diffusion with chemical reactions. *Chemical Engineering Science.* **22**(4), 525–530.

Chahal, D.S. (1983). Foundation in biochemical engineering kinetics and thermodynamics in biological systems. *American Chemical Society Symposium Series, American Chemical Society,* Washington, DC, **207**, 421.

Charles, M. (1978). Technical aspects of the rheological properties of microbial cultures. *Advances in Biochemical Engineering.* **8**, 1–62.

Durand, A., Pichon, P., and Desgranges, C. (1988). Approaches to kla measurements in solid state fermentation. *Biotechnology Techniques.* **2**(1), 11–16.

Garcia, O. (1983). A stochastic differential equation model for the height growth of forest stands. *Biometrics.* 1059–1072.

Gowthaman, M., Rao, K., Ghildyal, N., and Karanth, N. (1995). Estimation of KLa in solid-state fermentation using a packed-bed bioreactor. *Process Biochemistry.* **30**(1), 9–15.

Hasan, A. and Kabir, C. (1994). Aspects of wellbore heat transfer during two-phase flow (includes associated papers 30226 and 30970). *Old Production and Facilities.* **9**(3), 211–216.

Huang, A.A. (1975). Kinetic studies on insoluble cellulose–cellulase system. *Biotechnology and Bioengineering.* **17**(10), 1421–1433.

Humphrey, A., Moreira, A., Armiger, W., and Zabriskie, D. (1977). Production of single cell protein from cellulose wastes. *Biotechnology and Bioengineering Symposium* (United States), Univ. of Pennsylvania, Philadelphia.

Khanahmadia, M., Mitchell, D.A., Beheshti, M., Roostaazad, R., and Sanchez, L.R. (2006). Continuous solid-state fermentation as affected by substrate flow pattern. *Chemical Eng. Sci.* **61**(8), 2675–2687.

Knapp, J. and Howell, J. (1980). *Topics in Enzyme and Fermentation Biotechnology.* New York: John Wiley & Sons, vol. **4**.

Mandels, M., Hontz, L., and Nystrom, J. (2004). Enzymatic hydrolysis of waste cellulose. *Biotechnology and Bioengineering.* **16**(11), 1471–1493.

Metz, B., Kossen, N., and Van Suijdam, J. (1979). The rheology of mould suspensions. *Advances in Biochemical Engineering.* **11**, 103–156.

Mitchell, D.A., Do, D.D., Greenfield, P.F., and Doelle, H.W. (1991). A semi-mechanistic mathematical model for growth of *Rhizopus oligosporus* in a model solid-state fermentation system. *Biotechnology and Bioengineering.* **38**(4), 353–362.

Mitchell, D.A., Do, D.D., Greenfield, P.F., and Doelle, H.W. (2004). A semi-mechanistic mathematical model for growth of *Rhizopus oligosporus* in a model solid-state fermentation system. *Biotechnology and Bioengineering.* **38**(4), 353–362.

Mitchell, D.A., Von Meien, O.F., and Krieger, N. (2003). Recent developments in modeling of solid-state fermentation: Heat and mass transfer in bioreactors. *Biochemical Engineering Journal.* **13**(2), 137–147.

Moo-Young, M. and Blanch, H. (1981). Design of biochemical reactors mass transfer criteria for simple and complex systems. *Reactors and Reactions.* 1–69.

Nagel, F.J.J.I., Tramper, J., Bakker, M.S.N. and Rinzema, A. (2000a). Model for on-line moisture-content control during solid-state fermentation. *Biotechnology and Bioengineering.* **72**(2), 231–243.

Nagel, F.J.J.I., Tramper, J., Bakker, M.S.N., and Rinzema, A. (2000b). Temperature control in a continuously mixed bioreactor for solid-state fermentation. *Biotechnology and Bioengineering.* **72**(2), 219–230.

Nopharatana, M., Howes, T., and Mitchell, D. (1998). Modelling fungal growth on surfaces. *Biotechnology Techniques.* **12**(4), 313–318.

Olsson, S. and Jennings, D. (1991). Evidence for diffusion being the mechanism of translocation in the hyphae of three molds. *Experimental Mycology.* **15**(4), 302–309.

Oostra, J., Le Comte, E., Van Den Heuvel, J., Tramper, J., and Rinzema, A. (2001). Intra-particle oxygen diffusion limitation in solid-state fermentation. *Biotechnology and Bioengineering.* **75**(1), 13–24.

Pandey, A. (2003). Solid-state fermentation. *Biochemical Engineering Journal.* **13**(2), 81–84.

Peña Y Lillo, M., Pérez-Correa, R., Agosin, E., and Latrille, E. (2001). Indirect measurement of water content in an aseptic solid substrate cultivation pilot-scale bioreactor. *Biotechnology and Bioengineering.* **76**(1), 44–51.

Raghava, K. (1993). Biochemical engineering aspects of solid-state fermentation. *Advances in Applied Microbiology.* **38**, 99.

Rahardjo, Y.S., Weber, F.J., Le Comte, E.P., Tramper, J., and Rinzema, A. (2002). Contribution of aerial hyphae of *Aspergillus oryzae* to respiration in a model solid-state fermentation system. *Biotechnology and Bioengineering.* **78**(5), 539–544.

Rajagopalan, S. and Modak, J. (1995). Modeling of heat and mass transfer for solid state fermentation process in tray bioreactor. *Bioprocess and Biosystems Engineering.* **13**(3), 161–169.

Rajagopalan, S. and Modak, J.M. (1994). Heat and mass transfer simulation studies for solid-state fermentation processes. *Chemical Engineering Science.* **49**(13), 2187–2193.

Sangsurasak, P. and Mitchell, D.A. (1995). Incorporation of death kinetics into a 2-dimensional dynamic heat transfer model for solid state fermentation. *Journal of Chemical Technology and Biotechnology.* **64**(3), 253–260.

Sargantanis, J., Karim, M., Murphy, V., Ryoo, D., and Tengerdy, R. (1993). Effect of operating conditions on solid substrate fermentation. *Biotechnology and Bioengineering.* **42**(2), 149–158.

Saucedo-Castañeda, G., Gutiérrez-Rojas, M., Bacquet, G., Raimbault, M., and Viniegra-González, G. (1990). Heat transfer simulation in solid substrate fermentation. *Biotechnology and Bioengineering.* **35**(8), 802–808.

Smits, J., Van Sonsbeek, H., Tramper, J., Knol, W., Geelhoed, W., Peeters, M., and Rinzema, A. (1999). Modelling fungal solid-state fermentation: The role of inactivation kinetics. *Bioprocess and Biosystems Engineering.* **20**(5), 391–404.

Stuart, D., Mitchell, D., Johns, M., and Litster, J. (1999). Solid-state fermentation in rotating drum bioreactors: Operating variables affect performance through their effects on transport phenomena. *Biotechnology and Bioengineering.* **63**(4), 383–391.

Suga, K., Van Dedem, G., and Moo-Young, M. (1975). Enzymatic breakdown of water insoluble substrates. *Biotechnology and Bioengineering.* **17**(2), 185–201.

Van De Lagemaat, J., and Pyle, D. (2001). Solid-state fermentation and bio-remediation: Development of a continuous process for the production of fungal tannase. *Chemical Engineering Journal.* **84**(2), 115–123.

Viniegra-González, G., Favela-Torres, E., Aguilar, C.N., Rómero-Gomez, S.D.J., Díaz-Godínez, G., and Augur, C. (2003). Advantages of fungal enzyme production in solid state over liquid fermentation systems. *Biochemical Engineering Journal.* **13**(2), 157–167.

Von Meien, O.F. and Mitchell, D.A. (2002). A two-phase model for water and heat transfer within an intermittently-mixed solid-state fermentation bioreactor with forced aeration. *Biotechnology and Bioengineering.* **79**(4), 416–428.

Weber, F.J., Tramper, J., and Rinzema, A. (1999). A simplified material and energy balance approach for process development and scale-up of *Coniothyrium minitans* conidia production by solid-state cultivation in a packed-bed reactor. *Biotechnology and Bioengineering.* **65**(4), 447–458.

Weisz, P. and Prater, C. (1954). Interpretation of measurements in experimental catalysis. *Advances in Catalysis* **6**, 143.

Zhang, S., Ye, B.C., Chu, J., Zhuang, Y., and Guo, M. (2006). From multi-scale methodology to systems biology: To integrate strain improvement and fermentation optimization. *Journal of Chemical Technology and Biotechnology.* **81**(5), 734–745.

4

HISTORY OF SOLID-STATE FERMENTED FOODS AND BEVERAGES

SONG LIU, DONGXU ZHANG, JIAN CHEN

*Key Laboratory of Industrial Biotechnology, Ministry of Education,
School of Biotechnology, Jiangnan University, Wuxi, China*

YANG ZHU

Department of Biosciences, TNO Quality of Life, Ziest, The Netherlands

Contents

4.1 Introduction

The developments and changes in the food and beverage field is a microcosm of human history. Production of fermented foods and beverages is an age-old biotechnology and an important part of the history, culture, and science of human life. With the development of civilization, fermented foods and beverages became more and more indispensable to the entire population of many regions. They not only appease hunger but also provide delicious foods, have a probiotic function, and contribute to a people's cultural inheritance. Various fermented foods and beverages play important roles that are central to people's lives today.

During the fermentation of raw material, the processing of microorganisms (bacteria, yeast, and fungi) and enzymes cause many biochemical and physical changes. Generally, the style and flavor of the fermented foods and beverages in the world are the masterpieces of the microorganism's vital movement. The microorganisms needed for food fermentation have existed for billions of years. As humans and even animals evolved on Earth, they used microorganisms to preserve food components and yield attractive aromas and flavors, avoiding the food with unpleasant aromas or flavors. These were the beginning of fermented foods, including sour milk, cheeses, wines and beers, vinegar, lactic acid foods, such as sauerkraut, and hundreds of other fermented foods consumed today (Hui et al., 2004).

The origin of fermented food is lost in antiquity. We can only partly recover the stories about the development of the solid-state fermented foods and beverages. Table 4.1 shows the scheme of the known history of fermented foods (Farnworth, 2008).

Table 4.1   Milestones in the History of Fermented Foods

MILESTONE	DEVELOPMENT/LOCATION
ca. 10,000 BC to Middle Ages	Evolution of fermentation for salvaging surpluses, probably by pre-Aryans
ca. 7000 BC	Cheese and bread making practiced
ca. 6000 BC	Wine making in the Near East
ca. 5000 BC	Nutritional and health value of fermented milk and beverages described
ca. 3500 BC	Bread making in Egypt
ca. 1500 BC	Preparation of meat sausages by ancient Babylonians
2000 BC–1200 AD	Different types of fermented milks from different regions
ca. 300 BC	Preservation of vegetables by fermentation by the Chinese
500–1000 AD	Development of cereal–legume-based fermented foods
1881	Published literature on *koji* and *sake* brewing
1907	Publication of book *Prolongation of Life* by Eli Metchnikoff describing therapeutic benefits of fermented milks
1900–1930	Application of microbiology to fermentation, use of defined cultures
1970–present	Development of products containing probiotic cultures or friendly intestinal bacteria

Source: Data compiled from V.K. Joshi and A. Pandey. *Biotechnology: Food Fermentations,* Vol. 1. Educational Publishers and Distributors, New Delhi. 1999, 1–24; C.S. Pederson. *Microbiology of Food Fermentations,* AVI. Westport. CT. 1971. 1–274; IDF. Fermented Milks. *IDF Bull.,* No. 179, 16-32, 1984: E. Metchnikoff, *The Prolongation of Life,* G.P. Putnam's Sons, New York, 1908; K.H. Steinkraus, *Hand-book of Indigenous Fermented Foods,* Marcel Dekker, New York, 1983; C. Padmaja and M. George in *Biotechnology: Food Fermentations,* Vol. II, V. K. Joshi and A. Pandey, Eds., Educational Publishers and Distributors. New Delhi. 1999, 523–582.

4.2 Origin and Development of Solid-State Fermented Foods and Beverages

We cannot say with certainty where the first solid-state fermented food originated in the world, but the first consumer of fermented food actually may be animals. At the time fruit ripens, apes pick those that cannot be eaten in time and hide them in their lairs. The fruits are naturally fermented to a kind of wine with the effect of yeast. The apes were fascinated by the flavor of wine and all got drunk. It may be a mere accident when mankind first experienced the taste of fermented food. The first fermentation must have started with the storage of surplus milk, which resulted in a fermented product the next day (Farnworth, 2008). Except for drying, fermentation is one of the oldest food preservation methods. Fermentation became popular with the dawn of civilization because it could not only preserve food but also give food a variety of tastes, forms, and other sensory attributes. Slowly, people realized the nutritional and therapeutic value of fermented foods and drinks, which made fermented foods more and more popular (Farnworth, 2008)

4.2.1 Fermented Foods

4.2.1.1 Cheese   The first fermented food that can be found in archival documents would be the cheeses. The one-eyed cyclops of Homer's epic had learned the preparing method of cheese. The sign of cheese had been found in the archeology research of ancient Egypt and Mesopotamia. Cheese is also the first artificial food, a story even surviving about its origin. Back in 10,000 BC, when a shepherd in the Middle East crossed the desert on a long journey, milk was taken as a supply into a container made of a sheep's stomach. With the effect of high temperature in the desert and the chymosin on the stomach, the milk curdled. That is a story of accidental discovery. Actually, at about 9000 BC, cows were worshipped and milked in Libya. This was proved by the rock drawings discovered in the Libyan desert by archeology researchers (Pederson, 1979). In this warm climate, the growth of natural *Lactobacillus* caused the acidifying of milk that may be the origin of those acid-coagulated cheese varieties produced today, such as quark, cream cheese, and cottage cheese (Wood, 1998). After that, people found that some portion of the stomach of a calf can coagulate the milk more effectively, and various flavors occurred in the storage.

Most scholars agree that the art of cheese making traveled from Asia Minor to Europe, where it flourished in the hands of the Romans. It was the Roman culture that developed the art of cheese making as we know it today. The earliest record on cheese is from the development period of ancient Rome. At that period, cheese was a reward and a fixed-component diet of the heroes of ancient Rome. Like other areas of knowledge, Roman cheese-making expertise spread with their empire throughout Europe. With the collapse of the Roman Empire, many secret recipes were lost, and only a part of their method was known in some temples and distant countrysides. When the Mayflower navigated to America in 1620, cheese making spread in the new land.

By the 19th century, cheese had become popular, and production moved from the farm to the factory. Today, there are specific categories of cheese: hard cheese, like Parmesan; semihard cheese, like Swiss; semisoft cheese, like Muenster and Mozzarella; soft ripened cheese, like Brie and St. Andre; and soft cheese, like cream cheese and Chevre.

4.2.1.2 Bread   Bread has been one of the principal forms of food for man for a long time, dating back to the Neolithic era. Bread making was found by archeologists in ancient Egypt as early as 1000 BC (Wood, 1996). The Egyptians, in approximately 2600 BC, were lucky enough to have a sufficient amount of wild yeasts in the air from their beer brewing to accidentally discover its uses in leavening bread. Workers were often paid in loaves of bread. Paintings in the pyramids show that the dead were buried with loaves of bread to provide sustenance in the afterlife. The British Museum has one of these loaves—4000 years old (http://www.essortment.com/history-bread-41729.html). This kind of bread was made from an ancient wheat named "Emmer" (Farnworth, 2008). The workmanship of bread was taken back to Greece, where bread baking flourished, by Greek sailors and merchants. Rome took over the enterprise after their conquest of Greece, and in 150 BC formed the first baker's guilds. The Romans were probably the first to commercialize bread making by using yeasts separated from wine (Farnworth, 2008). The ancient Greeks and Romans knew bread both leavened and unleavened for a staple food. Even today, there is unleavened bread in some areas.

4.2.2 Fermented Condiments

4.2.2.1 Soy Sauce   Soy sauce is a liquid food condiment prepared from fermented rice or wheat and soybean with the help of molds, bacteria, and yeasts (Farnworth, 2008). Soy sauce dates back about 2,000 years, during the Zhu dynasty. In this period, meat and fish were the ingredients of a catsup. There is a record about the processing of soy sauce making in the book *Qi Min Yao Shu* from the period of Northern Wei. The word "soy sauce" first appeared in the book *Shang Jia Qin Gong* in the Song Dynasty. In AC 755, the processing of soy sauce making was introduced to Japan by Master Jianzhen. After that, soy sauce was introduced into Korea, Vietnam, Thailand, Malaysia, and the Philippines. Records of the Dutch East India Company list soy sauce as a commodity in 1737, when 75 large barrels were shipped from Dejima, Japan, to Batavia (present-day Jakarta) on the island of Java. Thirty-five barrels from that shipment were shipped to the Netherlands. This may be the first recorded trade of soy sauce from Asia to Europe. Today, the production of soy sauce in the world is about 8 million tons per year with 5 million tons being made on the China mainland alone.

4.2.2.2 Vinegar   People from many lands in the world have used vinegar in many different ways, for thousands of years. Around 5000 BC, the Babylonians were using the fruit of the date palm to make wine and vinegar to be used as food and a preservative or pickling agent. Vinegar residues were found in urns from ancient Egypt and have been traced to 3000 BC. The first written history of vinegar in China dates to 1200 BC. In ancient Rome different types of vinegar were made from wine, dates, figs, and other fruits and placed in bowls for the dunking of bread. The making of rice vinegar in China began 3000 years ago. During biblical times, vinegar was used to flavor foods, drunk as an energizing drink, and used as an herbal. In ancient

Greece, at around 400 BC, Hippocrates, who is considered the father of modern herbal medicine, prescribed apple cider vinegar mixed with honey for a variety of ailments, including coughs and colds. Besides its medicinal uses, vinegar was valued as a condiment.

The processing of vinegar making was likely introduced to other countries from China and the Middle East. Japanese obtained the knowledge about vinegar from China in 369–404 AC. Thereafter, it went to Britain in 100 BC. During the Middle Ages vinegar was well known to the European alchemists of the time. The vinegar industry in Europe flourished during the Renaissance (http://www.apple-cider-vinegar-benefits.com/vinegar-history.html) (as shown in Table 4.2).

Table 4.2 Milestones in the History of Vinegar

YEAR	EVENT	RESULT
794–1185	Samurai warriors of Japan used vinegar as a tonic	They believed the vinegar tonic gave them power and strength
1735-1826	John Adams, the second president of the United States, drank apple cider every morning for breakfast	This showed how the components of apple cider vinegar may support longer life (91 years old)
1909	Dr. Jarvis began studying herbal herbals and folk remedies after he started practicing herbal medicine	He helped to treat people with apple cider vinegar for a variety of ailments
1912	Dr. Alexis Carrel began an experiment that successfully kept cells of an embryo chicken heart alive for 30 years	This showed the importance of apple cider vinegar in health and longevity
1920s	Apple cider vinegar was made and drunk more than any other fruit juice in the United States	This implied that apple cider vinegar provides vim and vigor
1958	The book *Folk Herbal: A Vermont Doctor's Guide to Good Health* was published	The book chronicled the apple cider vinegar remedies studied by Dr. Jarvis
1968	The Vinegar Institute began on January 17th	This helped vinegar producers protect their rights
1973	Balsamic vinegar from the provinces of Modena and Reggio Emilia were introduced to America by Marcella Hazen, a cooking teacher and author	This showed Americans there are other kinds of vinegar besides apple cider vinegar
1990s	Self-help books published such as *Apple Cider Vinegar: Miracle Health System*	People educated on more ways to live healthier lives
1999	The International Vinegar Museum was opened on June 4 in Roslyn, South Dakota	New source of information available on the benefits of vinegar

4.2.3 Fermented Beverages

4.2.3.1 Wine Wine has a long history, and is closely intertwined with the history of agriculture, civilization, and trade. The archeological evidence in Iran shows that the origin of wine there dated from 6000 to 5000 BC. There is a story about the inventor of wine named Du Kang in the Xia Dynasty in China. Even today, the words "Du Kang" still mean wine in Chinese. Evidence from archeological sites in Greece, in the form of 6500 years old grape remnants, represents the earliest known appearance of wine production in Europe. In Egypt, wine played an important role in ancient ceremonial life. A thriving royal wine-making industry was established in the Nile Delta following the introduction of grape cultivation from the Levant to Egypt at 3000 BC.

At the beginning, wine was not distilled. The first evidence of distillation comes from Babylonia and dates from the 2nd millennium BC. Special shaped clay pots were used to extract small amounts of distilled alcohol through natural cooling for use in perfumes. By the 3rd century AD, alchemists in Alexandria, Egypt, may have used an early form of distillation to produce alcohol for sublimation or for coloring metal. In the 12th century distilled alcoholic beverages first appeared in Europe among alchemists who were more interested in brewing medical elixirs than in making gold from lead. The evidence of archeological sites in China shows that there is a distiller made in bronze in the Eastern Han Dynasty.

Now, there are many different distilled wines: brandy, vodka from Russia, whiskey, rum from Americas, gin from Holland, mezcal from Mexico, Chinese rice wine, and sake from Japan.

4.2.3.2 Fermented Tea Tea was discovered in 2737 BC by Shen Nong, also known as Yan Di or Shen Nong Shi in Chinese mythology. At the same time, he was a scholar, the father of agriculture, and the inventor of Chinese herbal herbal. In 479 AD, Turkish traders bargain for tea on the border of Mongolia. In 593 AD, Buddhism and tea journeyed from China to Japan. Japanese priests studying in China carried tea seeds and leaves back. And in the Tang Dynasty, tea became a popular drink in China for both its flavor and medicinal qualities. In 780 AD, Lu Yu wrote the first definitive book on tea, the *Ch'a Ching*.

Europeans learned about tea when a Venetian author credited the lengthy lives of Asians to their tea drinking in the 16th century. The Europeans imported tea from Asia by the Dutch East India Company in 1610 AD.

Tea is one of the most consumed beverages worldwide. In China and many Southeast Asian countries, green tea is preferred. In Western countries and the rest of the world, the consumption of black tea is the highest. Another kind of tea is *kombucha*, which is a sweet–sour tea beverage made actually from a tea extract supplemented with sugar and fermented with yeast and acetic acid bacteria. The fermentation process is necessary to produce black tea and kombucha. Black tea can be further subdivided into naturally oxidized tea (although the term *fermentation* is often used), and microbial fermented tea. The fermentation process for producing most black teas is actually an oxidation process catalyzed by the enzymes that are originally present in tea leaves (Mo et al., 2008).

Fermented tea can be divided into completely fermented tea and semifermented tea. Black tea or red tea is completely fermented tea and the result of complete oxidation of the leaf before being fired. Oolong tea is semifermented tea and gently rolled after picking, allowing its essential oils to slowly oxidize.

4.2.4 Fermented Vegetables and Meat

4.2.4.1 Fermented Vegetables   It is not clear when people began to conserve vegetables by fermentation. It seems that the development of fermented fruit and vegetable products took place from the time ancient people started collecting and storing food. It is said that vegetable fermentation started in China at the construction of the Great China Wall in the 3rd century (Farnworth, 2008).

Fermented vegetables in Europe go back to ancient Greece and the Roman Empire. And in the 13th century, Chinese sauerkraut was introduced into Europe by the Mongols. With the migration of the Jews, sauerkraut became accepted by the West Europeans. Sauerkraut from Germany is the most popular fermented vegetable. Today, sauerkraut is an industry that uses the knowledge of modern microbiology and fermentation technology. In East Asia, *kimchi* from Korea has a long tradition. In India, fermented cucumber is one of the oldest vegetables and has more than 3000 years' history (Farnworth, 2008).

4.2.4.2 Fermented Meat   Our ancestors managed to hunt even the largest of animals, but most of the meat was left to rot away. Meat preservation was always of primary concern. Meat that was pounded into thinner slices and then left in the sun for drying would be preserved for future use. Before 2200 BC, the Chinese discovered that salting meat was an effective way of preserving it. By 1500 BC, the Egyptians were using this technique, and the Greeks and Romans were salting meats before the Christian era. Plenty of salt was rubbed into hams, which were left out on racks for a few weeks. As the hams were left to dry out, more moisture was removed, which made them microbiologically stable. Of course, there was more to it, but basically Spanish hams were made that way for hundreds of years. And after that, people found that salts dug from different places can develop particular flavors because of the impurities, such as nitrate. Over time, some unique varieties of meat products developed in other regions of the world. All these fermentations are dependent on natural flora present in the raw materials, which decrease the pH and increase the shelf life of the meat. As with many other fermented foods, intensive research into the microbiology and the chemistry of sausage ripening was triggered when traditional empirical methods of manufacture no longer met the requirements of large-scale, consistent-quality, low-cost industrial production.

4.3 Development of SSF Food and Beverages with Modern Technology

4.3.1 Pasteurization

Pasteurization is a process that slows microbial growth in food. The process was named after its creator, the French chemist and micro-biologist Louis Pasteur. The first pasteurization test was completed by Louis Pasteur and Claude Bernard on April 20, 1864. The process was originally conceived as a way of preventing wine and beer from souring (Carlisle and American, 2005). Pasteurization is not intended to destroy all pathogenic microorganisms in the food or liquid. Instead, pasteurization aims to reduce the number of viable pathogens so they are unlikely to cause disease (assuming that the pasteurized product is stored as indicated and consumed before its expiration date). At first, pasteurization was used to prevent wine spoilage. Now, it's widely used in the production of milk, beer, butter, and some other foods. It has given a powerful boost to the food industry. Commercial-scale sterilization of food is not common because it adversely affects the taste and quality of the product. Certain food products are, however, processed to achieve a state of commercial sterility (Montville and Matthews, 2008).

4.3.2 Microorganism Breeding

Microorganism breeding technology has a very special role in modern biotechnology, especially in the fermentation industry. Microorganism breeding can lead to increased quality and production of fermented food and beverages. The first stage of microorganism breeding is natural selection. The desired strain is obtained through the selection of mutations. This is the only selected method in the long history of fermentation. In 1927, the mutation effect of x-rays was found by Mille. After that, many other methods causing mutations were found. There are many mutation methods, including chemical, physical, and biological. With the development of molecular biology in the latter half of the 20th century, genetic engineering was introduced into microorganism breeding, but the safety of this technology in food production needs further careful study.

4.3.3 Process Monitoring and the Bioreactor

Solid-state fermented food and beverages produced by manual labor has a very long history. In the 19th century, some fermented foods were industrialized, such as sauerkraut in Germany. With the development of sciences on the bioreactor, biological process, and computer technology, more and more new technologies have been used in solid-state fermentation. Many researchers focused on the substrate, process monitoring, and bioreactor design. In 1992, some showed that the size of the substrate has an effect on fermentation (Amin, 1992). The pretreatment of the substrate has a good effect on the fermentation processing. Research on the parameter of processing control has been on water activity, mass transfer, and air, temperature, and pH control (Chahal et al., 1992). Ventilation equipment is the key in aerobic solid-state fermentation. Some mathematical models of process have been established until 1995 (Sangurasakp and Mlttchelld, 1998). Based on the research of mathematical models, some bioreactors were created such as the tray fermentor, rotary bioreactor, packed-bed bioreactors, and column bioreactors. Because of the special requirements of the food industry, few bioreactors were used commercially.

4.3.4 Modern Testing Technologies

For the past 45–50 years, there was no testing technology except for the evaluation at the preliminary stage of solid-state fermented (SSF) foods and beverages manufacturing. With the development of instrument analysis technology, more and more compositions of SSF food and beverages were determined. Application of gas chromatography on the analysis of flavors in the SSF food and beverages were studied (Jeffrey et al., 2008). In 2010, high performance liquid chromatography was used to analyze the trace substrate (Marchand and De Revel, 2010). Recently, new methods, such as GC-MS, ICP-MS, spectrophotometer, NIR, AAS, and NMR were used to determine the sample qualitative and quantitative of SSF food and beverages. With modern testing technologies, many key compositions of SSF food and beverages were found, and some standards were consequently established. Production technologies of SSF food and beverages were optimized. Therefore, testing technologies are all very important to the industrialization of SSF food and beverages.

4.4 Concepts, Categories, and Characteristics of Solid-State Fermented Foods and Beverages

4.4.1 Introduction

Microorganisms have long played a major role in the production of food and beverages (Couto and Sanromán, 2006). Solid-state fermentation developed in the Orient is a very useful fermentation method that is presently used to produce a variety of foods, beverages, and related products (Vogel, 1983). All over the world, the preparation of solid-state fermented foods is a widespread tradition (Steinkraus, 1984). Owning to spatial variation in materials, climate, custom, and other environmental factors, unique SSF products have been developed in many parts of the world (Farnworth, 2008). Numerous kinds of SSF foods and beverages are being produced industrially in China, Japan, Korea, and other Oriental countries. Miso, shoyu, ontjam, kimchi, beet, tempeh, and fermented fish and meat are good examples.

Why are SSF food and beverages so important in our daily diet? The answer is clear—they have the potential for numerous benefits. First of all, solid-state fermentation often makes the food more nutritious, supplying added protein, minerals, and other nutrients that provide variety and nutritional fortification to otherwise starchy, bland diets (Steinkraus, 1984). It then tends to preserve the food, increasing its shelf life and lowering the energy costs involved in refrigeration. Besides, after fermentation, the food is more digestible (Farnworth, 2008). It reduces cooking time, which in turn reduces the energy consumption required for the preparation of foods (Steinkraus, 1984). In addition to its nutritive and preservative effects, solid-state fermentation enriches the diet through the production of a diversity of flavors, aromas, and textures (Farnworth, 2008).

With more and more alternative food preservation methods available today, fermentation (especially SSF) is no longer the most pressing requirement for preserving food in most countries. However, nutritional requirements are gaining in status on account of further advances in our understanding of nutrition and food science. A long time ago, there was no confirmed knowledge of the nutrition, technology, and quality of the ingenious solid-state fermented foods and beverages. On the contrary, plenty of articles and books related to SSF beverages and foods have rapidly increased in the last 20 years (Blandino et al., 2003). This has led to significant encouragement of and appreciation for SSF foods and beverages for their inherent health-promoting effects.

SSF foods are traditionally prepared in the home using relatively simple techniques, offering an economical way to preserve food materials, and producing a variety of tasty products (Farnworth, 2008). More specialized products are in development resulting from the rapid pace and complexity of modern life; the SSF system gradually has formed a standardized approach to making traditional fermented foods, such as using selected starter cultures, eliminating harmful toxic compounds, and preventing the growth of spoilage or food poisoning organisms (Steinkraus, 1984).

4.4.2 Categories of Solid-State Fermented Foods and Beverages

Solid-state fermented foods and beverages are produced worldwide using various manufacturing techniques, raw materials, and microorganisms (Table 4.3). Ten classes of solid-state fermented products may be recognized depending on the materials involved in the production, viz., dairy products, cereal products, beverages, legumes, fish products, meat products, fruit and vegetable products, additives and supplements, mushrooms, and many other mixed products, which emerged in very early times (Farnworth, 2008). Of these, cereal products, dairy products, legumes, and beverages are the most common.

Three popular cereals are wheat, rice, and maize. Bread is the most common type of fermented cereal product. Wheat dough is fermented by *S. cerevisiae* along with some lactic bacteria. Idli, dosa, vada, dhokla, and papadam are some common examples of Indian fermented cereal foods. These foods contain mixtures of wheat and legume flours, which are fermented by *Streptococcus*, *Pediococcus*, or *Leuconostoc*. Cheese constitutes one of the largest groups of fermented dairy products, besides yogurt. Yogurt is made from milk by lactic acid bacteria, often a mixture of *Lactobacillus bulgaricus* and *Streptococcus thermophiles* (Zhu et al., 2003). A great amount of food derived from legumes is made by SSF processes. Both China and India have great amounts of fermented legumes. Adjacent countries, such as Indonesia and Japan also have large numbers of consumers of fermented legumes, principally soybeans.

4.4.3 Variety of Flavors

One of the most important reasons that the solid-state fermented foods and beverages exist is that they can improve the qualities of foods and beverages by providing a pleasant taste and intense flavor. This is probably owing to their high content of free amino acids, peptides, calcium, peptides, and enzymes, resulting from the fermentation

Table 4.3   Common SSF Foods and Beverages

CATEGORY	PRODUCT NAME	CHARACTERISTICS AND USE	REGIONS
Dairy products	Cheese	Produced by coagulation of the milk protein casein. Cheese consists of proteins and fat from milk, usually the milk of cows, buffalo, goats, or sheep. Cheese is valued for its portability, long life, and high content of fat, protein, calcium, and phosphorus	Europe, Central Asia or the Middle East
Dairy products	Yogurt	Dairy product produced by bacterial fermentation of milk. Fermentation of lactose produces lactic acid, which acts on milk protein to give yogurt its texture and its characteristic tang	All through the world
Cereal products	Lao-chao	Paste, soft, juicy, glutinous—consumed as such, as dessert, or combined with eggs, seafood	China
Legumes	Miso	A traditional Japanese seasoning produced by fermenting rice, barley and/or soybeans, with salt and the fungus kōjikin, the most typical miso being made with soy	Japan
Legumes	Sufu	Sufu is a Chinese soybean cheese-like product obtained by solid-state fungal fermentation and ripening of tofu. This fermented product with its characteristic flavor has been widely consumed by Chinese people as a salty appetizer for many centuries	China
Legumes	Natto	It is a traditional Japanese food made from soybeans fermented with *Bacillus subtilis*. As a rich source of protein, natto formed a vital source of nutrition for breakfast	Japan
Beverages	Boza	It is a malt drink, made from wheat, millet, maize, and other cereals. Thick, sweet, slightly sour beverage	Albania, Turkey, Bulgaria, Romania
Beverages	Korean traditional wine	It is brewed in classical ways using nuruk, rice, flour, yeast, and some medicinal plants or herbs	Korea
Fruit and vegetable products	Kimchi	A traditional Korean fermented dish made of vegetables with varied seasonings	Korea

process. Solid-state fermentation processes often result in profound changes in flavor relative to the starting ingredients.

Taking soy-fermented products as an example, proteolysis is the most fundamental biochemical reaction occurring during the solid-state fermentation of soy substrates. The degradation products not only have a considerable influence on the nutritional values, but also contribute indirectly to the taste characteristics, in some cases serving indirectly as precursors of aromatic products (Zhang and Tao, 2009). In the case of rice wine, the wealth of compounds formed after solid-state fermentation provide pungent and solvent-like, sweet, fruity, buttery, and alcohol aromas. SSF is a key link within the production processes. It decides the flavor and quality, which are the essentially basic characters of high-quality alcoholic beverages.

4.4.4 Nutritional Values and Physiological Properties

Several means of processing including cooking, milling, sprouting, and fermentation can improve the nutritional properties of foods and beverages, and the best one probably is fermentation, which enhances nutritional quality through increased availability of vitamins, biosynthesis of essential amino acids and proteins, improved bioavailability of minerals, and the degradation of antinutritional or toxic factors (Steinkraus, 1984). Besides, solid-state fermentation results in a decrease in the content of poly- and oligosaccharides, which are nondigestible as well as carbohydrates (Chavan and Kadam, 1989).

Recently, more and more SSF foods have been found to have various benefits to health and physiological properties, in addition to their taste and basic nutritional values. Most Chinese fermented foods and beverages are good examples, such as sufu, douchi, red yeast rice, and rice wine, etc. (Farnworth, 2008).

The water-soluble protein content in sufu is distinctly higher than that of tofu (Liu and Chou, 1994; Wang, 2002). Many physiologically active components, such as soybean peptides, B vitamins, nucleic glycosides, and aromatic compounds which do not exist in unfermented soybeans, are produced in the fermentation of sufu. The vitamin B12 content, an essential nutrient for the nervous system, in *chou* ("strong smelling") sufu has been shown to be much higher (9.8 to 18.8 mg/100 g) than that in red sufu (0.42 to 0.78 mg/100 g), suggesting a high activity of microorganisms during fermentation of the sufu (Chou and Hwan, 1994).

Many researchers show that SSF foods and beverages are associated with reduced risks of cardiovascular diseases and some kinds of cancers. The antioxidative and antihypertensive effects have also been reported (Wang et al., 2003, 2004). There has been growing interest in the study of the physiological functionalities of Korean traditional wine (Yakju, 2005), due to the beneficial effects of prevention and treatment in blood circulatory trouble, thrombosis, and arteriosclerosis, etc., as a folk herbal (Yakju, 2005). The outstanding medicinal qualities of miso have been supported by scientific research. It is said that those eating miso soup daily suffer significantly less from stomach cancer and heart disease, and there are substances in miso that neutralize the effects of some carcinogens (Ito, 1972; Hirayama, 1981).

Many of the beneficial effects of physiological foods are due to the bioactive compounds found in the raw materials themselves. However, the solid-state fermentation used in the production imparts added benefits, which do not naturally exist in raw materials.

4.5 Materials and Methods of SSF Food and Beverages

4.5.1 Preparation of Materials

The materials of SSF food and beverages are mostly agricultural and farm-derived products, which greatly decreases the costs of fermentation compared with liquid-state fermentation. The major components of materials of SSF food and beverages can be identified as starch-bearing material, protein material, auxiliary materials, and plugging compounds, each playing different roles in the fermentation process (Macneil and Harvey, 1990).

Starch-bearing material contains cereals (including rice, corn, millet, and wheat) and tuber crops (including potatoes, sweet potatoes, and cassava). Besides, some fruits and vegetables rich in carbohydrates also can be used to ferment alcohol and vinegar, including apples, oranges, bananas, pears, jujubes, watermelons, hawthorns, and kelps. These materials provide a carbon source for the growth and metabolism of microorganisms. After a series of reactions (biochemical, physical, and chemical) in the course of fermentation, the materials are degraded into various kinds of saccharides, organic acids, alcohols, esters, and pigments. The starch-bearing material is always used in the production of SSF alcohol and vinegar.

People often use protein materials to ferment condiments such as soy sauce, lobster sauce, and preserved bean curd, which are rich in amino acids. The protein materials for SSF food and beverages include soybeans, horsebeans, peas, bean cakes, and tofu. They provide a nitrogen source for the growth and metabolism of microorganisms. During the fermentation, of complex biochemical reactions, the protein materials will be degraded to peptides, amino acids, alcohols, and pigments (Gu, 2008).

However, there are also negative factors in the SSF, for example, inferior aeration and low utilization of the substrate. Therefore, auxiliary materials are often added to improve the mass and heat transfer of mediums, such as bran, chaff, corn husk, straw, corn cobs, pumice, and porous glass fiber (Vogel and Tadaro, 1997).

4.5.2 Fermentation Processes

4.5.2.1 Initial Stage of Hydrolysis and Degradation of Macromolecular Compounds The materials in SSF are often rich in macromolecular compounds such as starch, cellulose, pectin, fat, and protein. Therefore, it is necessary to hydrolyze them from the macromolecular compounds to smaller molecular compounds, which are conveniently absorbed by microorganisms. The degradation needs various kinds of extracellular enzymes secreted by fermentation microorganisms, or is directly achieved by physical, chemical, and enzymatic methods (Tsao, 1983).

4.5.2.2 Intermediate Stage on Metabolic Products During this course, microorganisms in solid fermentation materials are gradually forming and secreting the final main products through further metabolism. For instance, acetic acid is produced greatly by acetic acid bacteria in the SSF of vinegar; ethanol is generated in a large quantity by yeast in the SSF of alcohols. Besides, other products in minor or trace quantities are accumulated in this stage, such as aromatic compounds, esters, organic acids, alcohols, and amino acids (Gu, 2008).

4.5.2.3 Later Stages of SSF Flavor Development The production process of SSF food and beverages often have a storage period, which is also called the later stage of ripening. In this stage, the unique flavor substances of fermented products are accumulated. This stage provide SSF food and beverages their own particular flavors and characters. There are significant differences in processing time of this stage for different fermented productions. For instance, the fermentation time of wine is over hundreds of days or years, while yogurt's is only several days (Cheng et al., 2008; Liang et al., 2009).

Actually, it may be difficult to sharply divide the three stages of SSF, for there is no crucial distinction among them. To try for such a distinction is convenient for us to understand more generally SSF food and beverages. Only a brief introduction about SSF food and beverages will be discussed; detailed processes will be described in later chapters (Tsao, 1983).

4.5.3 Key Points of Process Control in SSF Food and Beverages

4.5.3.1 Contamination Control   Sterilization is usually a vital part of the SSF. The contaminating microorganisms will compete with the production organisms for nutrients. They may even outgrow and displace the production organisms. Besides, contamination will also produce, in some, special products that will be poisonous or hard to separate from the final products. Therefore, contamination must be avoided, but as compared with liquid-state fermentation, the sterilization of SSF is difficult; that is why sterilization in SSF attracts much greater attention (Qiu, 2008).

The methods of SSF sterilization include moist heat, dry heat, filtration, chemical means, and radiation. For different materials of SSF, the suitable method should be chosen. Currently, moist heat is the most common sterilization option used for fermentation vessels and materials (Gu, 2008; Zhang, 2009). Filtration is the common choice for air supplement and very heat-sensitive medium constituents.

4.5.3.2 Production Strains   Production of SSF food and beverages depends on availability of the pure, stable microbial cultures. However, the deterioration of production strains would occur during the many steps (7–10) of the subculturing of inoculums. Mishandling of organisms at later stages may also cause the stain-deterioration. Therefore, a variety of techniques are used in the preservation of microorganisms to avoid strain–deterioration.

Several methods of preservation strains involve slowing down their metabolism. For example, desiccation involves removal of water from a culture and preventing its rehydration. During freezing, the water will be frozen that is unavailable to the cells, resulting in the freezing of microorganisms. In addition, the strains will be stored by a subculture on a suitable medium, using incubation to achieve growth and storage.

Actually, for some physiological reasons (including gene mutation, segregation of characters), the degradation of bacteria can hardly be avoided. Therefore, some methods of rejuvenation of production strains were developed, such as pure culture isolation, elimination of recession strains, and so on.

4.5.3.3 Granularity of Mediums   Generally speaking, the small particle size of the medium material can provide a bigger surface area than the large particle size for microbial growth, which can obviously improve biological reaction rate of the SSF. At the same time, small particle size can also lead to clotting in mediums. On the other hand, larger particle size can help to improve the transfer efficiency of mass and heat, and can also provide favorable conditions in breathing and ventilation. Therefore, it is necessary to choose an appropriate particle size in SSF (Wu, 2006; Qiu, 2008).

4.5.3.4 Temperature and Moisture Content of Mediums   During the SSF process, large quantities of heat are produced by microorganisms. In fermentation, the heat of metabolism is accumulated continuously, and the solid materials in the medium can be swelled up, which will lead to a decrease of the porosity of the solid materials and hinder the diffusion of the heat transfer. Over 3°C/cm of temperature gradient will then be formed in the medium. The metabolism of a microorganism is inhibited severely in SSF, which will finally lead to a decrease in the fermentation yields (Wu, 2006).

Moisture content also plays an important role in SSF, and the dosage of water will influence it significantly. A too low moisture content limits the growth of microorganisms. But a too high moisture content inhibits the dissolved oxygen levels in mediums, which can also affect the normal metabolism of microorganisms.

Forced ventilation is one of the main methods to cool temperature by evaporating water. This method can not only import oxygen and export carbon dioxide, but also take amounts of heat away from the mediums. However, the dry sterilized air can also influence the moisture content. To avoid drying of materials during the forced ventilation, sterilized water will be sprayed on the mediums regularly.

Therefore, coupling control of ventilation, temperature, and humidity is currently well used in large-scale SSF.

4.5.3.5 Maturity of SSF Flavor To prolong the time of fermentation is always applied in the SSF of soy sauce (Qin and Tian, 2001). It can fully exert all kinds of enzyme activities, which can improve the utilization of raw material and promote the color, aroma, and taste of soy sauce (Li and Luan, 2002). Besides, other technologies such as fermentation with multiple strains, the application of new bioreactors are also used to control the maturity of SSF.

Generally, these kinds of mature processes are used in the fermentation of vinegar. One process is aging of grains vinegar. In this process salt is added into grains vinegar, before it is moved to the pool or vinegar urn with compaction of fermented grains and salt layer inside (Lin, 2001). This SSF mixture will be stored more than one month. Another process is the concentration of raw vinegar via weather exposure for several months, preferably a year. The third way is the sealing of finished vinegar with a storage of more than one year (Li, 2009).

References

Amin, G. (1992). Conversion of sugar-beet particles to ethanol by the bacterium *zymomonas mobilis* in solid-state fermentation. *Biotechnology Letters.* **14**(6), 499–504.

Blandino, A., Al-Aseeri, M.E., Pandiella, S.S., Cantero, D., and Webb, C. (2003). Cereal-based fermented foods and beverages. *Food Research International.* **36**(6), 527–543.

Carlisle, R. and American, S. (2005). *Scientific American Inventions and Discoveries: All the Milestones in Ingenuity—from the Discovery of Fire to the Invention of the Microwave Oven.* New Jersey: Wiley.

Chahal, P.S., Chahal, D.S., and André, G. (1992). Cellulase production profile of *trichoderma reesei* on different cellulosic substrates at various pH levels. *Journal of Fermentation and Bioengineering.* **74**(2), 126–128.

Chavan, J.K. and Kadam, S.S. (1989). Critical reviews in food science and nutrition. *Food Science and Biotechnology.* **28**, 348–400.

Cheng, S.M., Shi, X.Y., and Xu, C. (2008). Exploration on optimization of the technology of making poria cocos yoghurt. *Journal of Anhui Agricultural Sciences.* **19**, 8297–8300.

Chou, C.C. and Hwan, C.H. (1994). Effect of ethanol on the hydrolysis of protein and lipid during the ageing of a Chinese fermented soya bean curd—sufu. *Journal of the Science of Food and Agriculture.* **66**(3), 393–398.

Couto, S.R. and Sanromán, M.A. (2006). Application of solid-state fermentation to food industry: A review. *Journal of Food Engineering.* **76**(3), 291–302.

Farnworth, E.R. (2008). *Handbook of Fermented Functional Foods.* Boca Raton: CRC Press.

Gu, G.X. (2008). *Wine Brewing Technology.* Beijing: China Light Industry Press.

Hirayama, T. (1981). Operational epidemiology of cancer. *Journal of Cancer Research and Clinical Oncology.* **99**(1–2), 15–28.

Hui, Y.H., Meunier-Goddik, L., Hansen, S., Josephsen, J., Nip, W.-K., Stanfield, P.S., and Toldra, F. (2004). *Handbook of Food and Beverage Fermentation Technology.* Boca Raton: CRC Press.

Ito, K. (1972). Achievements in the study of natural compounds in Japan. *Farmatsiia.* **21**, 70–75.

Jeffrey L, G., Timothy H.S., and Mary Anne, D. (2008). Characterization of volatile compounds contributing to naturally occurring fruity fermented flavor in peanuts. *Journal of Agricultural and Food Chemistry.* **56**(17), 8096–8102.

Li, J. and Luan, J. (2002). Enhancement of soy sauce flavor by improvement of post-ripening process. *China Brewing.* **2**, 28–29.

Li, X. (2009). Research of raising vinegar color, aroma, taste, body. *China Science and Technology Review.* **29**, 199–201.

Liang, Y.P., Liu, X.Z., Li, S.Y., Xu, L., Liang, X.J., and Li, J.-M. (2009). Effects of different aging methods on the quality of grape wine. *Liquor-Making Science and Technology.* **7**, 43–46.

Lin, Z. (2001). Research on improving low salt-solid state fermented soy sauce quality and flavor. *China Brewing.* **6**, 1–4.

Liu, Y.H. and Chou, C.C. (1994). Contents of various types of proteins and water soluble peptides in sufu during aging and the amino acid composition of tasty oligopeptides. *Journal-Chinese Agricultural Chemical Society.* **32**, 276–276.

Macneil, B. and Harvey, L. (1990). *Fermentation: A Practical Approach.* Oxford: Oxford University.

Marchand, S. and De Revel, G. (2010). A hplc fluorescence-based method for glutathione derivatives quantification in must and wine. *Analytica Chimica Acta.* **660**(1), 158–163.

Mo, H., Zhu, Y., and Chen, Z. (2008). Microbial fermented tea: A potential source of natural food preservatives. *Trends in Food Science and Technology.* **19**(3), 124–130.

Montville, T.J. and Matthews, K.R. (2008). *Food Microbiology: An Introduction.* Washington: American Society for Microbiology.

Pederson, C.S. (1979). *Microbiology of Food Fermentations.* Westport: AVI Publishing Company.

Qin, M. and Tian, F. (2001). Discussion on the development direction of soy sauce production technology. *Food Research and Development.* **22**, 21–22.

Qiu, L. (2008). *Principle and Application of Solid State Fermentation Engineering.* China Light Industry Press, Beijing.

Sangurasakp, K. and Mitchell, D. (1998). Validation of a model describing two dimensional dynamic heat transfer during solid-state fermentation in packed bed bioreactors. *Journal of Biotechnology and Bioengineering.* **60**, 793–794.

Steinkraus, K. (1984). Solid-state (solid-substrate) food/beverage fermentations involving fungi. *Acta Biotechnologica.* **4**(2), 83–88.

Tsao, G.T. (1983). *Annual Reports on Fermentation Processes.* Waltham: Academic Press.

Vogel, H.C. (1983). *Fermentation and Biochemical Engineering Handbook: Principles, Process Design, and Equipment.* Park Ridge: Noyes Publications.

Vogel, H.C. and Tadaro, C.L. (1997). *Fermentation and Biochemical Engineering Handbook—Principles, Process Design, and Equipment* (2nd edition). New York: William Andrew Publishing.

Wang, J.H. (2002). The nutrition and health function of sufu: A fermented bean curd. *China Brewing.* **4**, 4–6.

Wang, L., Li, L., Fan, J., Saito, M., and Tatsumi, E. (2004). Radical-scavenging activity and isoflavone content of sufu (fermented tofu) extracts from various regions in China. *Food Science and Technology Research.* **10**(3), 324–327.

Wang, L., Saito, M., Tatsumi, E., and Li, L. (2003). Antioxidative and angiotensin i-converting enzyme inhibitory activities of sufu (fermented tofu) extracts. *Japan Agricultural Research Quarterly.* **37**(2), 129–132.

Wood, B.J. (1998). *Microbiology of Fermented Foods.* London: Blackie Academic & Professional Londres.

Wood, E. (1996). *World Sourdoughs from Antiquity.* Berkeley: Ten Speed Press.

Wu, Z. (2006). *Solid State Fermentation Technology and Application.* Beijing: Chemical Industry Press.

Yakju (2005). Physiological functionalities of Korean traditional wine. *Annual Meeting and International Symposium Microorganisms and Human Well-Being.*

Zhang, L. (2009). Research on Control System of Pure Breed Solid State Fermentation. Nanjing, Nanjing Forestry University. Master: 72.

Zhang, Y.F. and Tao, W.Y. (2009). Flavor and taste compounds analysis in Chinese solid fermented soy sauce. *African Journal of Biotechnology.* **8**(4), 673–681.

Zhu, Q.J., Wang, D., Chen, T.C., Lin, G.H., and Long, A.X. (2003). Studies on physical qualities during different ages of manufacture and storage of yogurt. *Food Science.* **24**(2), 44–48.

5

SOLID-STATE FERMENTED FOOD OF ANIMAL ORIGIN

PENG ZHOU, DASONG LIU, YINGJIA CHEN, AND TIANCHENG LI

State Key Laboratory of Food Science and Technology,
School of Food Science and Technology, Jiangnan
University, Wuxi, Jiangsu Province, China

Contents

5.1 Introduction

Solid-state fermentation has long played an important role in the processing and preservation of food of animal origin, such as dry-cured ham, sausage, and fermented aquatic food products. The processing of solid-state fermented food products is very complex and involves numerous physical, chemical, and biochemical changes that contribute to their characteristic sensory properties, and nutritional and functional values. The quality of these products depends largely on the properties of raw material, especially the quantity and quality of proteins and lipids, and the activities of endogenous enzymes during processing. In addition, microbes and processing parameters such as temperature and humidity also play a key role in determining the quality of these products. Solid-state fermented food of animal origin has gained large popularity all over the world and is of great economic importance to the food industry.

5.2 Ham

5.2.1 Introduction

Dry-cured ham is a traditional meat product made from pig's hind leg after a long process of curing and fermentation involving complicated physical, chemical, and biochemical changes that determine the final quality of the product (Zhang and Chen, 2002; Zu, 2004; Zu and Zhang, 2004; Dang et al., 2008). Lipids undergo an intense lipolysis, resulting in a large number of free fatty acids that are further subjected to oxidation (Huan et al., 2003; Gilles, 2009). Proteins will also go through intense proteolysis, generating high amounts of peptides and free amino acids, which may experience a further degradation (Zhang and Chen, 2002; Jiménez-Colmenero et al., 2010). Moreover, with the loss of moisture and decrease of water activity, the Maillard reaction between reducing sugars and free amino acids is more likely to occur, giving rise to the formation of characteristic flavor compounds (Huan et al., 2003).

The final quality of dry-cured hams such as flavor, texture, and color depends largely on the properties of raw ham, including the quantity, quality, and distribution of lipids, as well as endogenous enzymes, mainly lipases and proteases (Toldrá and Flores, 1998; Huan et al., 2003; Jiménez-Colmenero et al., 2010). Microbes also play a significant role during the process, and the flora is very complex and mainly involves molds, yeasts, lactic acid bacteria, and staphylococci (Lücke, 2000; Liu, 2005; Leroy et al., 2006; He et al., 2008; Jiao, 2008; Fadda et al., 2010). Furthermore, the processing parameters, including temperature, salt content, and time, also contribute largely to the quality traits (Toldrá and Flores, 1998; Huan et al., 2003; She, 2005; Jiménez-Colmenero et al., 2010). Although the flavor varies significantly among different hams all over the world, common flavor compounds are identified, including alcohols, aldehydes, ketones, esters, sulfocompounds, alkyls, and alkenes (Huan et al., 2003; Huan et al., 2006; Dang et al., 2008).

Originally, the processing of dry-cured ham was used as a method to preserve meat; however, it has gained large popularity all over the world and brought enormous economic value to the meat industry (Toldrá and Flores, 1998; Jiménez-Colmenero et al., 2010). There are various kinds of hams presenting distinctive characteristics in markets. With a history of centuries, the traditional Chinese hams include Jinhua ham, Rugao ham, and Xuanwei ham (Zhang, 2007; Lu, 2008). Traditional western hams originate from the Mediterranean area and are ranked as the top meat products for their unique flavor, such as Parma in Italy, Iberian and Serrano in Spain, Bayonne and Corsican in France, and Westphalian and the Black Forest in Germany (Zhang and Chen, 2002; Zu, 2004; Dang et al., 2008; Jiménez-Colmenero et al., 2010). In America, country ham, such as Smithfield, is of great economic importance for the meat industry (Zu and Zhang, 2004).

5.2.2 Dry-Cured Ham Components and Nutritional Properties

A typical process for dry-cured ham making involves three steps: salting, postsalting, and ripening–drying (Gilles, 2009). In fact, some hams are presalted by a mixture of curing agent, including salt, nitrate and nitrite, and some are even smoked before the ripening–drying process (Toldrá and Flores, 1998). Ripening–drying is the most complicated and also the most crucial step for determining the final quality of dry-cured hams (Toldrá and Flores, 1998). Intense changes during these processes include the degradation of lipids and proteins, and the generation of nutritional and flavor compounds (Zhang et al., 2010). The sensory properties and microbial stability of dry-cured hams are mainly attributed to endogenous enzymes and curing agents, with a lesser contribution from microorganisms (Lücke, 2000).

5.2.2.1 Lipids and Fatty Acids   Lipids play a key role in determining the sensory properties of dry-cured ham. Both the lipids of muscle and adipose tissue of fresh meat and their changes during processing may affect the quality of the final dry products (Toldrá and Flores, 1998; Lücke, 2000). Lipid traits of fresh meat such as quantity, quality, and distribution are strongly related with the sensory properties, including marbling, oily aspect, texture, and color (Gilles, 2009). There are mainly two kinds of lipids in meat: phospholipids and triacylglycerols. The former are located in the membrane of muscle cell, serving as the precursor of many aroma compounds, while the latter are located in adipose tissues and intermuscular adipose cells, serving as the solvent for many aroma compounds (Gilles, 2009). During processing, lipids are subjected to lipolysis and oxidation. Lipolysis generates free polyunsaturated fatty acids, which are more liable to undergo oxidation and hence form many volatile components (Huan et al., 2003; Decker and Park, 2010). Most of the volatiles are generated by lipid oxidation regardless of the type of dry-cured hams. Phospholipids contain higher proportion of polyunsaturated fatty acids, one third of which has 4, 5, or 6 double bounds. This could explain why phospholipids are the main substrates for lipid oxidation in muscle (Gandemer, 1999). The extent of lipolysis and oxidation is based on many parameters of the process, such as the way and the level of salting and the duration of processing (Dang et al., 2008). Former studies suggested that dry-cured hams subjected to a longer ripening process contain a larger amount of volatiles generated by lipid degradation, and hence exhibit higher aroma intensity (Gilles, 2009).

Endogenous enzymes play a significant role in the generating of fatty acids. There are mainly three enzymes engaged in the lypolysis of triacylglycerols: basic lipase (lipoproteins lipase), neutral lipase (hormone sensitive lipases), and acid lipase (Toldrá and Flores, 1998; Coutron-Gambotti and Gandemer, 1999; Lu, 2008). On the other hand, phospholipids are mainly hydrolyzed by three other enzymes: phospholipases A1, phospholipase A2, and lysophospho-lipase (Coutron-Gambotti and Gandemer, 1999; Zhang et al., 2010). Numerous studies have been done to investigate the postmortem activities of these enzymes in fresh muscle and their changes during the processing (Gilles, 2009). The generation of free fatty acids in dry-cured hams promotes further lipid oxidation. However, former studies also suggested that lipolysis of phospholipids during process-ing may prevent the long-chain polyunsaturated fatty acids from fur-ther oxidation, but the underlying mechanism remains to be further studied (Gilles, 2009).

Although lipid oxidation contributes to the deterioration in meat quality during processing and storage, it is essential to the forma-tion of the characteristic aroma of dry-cured ham during long-term processing. The main mechanism for lipid oxidation is autoxida-tion, a process involving initiation, propagation, and termination (Jiménez-Colmenero et al., 2010; Zhang et al., 2010). A large amount of volatiles are generated through this process, such as aldehydes, alkanes, alcohols, esters, and carboxylic acids (Huan et al., 2003; Lu, 2008; Jiménez-Colmenero et al., 2010). Aldehydes exhibit low odor threshold and hence contribute largely to the overall aroma of the final dry products (Lu, 2008; Zhang et al., 2010). Based on the length of the chains, they may present pleasant or unpleasant odors, such as rancid, deep fried, oily, and fruity (Huan et al., 2003; Zhang et al., 2010). Ketones originating from lipid oxidation may present fruity, fatty, or blue cheese aroma notes (Gilles, 2009). Linear alcohols with short chains exhibit a slight aroma, however, with the increases in length of chains, the intensity of the aroma increases (Huan et al., 2003). Given the fact that phospholipids and triacylglycerols in most raw hams contain a similar composition of fatty acids, many common flavor compounds are identified from different dry-cured hams and the major difference lies in the quantity of each volatile (Huan et al., 2003; Gilles, 2009).

5.2.2.2 Proteins, Peptides, and Amino Acids Degradation of proteins during the process also plays a significant role in determining the quality of dry-cured ham. Although the progress may vary according to specific conditions including salt content and processing time, proteolysis mainly contains the three following steps: the initial degradation of major myofibrillar proteins, generating intermediate-size polypeptides and leading to changes in texture; the production of small peptides and free amino acids through further degradation of polypeptides, directly delivering taste properties to the final dry products; and further participation in the Maillard reaction and Strecker degradation for free amino acids, indirectly contributing to the flavor of dry-cured ham (Toldrá and Flores, 1998; Huan et al., 2003; Zhang et al., 2010). Moreover, proteolysis during this process is one of the key factors that lead to the improvement of functional value of meat and meat products (Zhang et al., 2010).

An intense degree of proteolysis has been reported in the final dry-cured hams, and these changes are mainly caused by endogenous proteinase, including endopeptidases (cathepsins B, D, H, and L, and calpains) and exopeptidases (peptidases and aminopeptidases) (Toldrá and Flores, 1998; Zhang and Chen, 2002). It suggested that cathepsins and calpains may play a more important role in muscle proteolysis, while much less attention has been put on peptidases and aminopeptidases, even these enzymes are engaged in the latter steps of proteolysis (Toldrá and Flores, 1998). However, it was reported that calpains were so instable that no activity related to them could be detected after curing; therefore, the possible role of calpains in proteolytic degradation is negligible (Zhang and Chen, 2002).

Free amino acids in dry-cured ham are mainly generated from extensive proteolysis, including alanine, leucine, arginine, lysine, valine, aspartic, and glutamic acids (Zhang and Chen, 2002). The extent of free amino acids release is largely determined by processing time, aminopeptidase activity, and properties of raw hams (Jiménez-Colmenero et al., 2010). Free amino acids are highly bioavailable and could be absorbed by intestinal mucosa. Moreover, these free amino acids can directly present flavor characteristics and taste properties: Alanine and glycine are related with a sweet taste, hydrophobic amino acids are associated with a bitter taste, and sodium salt of aspartic and glutamic acids enhance taste (Huan et al., 2003; Zhang et al., 2010). Free amino acids are also precursors of many water-soluble flavor compounds. Among the flavor compounds in dry-cured ham, sulfocompounds play a significant role in dry-cured ham flavor formation and are mainly formed through Strecker degradation of cysteine, cystine, and methionine (Huan et al., 2003). On the other hand, further reactions of these precursors with reducing sugars can generate Maillard products further contributing to the dry-cured ham flavor profile.

Food-based strategies would be an excellent option to maintain health and wellness and decrease health care costs. Some free amino acids in dry-cured ham present bioactive properties. It was reported that tryptophan is related to recovery from mental fatigue and leucine is associated with recovery from physically exhausting activity (Jiménez-Colmenero et al., 2010). In addition, the hydrolysis of proteins during the processing of dry-cured ham also generates many bioactive peptides (Decker and Park, 2010; Zhang et al., 2010). A number of histidine-based dipeptides exhibiting antioxidant activity, such as carnosine and anserine, have been discovered in dry-cured ham (Jiménez-Colmenero et al., 2010). Some of the bioactive peptides can regulate blood pressure, owing to their inhibitory activities against angiotensin I-converting enzyme which could indirectly cause contraction of the artery and hence increase blood pressure (Decker and Park, 2010; Jiménez-Colmenero et al., 2010).

5.2.2.3 Other Components   During postmortem storage, an intensive enzymatic hydrolysis of ribonucleotides generates guanosine monophosphate, adenosine monophosphate, and inosine monophosphate, which may provide *umami*-taste characteristics or enhance flavor properties, including meaty, mouth-filling, and brothy qualities (Zhang et al., 2010). These compounds can work in synergy with some of the free amino acids forming during the processing of dry-cured ham to enhance the intensity of flavor (Zhang and Chen, 2002).

Nitrates can be added to dry-cured ham as a reservoir for slow generation of nitrites, which would enhance the preserving effect of salting by preventing the growth of spoilage bacteria and pathogenic bacteria (Toldrá and Flores, 1998; Jiménez-Colmenero et al., 2010). Moreover, nitrate and nitrite also play an important role in the development of characteristic cured flavor, since nitrite can react with a large number of meat components and hence affect the reactions occurring in meat during processing (Toldrá et al., 2009). One disadvantage of using nitrate and nitrite is that they can form N-nitrosamines, which are potential carcinogens. The key point in using nitrate and nitrite as preservatives in dry-cured ham is to strike a balance between the assurance of microbiological safety and the diminution in forming of nitrosamines (Toldrá et al., 2009).

Dry-cured ham is also a good source of B-group vitamins: riboflavin, thiamine, niacin, vitamin B_{12}, and B_6, and it also contains a small amount of vitamins E, A, C, D, and K (Jiménez-Colmenero et al., 2010). It is reported that the content of vitamin E could be improved by meat fortification or dietary supplementation (Decker and Park, 2010). Vitamin E can improve the quality of meat and meat products, through improving meat color, limiting lipid oxidation, and the increasing the water-holding capacity of the final product (Zhang et al., 2010).

Dietary minerals are critically important for health and well-being in that they are highly involved in bone health, muscle and nerve function, and development of disease (Decker and Park, 2010). Dry-cured ham is a valuable source of minerals such as iron, zinc, magnesium, and selenium, considering their amounts and high bio-availability (Jiménez-Colmenero et al., 2010). Selenium is a fundamental trace mineral for humans, and it plays an important role in the prevention of cancer and cardiovascular diseases (Zhang et al., 2010). Sodium is added in the raw ham as a curing agent to prevent deterioration of the product and to generate the characteristic dry-cured ham flavor (Zu and Zhang, 2004). However, high sodium consumption is related with high blood pressure, and it is suggested that the amount of sodium added in raw ham need to be reduced without affecting the quality of final products (Jiménez-Colmenero et al., 2010).

High levels of cholesterol in meat and meat products have a negative bearing on meat's health image for consumer (Fernández-Ginés et al., 2005). According to the nutritional recommendations, the upper limitation for cholesterol intake is 300 mg/day (Jiménez-Colmenero et al., 2010). However, cholesterol may be subjected to oxidation during the processing of dry-cured hams, consequently reducing the health-related concerns about cholesterol levels in meat (Fernández-Ginés et al., 2005).

5.2.3 Major Types of Hams

Dry-cured hams are widely accepted all over the word for their sensory properties and nutritional and functional properties. However, the qualities and characteristics such as flavor, texture, and color vary largely among different kinds of dry-cured hams in that the quality of raw hams and processing parameters differ greatly all over the world (Huan et al., 2003; Zu, 2004).

5.2.3.1 Hams in European Countries and America   In European countries, dry-cured hams are very popular, and some of them are ranked as the top meat products for their unique sensory properties, such as Parma and Light Italian Country in Italy, Iberian and Serrano in Spain, Bayonne and Corsican in France, and Westphalia and the Black Forest in Germany.

The dry-cured Parma ham originates from the mountainous Parma region of Italy, which is also famous for its cheese production (Zu, 2004). Its characteristic quality can not only be attributed to the well-developed and complicated processing technology, but also to the distinctive local geographic terrain and climate. Produced in Langhirano of the Parma region, every single Parma ham benefits from the breeze from the Apennines and also the favorable dry climate from northern areas (Fang, 2008; Guo, 2009). The unique Parma pork hind quarter (also known as pork leg) comes from the pigs being fed with whey, the by-product of cheese manufacture, and other abundant local agricultural products, including chestnut, corn, and oats (Zu, 2004). The maturing period should be no less than 9 months and weight per animal should be more than 150 kg, ensuring the production of firm muscle that is low in moisture content. Such kind of muscle will absorb less salt than normal muscle during the curing process, which is desirable for the ham flavor formation during the subsequent ripening stage (Li, 2009). The curing process is carried out in a chill room through a series of complex regulations of both temperature and humidity (Zu, 2004). Then, the cured ham is transferred to an aging room with good ventilation; this process depends totally on the natural climate and weather conditions; however, the ripening process is accomplished in a basement with faint light (Zu, 2004). In general, it will take more than 12 months to form the final dry-cured Parma ham renowned for its bright red color and characteristic flavor and texture (Jens et al., 2003).

The Iberian ham is representative of traditional Spanish meat products and is also one of the most famous dry-cured hams in the world (Zu, 2004). Legend has it that its development is based on the processing technology of the Jinhua ham brought to European countries by Marco Polo (Li, 2009). The raw material is from a native black-footed Spanish pig fed with acorns, which is rich in oleic acid and vitamin E (Zu, 2004; Wang, 2008). Consequently, the content of oleic acid and vitamin E in the muscle is increased, which is favorable for the formation of ham flavor and the increase in oxidative stability of meat (Zu, 2004). During the curing process, the temperature is changed circularly between 2°C and 5°C at a pace of 2°C every 2 h for a week to promote salt penetration (Li, 2009). On the other hand, at the beginning of the fermentation, a layer of lard is brushed onto the surface of the ham in order to prevent excessive lipid oxidation (Li, 2009). It may take more than 18 months to form the final dry product with characteristic marbling, tender texture, and delicate flavor (Wang, 2008).

In America, the Country ham is of great economic importance and refers to those hams produced in rural areas; however, if it is not produced in a rural area, then only "Country-style ham" can be used on the label. Country ham is mainly produced in Virginia, Georgia, and Missouri; the famous Smithfield ham is manufactured in Virginia (Zu, 2004). Homogenous distribution of salt before aging is fairly important for the formation of ham flavor and storage stability. As the aging process advances, the texture will become harder and the intensity of the flavor will get stronger. After the aging process, smoking will be applied to certain kinds of hams (Zu, 2004).

5.2.3.2 Hams in China   In China, the Jinhua ham, Xuanwei ham, and Rugao ham are widely known as the "Famous Three" and are representative of traditional Chinese meat products (Zhang, 2007; Dang et al., 2008).

Originated and produced in the Jinhua area of Zhejiang province in China, Jinhua ham is one of the most famous traditional meat products and is widely accepted in China and Southeast Asia (Huan et al., 2005). Legend has it that the processing of the Jinhua ham was formed as early as the Tang Dynasty, 618–907 AD; however, the name Jinhua ham was granted by the first emperor of the South Song Dynasty, approximately 800 years ago (Huan et al., 2005; Zhou and Zhao, 2007). The processing of the Jinhua ham was brought to European countries by Marco Polo during the 13th and 14th centuries and produced a worldwide impact on the development of dry-cured ham processing technology (Zhou and Zhao, 2007). It is the most famed dry-cured ham in China and has extraordinary qualities, such as a bamboo leaf-like shape, golden yellow skin, pure white fat, and rose-like muscle (Zhao et al., 2005; Zhang, 2007). The amount of free amino acids in dry-cured hams is 14–16 times that of raw hams, and numerous volatile components have been discovered in the final ham products, making a major contribution to the unique flavor of Jinhua ham (Zhou and Zhao, 2007).

Traditionally, the production of the Jinhua ham totally depends on the climate and weather conditions. The processing starts in winter and then changes according to the alteration of climate. Jinhua area, the place of origin for the Jinhua ham, is mountainous with well-defined seasons and regularly changing temperature with small fluctuations (Zhang, 2007). In winter, the temperature varies between 0°C and 10°C, and it is desirable for curing during that period of time. As the rainy spring comes, the ambient humidity increases and the temperature goes up to 20°C; while, in summer, the ambient humidity decreases and the temperature goes up further to 30°C (Zhao, 2004). Such a trend in changes of temperature and ambient humidity during spring and summer is favorable for fermentation (Chen, 2007; Lu, 2008). When the autumn comes, the temperature drop and correspondingly the ripening stage begin (Lu, 2008). Therefore, the whole process for the production of the Jinhua ham, starting in one winter and finishing the subsequent winter, lasts about 10 months.

With the increasing demand for the Jinhua ham, the traditional processing is no longer applicable for industrial mass production. In model processing, the alterations of temperature and ambient humidity are controlled by instruments instead of nature, thus, the overall processing time can be reduced and the undesirable changes in quality resulting from unfavorable changes in climate can also be avoided (Toldrá and Flores, 1998; Lu, 2008). Compared with traditional processing, curing is performed in a chill room set at 0–5°C; drying is accelerated by aeration, and ripening is achieved at high temperature during the modern processing of the Jinhua ham (Fang, 2006; Lu, 2008). However, there are still many technical problems that remain to be solved during the processing of the Jinhua ham. The high salt content needs to be reduced without affecting the final quality of the products and excessive lipid oxidation, resulting from direct and long exposure to air, also needs to be prevented (Dang et al., 2008).

The processing of the Jinhua ham mainly contains six steps: raw ham fabricating, salting, soaking and washing, sun-drying and shaping, ripening, and postripening (Huan, 2005; Jiang, 2005). The best raw ham needs to be fresh, and also have a thin skin and slim shank bone (Zhao et al., 2007). Traditionally, the hind quarter from the "Shuangtouwu" pig or its crossbreed is used as the raw material in that it is rich in intermuscular fat but low in subcutaneous fat, which is favorable for the formation of the unique characteristics of the Jinhua ham during the ripening stage (Zhang, 2007; Zhao et al., 2007). Salting is a critical step for Jinhua ham processing, and spoilage may occur if it is not well controlled. If the temperature is too high, microbial propagation can lead to the deterioration of the final product; however, if the temperature was too low, penetration of the curing agent would be affected and also ice crystals could form within muscle, which is unfavorable for the flavor formation (Zeng, 2006; Chen, 2007). Moreover, if the ambient humidity falls below 70% RH, loss of moisture will occur so fast that it may result in insufficient salt penetration; however, if the ambient humidity goes up to 90% RH, too much salt would be carried away in the form of brine (Zu and Zhang, 2004; Zhou and Zhao, 2007).

The main purpose of soaking and washing is to get rid of excessive salt and any dirty substances on the surface of the leg, while subsequent sun-drying is performed to achieve appropriate dehydration in case of microbial spoilage (Zhu, 2005; Chen, 2007). Ripening determines the final quality of the Jinhua ham and depends on temperature and ambient humidity. Low temperature would slow down flavor formation; however, high temperature with low ambient humidity may promote moisture loss and also fat oxidation (Zeng, 2006; Zhou and Zhao, 2007). On the other hand, high ambient humidity may cause ham spoilage (Chen, 2007). During this stage, skin and muscle may shrink due to moisture loss, and further shaping is needed to cut off protruding bones, superfluous skin, and fat (Zhou and Zhao, 2007). Before postripening, the hams are further cleaned to remove mold spores and dust on the surface and then a thin layer of vegetable oil is applied to soften muscle and prevent excessive lipid oxidation (Zhao, 2004; Zhu, 2005). The purpose of postripening is to stabilize and strengthen ham flavor (Zhao, 2004; Zhu, 2005; Chen, 2007). Once the Jinhua ham is fully ripened after such a complex and time-consuming process, it will be graded mainly according to the aroma intensity (Zhou and Zhao, 2007). In general, the Jinhua ham can be stored for years.

The Xuanwei ham originates from the Xuanwei area of Yunnan province in China and has a history of more than 1000 years; however, the name Xuanwei ham formally came into being during the reign of the Qing dynasty, about 270 years ago (Yang, 2006; Wang et al., 2006a). It has a yield of more than 10,000 tons per year (Wang et al., 2006a). Thanks to the unique local geographic terrain and climate, both the changes in temperature and ambient humidity in the Xuanwei area are favorable for the processing of dry-cured ham (Xiao and Xu, 1998; Qiao and Ma, 2004). The hind leg is from a local breed named Wujin pig and its high intermuscular lipid content makes it a desirable material for the generation of characteristic ham flavor (Qiao and Ma, 2004; Liu et al., 2009). Nitrate and nitrite are not added during the curing processing (Liu et al., 2007).

The Rugao ham is formed and produced in the Rugao area of the Zhejiang province in China and the earliest legend with reference to the processing of Rugao ham can be traced back to the reign of Xianfeng, emperor of the Qing dynasty, about 150 years ago (Pan, 2007; Zhang, 2008; Wei et al., 2009). The raw ham is also from a distinctive local pig with thin skin, slim shank bone, and tender meat. Based on the period when the processing starts, Rugao ham can be divided into two categories: "winter ham," the curing of which starts between November and December of the traditional Chinese calendar, and "spring ham" with curing started between January and February of the traditional Chinese calendar (Qiu and Zhang, 2007).

5.2.4 Perspectives

In recent years, the demand for meat products with health-promoting and disease-preventing ingredients has been rapidly growing. In response to this increasing interest in the effect of food-based strategies on the prevention of certain diseases, many studies were conducted to investigate the relationship between nutritional factors and various pathologies, such as high blood pressure, cardiovascular disorders, cancers, obesity, diabetes, or osteoporosis (Jiménez-Colmenero et al., 2010). However, increasing the nutrition and health benefits of conventional meat products is not always easily achieved, since many factors such as taste, cost, and regulatory hurdles must be taken into account (Fernández-Ginés et al., 2005; Leroy et al., 2006; Decker and Park, 2010; Fadda et al., 2010). Improvement of nutritional and functional value can be achieved by feeding management, such as adding omega-3 fatty acid, selenium, and vitamin E to the animal diets to increase animal production and meat composition and quality (Decker and Park, 2010; Zhang et al., 2010). However, one of the drawbacks of this practice is that oxidative stability of meat products can be weakened by increasing levels of unsaturated fatty acids (Decker and Park, 2010). Moreover, nutritional and functional compounds including dietary fibers, vegetable and dairy proteins, and lactic acid bacteria can be directly added to the meat and meat products during processing (Zhang et al., 2010). Dry-cured ham has a large presence in markets and has great potential for delivering functional compounds, especially bioactive peptides, during the process of curing, fermentation, and aging (Lu, 2008; Jiménez-Colmenero et al., 2010; Zhang et al., 2010). Moreover, dry-cured ham is also an excellent source of fatty acids, amino acids, dietary fiber, minerals, and vitamins (Decker and Park, 2010; Zhang et al., 2010).

Although dry-cured ham presents eminent nutritional and functional properties, there are still many aspects that remain to be improved. Genetic selection as well as management and feeding practices can be applied to improve the quality of raw ham, such as fat content and fatty acid composition (Dang et al., 2008; Lu, 2008; Gilles, 2009; Jiménez-Colmenero et al., 2010). There are also various measures that can be taken to optimize the processing systems, and consequently obtain an optimal concentration of compounds with possible heath implications without affecting the overall quality of final products. Such strategies include salt reduction, generation of bioactive peptides, and reduction of nitrates and nitrites (Dang et al., 2008; Toldrá et al., 2009; Jiménez-Colmenero et al., 2010; Zhang et al., 2010). Moreover, the traditional processing of dry-cured ham is always so complex and time consuming that it would reduce the overall production and increase manufacturing cost. Therefore, optimization should also be aimed at accelerating the processing and hence reducing the cost of production without changing the original taste and flavor (Zhang and Chen, 2002; Zu and Zhang, 2004).

5.2.5 Summary

Dry-cured ham is widely appreciated for its distinctive sensory properties and nutritional and functional value. The processing of dry-cured ham is complicated and numerous biochemical changes occur intensively, primarily related to lipids and proteins. Endogenous enzymes, mainly lipases and protease, play a vital role during this process. The final quality of the dry-cured ham largely depends on the properties of raw materials and the processing parameters.

5.3 Sausage

5.3.1 Introduction

Fermented sausages are made from raw meat material and additional ingredients. The main raw material can be one or more of meats including pork, beef, mutton, venison, and turkey, among others. Sometimes, offal like blood and liver will be added. The additional ingredients include salt, spice, wine, and fermentation starter. After mincing and chopping, the mixed materials will be poured into an edible casing and undergo the course of microbial fermentation to eventually form a sausage type of product with characteristic fermented flavor, color, and texture, and long shelf life without refrigeration.

As one of the most important traditional processed meat products, fermented sausage has a long history. Italians started making natural fermented dry sausage about 2000 years ago, while it has a history of around 3500 years in China. In the 1850s, the manufacturing technology was brought from the Mediterranean regions to Hungary and then to America by the Central European immigrants. Nowadays, the scale of enterprise for fermented sausage production in Europe is small or medium, while in America, large-scale industrial producing lines were established as early as the beginning of 20th century. Nowadays, fermented sausage occupies an important part of the meat consumption (Ordonez et al., 1999).

The classification of fermented sausage is diverse, depending on various criteria. One of the most popular ways is to categorize according to the origin, for instance, Italian salami, German Mettwurst, Bologna sausage, Polish sausage, Spanish chorizo, Chinese sausage, and Hungarian salami. Besides, they can also be named after the raw meat materials used, such as pork sausage, mutton sausage, fish sausage, and venison sausage. Depending on the drying speed, fermented sausages can be divided into fast-fermented sausage, medium-fermented sausage, and slow-fermented sausage. If taking final water content into consideration, they are grouped as dry sausage and semidry sausage. Dry sausages undergo a long period of air drying at a low drying speed leading to a total moisture loss of 20%–50% and the pH ranges from 4.7 to 5.0. Semidry sausages undergo a rapid air drying with a comparatively high drying speed resulting in a final moisture loss of 15% and a slightly lower pH than dry sausages. Most semidry sausages are cooked or smoked. In addition, the low acid fermented sausage and high acid fermented sausage are distinguished by the degree of fermentation.

Fermented sausage is nutritious. The meat, as major raw material, contains a lot of nutrients that humans need for everyday life, including proteins, lipids, carbohydrates, nitrogenous extract, minerals, and vitamins. Intensive attention has been focused on meat products, including fermented sausages, with physiological functions to enhance health conditions (Decker and Park, 2010; Zhang et al., 2010). During fermentation, the biochemical substances with a high molecule weight such as proteins, lipids, and carbohydrates will be broken down by the action of microorganisms to form compounds with small molecular weight such as peptides, free amino acids, and fatty acids, which are easier to absorb. In addition, the universal existence of lactic acid bacteria enhanced its nutritional value. Lactic acid bacteria can diminish not only the amount of precursors of carcinogens but also the activity of enzymes, which catalyze the transition of the precursors to carcinogens, therefore lowering the risk of contamination with carcinogenic substances. Lactic acid bacteria are also regarded as a natural food protection so that no chemical preservatives are needed during the processing of fermented sausage.

In addition to the highly nutritional value, the attractive color, the texture, and the distinctive flavor and aroma of fermented sausage all contribute to its popularity all over the world. Another important contribution is that it can be stored safely for a year or even longer without refrigeration.

Current studies on fermented sausage mainly focused on three aspects (Wei et al., 2001; Wang, 2006). The first is to improve the processing technology. Fermented sausage owes its popularity to color, flavor, and aroma, but the mechanisms and factors involved remain unclear. The development of processing technology is necessary for industrial production. The second aspect focuses on the strains and starter culture. Naturally fermented sausage has the drawbacks of unstable quality, long producing period, and producing seasonality. In order to eliminate the disadvantages and achieve better control of the process of manufacture, research has focused on the isolation of pure strains from the spontaneously fermented meat products for commercial use. The other field is related to the additives, with the purpose to improve the sensory and quality of the sausages and to make the products more beneficial for human health, which intensively follow the trend of functional foods.

5.3.2 Major Types of Sausage

Despite the various rules established for sausage classification, we would like to give a brief introduction according to their origin.

5.3.2.1 Chinese Sausage Fermented sausage has a history of around 3500 years in China. In the spring and autumn, people mixed meat with salt and spice, poured them into sheep intestine, and then air dried. At that time, the process was merely applied as a means of meat storage. It is not until the Northern Wei Dynasty that the detailed written record of the producing process of Chinese sausage appeared.

Table 5.1   Typical Ingredients of Chinese Sausages

NAME	INGREDIENTS (G/1000 G SAUSAGE)
Cantonese sausage	Lean meat, 596; fat, 256; salt, 19; sugar, 65; white wine, 21; white soy sauce, 43; sodium nitrate, 0.4
Sichuan sausage	Lean meat, 739; fat, 185; salt, 28; sugar, 9; liquor, 9; soy sauce, 28; sodium nitrate, 0.05; pepper, 0.9; spice (mixture of anise, kaempferol, cinnamon, hay, piper longum linn) 1

Chinese sausage can be approximately classified into two categories, namely Cantonese sausage and Sichuan sausage. The predominant ingredients of Cantonese sausage are pork, spice, sugar, salt, nitrite, and nitrate, while Sichuan sausage also contains chili and pepper to render a distinctive spicy flavor. Although different types of Chinese sausage have different names and flavors, the manufacturing process is almost the same, including the selection and trimming of the raw meat, curing, stuffing, rinsing, drying, or roasting (Wang et al., 2006b).

Pork is the preferred raw meat material for Chinese sausage, and the other ingredients usually vary according to the region as shown in Table 5.1 (Wang et al., 2006b). As for meat processing, traditionally three ways—cutting into slices, cutting into strips, and chopping—can be adopted. A previous study (Wang, 2006) shows that cutting the pork into slices is the best option, while cutting into strips displays the highest water content and pH value, and chopping leads to an improper color and a comparatively soft texture.

Curing occurs after the thorough mixing of the ingredients. Salt and nitrite play a significant role during this course. The application of salt gives products the appropriate flavor and also helps to prevent the growth of spoilage microorganisms. The nitrite can combine with the myoglobin to form nitrosomyoglobin, which displays the fresh red color of salted meat. Nitrite can only be reduced under acidic conditions, so the meat exhibits a faint red color when the pH is around 7.0. As a result, sugars will be added regularly during early fermentation in order to lower the pH. The color of salted meat is also related to the amount of nitrite, temperature, the additives, the strains of microbe, and some other factors. The lactic acid could lower the water-holding capacity of the meat and also make the meat an unfriendly matrix for other spoilage bacteria. After curing, the shaped sausage should be rinsed to remove contaminations on the surface, and then subjected to either air-drying or roasting. The air-drying process includes exposing the meat under sunlight for 2–3 days, and then hanging in the natural environment with flowing air for 10–15 days. However, roasting only takes about 3 days at a temperature between 40°C–60°C.

5.3.2.2 European Sausage The Mediterranean region is the home of salami, a fermented sausage famous worldwide. About 2000 years ago, the Romans had learned the technology to make naturally fermented dry sausage, which is still prevalent today. The traditional ingredients in salami usually contain meat, including pork and beef, fat, salt, spices, garlic, wine, herbs, and vinegar. Salt is an indispensable ingredient; the word *salami* originates from the word *sale*, which means salt.

Salami is a generic term of cured, air-dried fermented sausages prevailing in European countries, including Italy, France, Germany, Spain, Hungary, Slovenia, Turkey, Greece, Czech Republic, Belgium, Luxembourg, Romania, and Bulgaria. Italy alone produces several types of salami. Table 5.2 shows some named salamis and their origins.

The manufacturing process of salami is somewhat the same as Chinese sausage with only slight differences. First of all, the raw meat materials can be pork, beef, mutton, venison, fish meat, or their mixture. In addition, blood and liver can also be added. One group

Table 5.2 Origin of Various Salami

ORIGIN		SAUSAGE NAME
Italy	Milan	Milanese
	Genova	Genovese
	Tuscany	Fegatelli
		Finocchiona
	Parma	Felino
	Calabria	Sopressata
		Nduja
	Marche	Ciauscolo
France		Saucisson
Hungary		Winter salami
Germany		German salami
Spain		Longaniza
		Salchichon
		Catalan fuet
Denmark		Danish

of sausages named "special sausages" include liver sausages, head cheeses, and blood sausages. These sausages are time consuming to prepare because the meat should be cooked in water before it is stuffed into casings, and then cooked again. They are generally made on the farm after the pigs are slaughtered. The casings can be both natural and artificial. Even an inedible cellulose casing can be employed for curing. As for a natural casing, bovine cecum and pig intestine are preferred. As a result, the diameter of salami is generally larger than Chinese sausages. Thirdly, the course of curing and drying is determined by factors such as climate and the characteristics of a casing. The curing of salami usually takes about 36 weeks, but in some special conditions it can also be as short as 24 weeks. After fermentation, the sausages need to be dried. It demands cooking in a smoker or water. The cooking temperature should remain between 80°C and 85°C, and the internal temperature should reach 72°C.

5.3.3 Starter Cultures

In the past, sausages were fermented by the microbes naturally present in the matrix. Such kinds of fermentation are still adopted by the home sausage makers today. However, spontaneous fermentation is unsuitable for industrial production because of the potential risks of inconstant quality. As commercial food products, the total number of microbes should be strictly controlled and the production per day can reach as high as thousands of pounds, so natural bacteria are no longer reliable and starter cultures come into play. Starter cultures contain live or dormant microorganisms, which can metabolize smoothly in the fermentation matrix. Pure microbes were successfully inoculated to fermented sausage in 1921. Niomovaara first applied pure fermentation starter in large-scale industrial manufacture in 1955. The microorganisms in the starter cultures for fermented sausages primarily involved three groups of microorganisms, namely bacteria, molds, and yeast, and the mixed types of starter cultures are more common in regular industrial production.

5.3.3.1 Bacteria   Lactic acid bacteria are widely used in fermented sausage. The major species include *Lactobacillus plantarum*, *Lactobacillus*

acidophilus, Lactobacillus lactis, Lactobacillus casei, Lactobacillus sake, Lactobacillus brevis, Lactobacillus curvatus, Lactobacillus büchneri, Lactobacillus farciminis, Lactobacillus gayonii, Lactobacillus pentosus, Lactobacillus fermentum, Lactobacillus paracasei, and Lactobacillus versmoldensis (Rebecchi et al., 1998; Aymerich et al., 2003; Krockel et al., 2003; Leroy et al., 2006; Zhao, 2008). It was reported that *Lactobacillus sake* and *Lactobacillus curvatus* dominated the fermentation process (Rebecchi et al., 1998; Aymerich et al., 2003; Papamanoli et al., 2003; Rantsiou et al., 2005; Leroy et al., 2006; Zhao, 2008).

The lactic acid bacteria are essential for high product quality. First, they shorten the fermentation time, improve the color stability, give the sausage a unique fermented flavor, and also lengthen the shelf life without refrigeration. Besides, the lactic acid bacteria also contribute to the nutritional qualities of the fermented sausage. Lactic acid bacteria consume carbohydrates to generate lactic acid, which is effective in lowering the pH of the raw materials, and inhibits the growth of the spoilage microbes. In addition, lactic acid bacteria may produce bacteriocins to prevent the growth of spoilage bacteria, so they are regarded as a natural food protectant so that no chemical additives are needed during processing. *Lactobacillus sake* is reported to release a bacteriocin that inhibits the growth of *Listeria monocytogenes* (Schillinger et al., 1991). Further, some lactic acid bacteria can survive in the human gut and help maintain the balance of the human intestinal flora.

Micrococcus and *Staphylococcus* both belong to the Micrococcaceae, and have similar functions during the ripening of the fermented sausage. *Micrococcus* and *Staphylococcus* promote the curing through breaking down proteins and lipids, and reducing nitrate and hydroperoxide. The selection of *Micrococcus* and *Staphylococcus* should be based on the following criteria: First of all, they should be safe to health without generation of any endotoxin or biogenic amine. Their tolerance to salt and nitrite must be high enough for their survival. They should not produce carbon dioxide, mucus, hydrogen sulfide, and ammonia, because carbon dioxide and mucus destroy the texture while hydrogen sulfide and ammonia have negative effect on the flavor. Besides, they should have a high capacity to reduce nitrate and produce hydroperoxide reductase. For other bacteria not discussed in this chapter, the reader can turn to the review of Leroy for more detailed information (Leroy et al., 2006).

5.3.3.2 Molds Molds, mainly *Penicillum nalgiovense* and *Penicillium chrysogenum*, grow on the surface to give the sausage a white or gray appearance, and also regulate the water content. Like lactic acid bacteria, they can inhibit the growth of spoilage microbes due to their higher competitive growth rate. Through consuming oxidants, they also diminish the fading of color caused by oxidants. Moreover, they contribute to the formation of characteristic flavor and aroma by breaking down lipids and proteins. A good review has been done on molds (Sunesen and Stahnke, 2003).

5.3.3.3 Yeast Yeasts, mainly *Debaryomyces hansenii* and *Candida famata*, are also frequently used (Encinas et al., 2000; Olesen and Stahnke, 2000). They can survive on the surface or near the surface because of their high tolerance to salt. Their roles during ripening may be similar to the molds. The primary functions of yeast are to diminish the oxidant remained in the casing and inhibit the growth of *Staphylococcus aureus*. In addition, they also improve the flavor due to their ability to break down lipids and proteins. However, they lack the ability to deoxidize nitrate; therefore, they are often accompanied by micrococci and lactic acid bacteria.

5.3.4 Trends

As a kind of fermented meat products, fermented sausages are widely accepted by people all over the world because of their long shelf life, unique flavor, and wholesomeness. However, traditional naturally fermented sausages need a long period for ripening, and their quality fluctuates considerably. Recent studies of fermented sausage focus mainly on starter cultures and processing technology.

   Previous studies have been focused on isolation of useful strains from the naturally fermented meats, as well as their action on physical, chemical, and biological changes of the sausages. Some researchers showed that the house flora are superior to commercial ones (Leroy et al., 2006). Our knowledge on fermentation agents is very limited. Currently, the modern genetic engineering techniques, molecular biology methods, and metabolic engineering techniques are show-ing their advantages in the production of starter culture (Hentiksen et al., 1999; Kuipers et al., 2000; Baruzzi et al., 2006). In addition, the processing technology should be optimized to ensure the product safety and quality, enhance nutritional and functional properties, and prolong the shelf life.

5.4 Fermented Aquatic Food Products

5.4.1 Introduction

Due to the unique taste and flavor, solid-state fermented aquatic food products, such as fish sauce and shrimp paste, are popular as cook-ing condiments all over the world, especially in Thailand, Indonesia, China, Malaysia, Japan, Korea, and other Asian countries.

The application of fermentation to fish can date back to ancient times. Roman fermented products *Liquamen* and *Garum* played a critical role in their daily cooking. In the 12th century, fermented fish was a vital seasoning in Northern Europe. Additionally, fermented mackerel was the essential export commodity in the Baltic Sea area. Many fermented fish products are available in the worldwide market, for instance, *yulu* in China, *patis* in the Philippines, *kapi* in Thailand, *belacan* in Malaysia, *myelochi-jeot* in Korea, and *jaadi* in Sri Lanka. However, these traditional fermented fish products are produced by spontaneous fermentation by bacteria originating from the materials themselves under natural conditions. Although a pleasant flavor can be obtained with these traditional methods, constant and stable qual-ity can not be guaranteed, and the production cycle is always very long, which restricts their application to industry.

Fermented aquatic food products can be classified into three groups: (1) fermentation based on endogenous enzymes in the fish meat and viscera; (2) fermentation depending on enzymes produced by bacteria together with endogenous enzymes; and (3) rapid fermentation prod-ucts and others.

Fermented fish products contain a variety of amino acids and pep-tides formed by hydrolysis of salt-soluble proteins in raw materials through fermentation. These products have high nutritional value due to their high contents of all the essential amino acids, vitamins, and minerals. A high content of glutamate was detected in ancient fish sauce (*garum*; Miro et al., 2010).

5.4.2 Major Fermented Aquatic Food Products

5.4.2.1 Fish Sauce   Fish sauce is a pellucid liquid with a dark color, salty taste, and mild fishy smell. In China, it is called "yulu." The history of Chinese fish sauce dates back to 3000 BC. Nowadays, fish sauces are produced in many parts of China, with those produced in Fuzhou and Shantou being most famous.

Generally, a traditional approach to produce fish sauce in China is similar to that in Thailand, including pickling and autolysis (prefermentation), middle-state fermentation, filtering, basking, final-state fermentation, dilution and sterilization, and packing. Pickling and autolysis generally need 7 to 8 months. Middle-state fermentation can start when the fish meat becomes soft and red, or the meat and bone are tending to separate. At that time, the partially fermented fish is transferred to ceramic containers in the open air to continue fermentation with frequent stirring. To stabilize quality and homogenous flavor, partially fermented fish of different ages and species should be mixed. The final fermentation needs an extra 1 to 3 months. Fully ripened fish sauce contains very few bacteria. The traditional Chinese production cycle of fish sauce may take several months or even a year.

In Thailand, fish sauce is usually produced in a container with two layers of bamboo. Clean fish flesh mixed with salt is transferred to the fermentation container, and placed between the bamboo layers with something heavy loaded on the top bamboo layer to keep fish flesh soaking in the brine. After 3 to 5 months of fermentation and 0.5 to 3 months of ripening, the first-grade fish sauce with highest quality is obtained. Hot brine is added into fermentation containers after first-grade fish sauce is extracted, to obtain more fish sauce regarded as second-grade fish sauce with a lower quality and shorter shelf life. To overcome such flaws, additives are added to enhance the color or flavor of the second-grade products. Sometimes, first-grade products can be mixed with second-grade products. The details of fish sauce production in Thailand, Malaysia, Korea, and some other Asian countries has been reviewed (Lopetcharat et al., 2001).

The flavor of fish sauce is due to a combination of volatile compounds. Fish species, the type of salt, the ratio of fish and salt, minor ingredients, and fermentation conditions are the five critical factors that determine the quality of fish sauce (Lopetcharat et al., 2001). In a recent study, 16 major odor-active compounds were found to contribute to the characteristics of fish sauce products. Microorganisms found in traditional fermented fish products include lactobacilli, micrococci, molds, and yeasts (Wu et al., 2000; Fukami et al., 2004). Some scientists reported that *Staphylococcus xylosus*, *S. saprophyticus*, *S. carnosus*, and *S. cohni*, which contribute to the flavor of fermented products, are detected in the fish sauce products of southeastern Asian countries (Ijory and Ohta, 1996; Mura et al., 2000). In addition, *Pediococcus pentosaceus*, *Weisella confusa*, *Lactobacillus plantarum*, *Lactococcus garviae*, and *Zygosaccharomyces rouxii* were also detected in Thailand fermented products called *plaa-som*. Most traditional fermented products depend on the biochemical action generated by the bacteria of the raw material. The types and amounts of microorganisms may vary, depending upon many factors, such as seasons, species, and storage methods. Thus, it is difficult to control the quality and safety of traditional fermented fish products.

Because of the different microorganisms and technology that are used in the fermentation, the change in pH value of fermented fish products varies. Usually, fish sauce is obtained from natural fermentation, and quite a large amount of salts are added to avoid excessive microbial growth. In general, the final pH of matured products is higher than 5.0. It was reported that the final pH of fish sauce produced from Caspian Kilka ranged from 6.5 to 7.0 (Koochekian and Moini, 2009).

Proteins are hydrolyzed into peptides and amino acids during fish sauce fermentation. Some amino acids are degraded to aldehydes and ketones through decarboxylation and deamination. These peptides and amino acids, along with some organic acids and nucleotides, contribute to the unique flavor of the fermented products. Fish fats are also hydrolyzed during fermentation. In acidic conditions, the increase of fat-decomposing bacteria results in the increase of free fatty acid, which promote the formation of the unique flavor.

5.4.2.2 Fermented Fish   Fermented fish is a traditional fish product in China, and it has a great market demand both in China and Japan. Local citizens living in the Yangtze valley always produced cured fish in winter. There are many types fermented fish products with various flavors, for instance, smoked fish in the Sichuan and Hunan provinces, and "drunk fish" in the Jiangxi and Fujian provinces. Drunk fish is a traditional vinasse fermented food in Jiangxi Province with a unique feature. It is a product of two-stage fermentation of salty fish with a mixed full-bodied flavor of wine and rice. The quality of traditional fermented fish products based on natural fermentation may vary between different batches of raw material and also there may be some problems such as a high level of salt and lipid oxidation.

Although methods to produce fermented fish vary in different areas, the principal processes include pickling and drying. Manufacture parameters in pickling such as salt dosage, pickling time, and temperature may differ among manufacturers. *Plaa-som*, the Thai fermented fish, was prepared with different salt concentrations (Paludan-Müller et al., 2002). It was found that a salt concentration ranging from 6%–11% caused a suppression of lactic acid bacteria growth together with the formation process of *plaa-som*. The predominant lactic acid bacteria and yeast species during fermentation of *plaa-som* were identified as *P. pentosaceus* and *Z. rouxii*, respectively. It is easy to understand that different parameters used in pickling result in different flavors and taste. With the increase of demand for healthy food, the study of low-salt and low-temperature fermentation technology becomes a crucial project in the traditional fermented fish producing industry.

Some lactic acid bacteria, micrococci, and staphylococci can produce protease, lipase, nitrate reductase, amino acid desaminase, and amino acid decarboxylase, which may contribute to the formation of certain textures, colors, and flavors in fermented fish. In addition, these microorganisms can also inhibit the growth of pathogenic bacteria and putrefying bacteria by excreting acids and antibiotics to improve the product safety (Corbiere et al., 2006). Some bacteria have been isolated from the fermented fish product, including *Lactobacillus plantarum*, *L. alimentarius*, *L. curvatus*, *Pedicoccus acidilactici*, *P. pentosaceus*, *Staphylococcus sciuri*, *S. epidermidis*, *S. simulans*, *S. warneri*, *S. xylosus*, *S. chromogenes*, and *S. saprophyticus*. Lactic acid bacteria tolerate salt concentrations up to 8% and for staphylococci 11%.

Diverse microorganisms were identified from traditional fermented fish in northeastern India, ranging from species of lactic acid bacteria to species of yeasts, as well as certain pathogenic bacteria (Namrata et al., 2004). Lactic acid bacteria (*L. farciminis, L. pentosus, Leuconostoc* sp.) were also found in fermented fish in Thailand (Somboon et al., 1998).

Acids, aldehydes, esters, alcohols, and ketones are predominant substances that contribute to the flavor of fermented products. Histidine, glycine, arginine, and alanine are the main free amino acids in fermented fish. The characteristic volatile chemical substances of fermented fish are 1-limonene, sabinene, 3-pinene, linalool, phenethyl alcohol, acetaldehyde, heptanal, pelargonaldehyde, and carprylaldehyde. Acetaldehyde can be an index for flavor estimation. Aldehydes, ketones, alcohols, and esters were reported as products of lipid oxidation and enzyme-mediated protein breakdown during fermentation (Cha and Cadwallader, 1995).

5.4.2.3 Shrimp Paste Shrimps are a rich source of protein and vitamins, and also contain potassium, iodine, magnesium, phosphorus, and vitamin A. It was reported that traditional shrimp paste in the Philippines contains measurable quantities of polyunsaturated fatty acid, including DHA (Nemesio et al., 2001). After pickling and fermentation, an increase of soluble Ca was observed in shrimp paste, and digestibility had been improved.

Shrimp paste has a mixed flavor of sweet, sour, spicy, and salty with a tender mouth feel. The manufacturing procedure for shrimp paste is shown in Figure 5.1. To make shrimp paste, one should first choose mature and intact shrimps and transfer them to clear water for 2 to 3 days without feeding. This procedure is critical to avoid the existence of unpleasant flavor in the final products. Glutinous rice should also be chosen and soaked in hot water for 8 to 10 h, and then steamed. After that, clean shrimps were mixed with glutinous rice and additives. The tank used to store the shrimp paste must be unglazed ceramic. Moreover, the container should be washed with hot water and white wine (45%, v/v) before use. After the shrimp–rice mixture is transferred to the container, it should be sealed with preservative films or plates with the same diameter of the container.

Figure 5.1 Manufacturing procedure for shrimp paste.

The time of fermentation is 15 to 20 days in summer, 20 to 25 days in spring and autumn, or 30 to 35 days in winter.

Shrimp paste, as a traditional food or condiment, also has some other functions or utilities. Recently, some scholars reported the antioxidant activity of shrimp paste or extracts of shrimp paste (Peralta et al., 2008); Binsan et al., 2008). Prolonged fermentation would improve antioxidant ability and free amino acid in the salt-fermented shrimp paste. It was reported that lipid-derived compounds in shrimp paste were lower compared to fermented fish, while the nitrogen-containing substances mainly contribute to the characteristic flavor of shrimp paste (Cha and Cadwallader, 1995).

5.4.2.4 Fermented Shrimp Fermented shrimp is characterized by a texture of tenderness and brittleness along with an agreeable white color. The first step to produce fermented shrimp is the selection of raw shrimp. Dead and damaged ones must be excluded. After that, a cleaning step using running water is necessary. This is the prerequisite for obtaining an agreeable white product. The washed shrimps are put into a container with another twenty minutes of flow-water cleaning which guarantees the shrimp bodies to absorb enough moisture. The shrimp bodies are then dried and mixed with salt, alcohol, monosodium glutamate, ginger sauce, green bean starch, and egg white. Among all the ingredients added to the shrimp, the starch is the most important one. Green bean starch is much better for a white color. Some additives, such as borax, are added for better water-holding capacity and texture.

5.4.2.5 Scallop Sauce   Scallop sauce is a popular fermented food or condiment with characteristic flavor and taste. For many years, lots of scallop skirts were produced as the waste of scallop manufacturing, which might lead to environmental problems. However, a scallop skirt is a rich source of proteins, fats, vitamin, and minerals. It has a remarkable high level of Gly, Glu, Asp, and Arg. More than 18 kinds of amino acid are included, and 30% of them are essential amino acids. Thus, using a scallop skirt to produce scallop sauce through fermentation is an alternative approach to reuse the waste and assuage the environmental problems.

Figure 5.2 shows the manufacturing procedure for scallop sauce in China. To make scallop sauce with good quality, scallops should be chosen in early spring or late autumn. The scallops are cleaned and then put into boiled water until the shell is open. The scallop skirt is separated from the meat and then readied to produce scallop sauce. Scallops are mixed with *Aspergillus oryzae* and other ingredients such

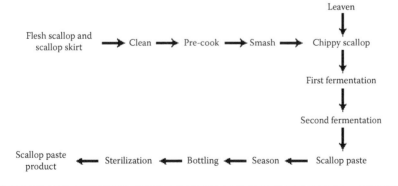

Figure 5.2   Manufacturing procedure for scallop paste in China.

as bean pulp and wheat bran to make *koji*. The temperature is set at 30°C for the first 8 hours, 32°C at the next 22 hours, and 28°C for the last 15 hours. After koji making, there is another 30 days for low-salt solid fermentation until the scallop sauce is obtained.

5.4.3 Trends

Traditional methods of producing solid-state fermented fish products are based mainly on small-scale and manual operations. Most of the products are made by individuals and families for domestic consumption. Furthermore, the microbes contributing to the fermentation originate from the raw material, air, ingredients, container, or the workers. The process of fermentation is not controlled scientifically and the quality of the products largely depends on the experience of the workers. In future, industrial standards for solid-state fermented fish products should be developed. Time, temperature, microorganisms, container, ingredients, and the raw materials used in the fermentation process should be standardized to obtain a product with high quality and safety to meet the demand of a larger market.

References

Aymerich, T., Martin, B., Garriga, M., and Hugas, M. (2003). Microbial quality and direct PCR identification of lactic acid bacteria and non-pathogenic staphylococci from artisanal low-acids sausages. *Applied and Environmental Microbiology.* **69**, 4583–4594.

Baruzzi, F., Matarante, A., Caputo, L., and Morea, M. (2006). Molecular and physiological characterization of natural microbial communities isolated from a traditional southern Italian processed sausage. *Meat Science.* **72**, 261–269.

Cha, Y.J. and Cadwallader, K.R. (1995). Volatile compounds in salt-fermented fish and shrimp pastes. *Journal of Food Science.* **60**, 19–24.

Chen, J. (2007). *Studies on the Moulds Population during Ripening and Their Effects on the Safety and Quality of Jinhua Ham.* Hangzhou, Zhejiang, China: Zhejiang Gongshang Univ: 86 p.

Corbiere, M., Leroy, S., and Talon, R. (2006). Staphylococcal community of a small unit manufacturing traditional dry fermented sausage. *International Journal of Food Microbiology.* **108**, 210–217.

Coutron-Gambotti, C. and Gandemer, G. (1999). Lipolysis and oxidation in subcutaneous adipose tissue during dry-cured ham processing. *Food Chemistry.* **64**, 95–101.

Dang, Y.L., Wang, Z., and Xu, S.Y. (2008). The analyze and improving method of quality of dry-cured ham. *Science and Technology of Food Industry (China).* **2**, 308–312.

Decker, E.A. and Park, Y. (2010). Healthier meat products as functional foods. *Meat Science.* **86**, 49–55.

Encinas, J.P., Lopez-Diaz, T.M., Garcia-Lopez, M.L., Otero, A., and Moreno, B. (2000). Yeast populations on Spanish fermented sausages. *Meat Science.* **54**, 203–208.

Fadda, S., Lopez, C., and Vignolo, G. (2010). Role of lactic acid bacteria during meat conditioning and fermentation: Peptides generated as sensorial and hygienic biomarkers. *Meat Science.* **86**, 66–79.

Fang, W. (2008). Parma ham. *Meat Research (China).* **1**, 1.

Fang, Y.Q. (2006). *Research on the Jinhua Dry-Cured Ham New Product.* Hangzhou, Zhejiang, China: Zhejiang Gongshang Univ: 50 p.

Fernández-Ginés, J.M., Fernndez-Lpez, J., Sayas-Barber, E., and Prez-Alvarez, J.A. (2005). Meat products as functional foods: A review. *Journal of Food Science.* **70**, 37–43.

Fukami, K., Funatsu, Y., and Kwasaki, K. (2004). Improvement of fish-sauce (Moromi) made from frigate mackerel. *Journal of Food Science.* **69**, 45–49.

Gandemer, G. (1999). Lipids and meat quality: Lipolysis, oxidation, Maillard reaction and flavour. *Sciences des Aliments.* **19**, 439–458.

Gilles, G. (2009). Dry cured ham quality as related to lipid quality of raw material and lipid changes during processing: A review. *Grasas Y Aceites.* **60**, 297–307.

Guo, X.D. (2009). Processing and RFID technology of Italian ham. *Meat Industry (China).* **7**, 7–12.

He, Z.F., Zhen, Z.Y., Li, H.J., Zhou, G.H., and Zhang, J.H. (2008). Microorganisms flora study on Jinhua ham fermentation. *Food Science (China).* **29**, 190–195.

Hentiksen, C.M., Nilsson, D., Hansen, S., and Johansen, E. (1999). Industrial applications of genetically modified microorganisms: Gene technology at Chr. Hansen A/S. *International Dairy Journal.* **9**, 17–23.

Huan, Y.J. (2005). *Studying on the Changes of Lipid Flavor Compounds during Processing of Jinhua Ham.* Nanjing, Jiangsu, China: Nanjing Agricultural Univ: 165 p.

Huan, Y.J., Zhou, G.H., and Xu, X.L. (2003). Flavor comparison and formation of mechanism analysis of dry-cured ham from China and west countries. *Food and Fermentation Industries (China).* **29**, 81–88.

Huan, Y.J., Zhou, G.H., Xu, X.L., Liu, Y.M., and Wang, L.P. (2006). Studying on flavor characteristics of different grades of Jinhua ham. *Food Science (China).* **27**(39–45).

Huan, Y.J., Zhou, G.H., Zhao, G.M., Xu, X.L., and Peng, Z.Q. (2005). Changes in flavor compounds of dry-cured Chinese Jinhua ham during processing. *Meat Science.* **71**, 291–299.

Ijory, F.G. and Ohta, Y. (1996). Physicochemical and microbiological change associated with bakasang processing: A traditional Indonesian fermented fish sauce. *Journal of the Science of Food and Agriculture.* **71**, 69–74.

Jens, K.S.M., Giovanni, P., Laura, G., Jakob, C., and Leif, H.S. (2003). Monitoring chemical changes of dry-cured Parma ham during processing by surface autofluorescence spectroscopy. *Journal of Agricultural and Food Chemistry.* **51**, 1224–1230.

Jiang, Y.X. (2005). *Studies on the Changes of Protein during the Processing of Jinhua Dry-Cured Ham.* Beijing, China: China Agricultural Univ: 51 p.

Jiao, Y. (2008). *Isolation and Identification of Aroma Production Staphylococcus in Ham and Their Effect on Fermented Sausage.* Yangzhou, Jiangsu, China: Yangzhou Univ: 77 p.

Jiménez-Colmenero, F., Ventanas, J., and Toldrá, F. (2010). Nutritional composition of dry-cured ham and its role in a healthy diet. *Meat Science.* **84**, 585–593.

Koochekian, S.A. and Moini, S. (2009). Producing fish sauce from Caspian kilka. *Iranian Journal of Fisheries Sciences.* **8**, 155–162.

Krockel, L., Schilliger, U., Franz, C.M.A.P., Bantleon, A., and Ludwig, W. (2003). *Lactobacillus versmoldensis* sp. Nov., isolated from raw fermented sausage. *International Journal of Systematic and Evolutionary Microbiology.* **53**, 513–517.

Kuipers, O.P., Buist, G., and Kok, J. (2000). Current strategies for improving food bacteria. *Research in Microbiology.* **151**, 815–822.

Leroy, F., Verluyten, J., and Vuyst, L.D. (2006). Functional meat starter cultures for improved sausage fermentation. *International Journal of Food Microbiology.* **106**, 270–285.

Li, J.R. (2009). The comparison of technology between Jinhua ham, Parma ham, Iberian ham and Serrano ham. *China Condiment (China).* **2**, 36–39.

Liu, G.Y. (2005). *Studies on Fermented Western Ham.* Zhenjiang, Jiangsu, China: Jiangsu Univ: 69 p.

Liu, H., Liu, Y., and Ma, C.W. (2009). Generation mechanism of white spots in Xuanwei ham's muscle. *Food Science and Technology (China).* **34**, 118–121.

Liu, N., Sun, X.Y., and Ge, C.R. (2007). Processing and its characteristics of Yunnan ham. *Farm Products Processing (China).* **6**, 18–19.

Lopetcharat, K., Yeurg, J.C., and Jae, W.P. (2001). Fish sauce products and manufacturing: A review. *Food Reviews International.* **17**, 65–88.

Lu, R.Q. (2008). *Changing of Lipids and Flavor Compounds during Modern Processing of Jinhua Ham.* Wuxi, Jiangsu, China: Jiangnan Univ: 71 p.

Lücke, F.K. (2000). Utilization of microbes to process and preserve meat. *Meat Science.* **56**, 105–115.

Miro, S., Toshimi, M., Daigo, I., Sachis, E., Hiroshi, M., Takeshi, K., and Robert, I.C. (2010). Amino acid and minerals in ancient remnants of fish sauce (parum) sampled in the "garum shop" of Pompeii, Italy. *Journal of Food Composition and Analysis.* **23**, 442–446.

Mura, K., Maeda, H., Tanaka, A., Koizumi, Y., and Yanagida, F. (2000). Isolation of protease-productive bacteria from fish sauce collected in Vietnam. *Journal of Food Processing and Preservation*. **26**, 263–271.

Namrata, T., Joydeb, P., and Jyoti, P.T. (2004). Microbial diversity in *Ngari*, *Hentak* and *Turgatap*, fermented fish products of North-East India. *World Journal of Microbiology and Biotechnology*. **20**, 599–607.

Nemesio, M., Grace, G., and Victor, C.G. (2001). Polyunsaturated fatty acid contents of some traditional fish and shrimp paste condiments of the Philippines. *Food Chemistry*. **75**, 155–158.

Olesen, P.T. and Stahnke, L.H. (2000). The influence of *Debaryomyces hansenii* and *Candida utilis* on the aroma formation in garlic spiced fermented sausages and model minces. *Meat Science*. **56**, 357–358.

Ordonez, J.A., Hierro, E.M., Bruua, J.M., and De La Hoz, L. (1999). Changes in the components of dry-fermented sausage during ripening. *Critical Reviews in Food Science and Nutrition*. **39**, 329–367.

Pan, M. (2007). *Studies on Relationship between Microorganism Flora and Quality of Rugao Ham*. Yangzhou, Jiangsu, China: Yangzhou Univ: 81 p.

Papamanoli, E., Tzanetakis, N., Litopoulou-Tzanetaki, E., and Kotzekidou, P. (2003). Characterization of lactic acid bacteria isolated from a Greek dry-fermented sausage in respect of their technological and probiotic properties. *Meat Science*. **65**, 859–867.

Peralta, E.M., Hatate, H., Kawabe, D., Kuwahara, R., Wakamatsu, S., Yuki, T., and Murata, H. (2008). Improving antioxidant activity and nutritional compounds of Philippine salt-fermented shrimp paste through prolonged fermentation. *Food Chemistry*. **111**, 72–77.

Qiao, F.D. and Ma, C.W. (2004). Characteristics analysis of the specific quality and forming reasons of traditional Xuanwei ham. *Food Science (China)*. **25**, 55–61.

Qiu, H.B. and Zhang, W.H. (2007). Processing of typical Chinese "north ham": Rugao ham. *Meat Research (China)*. **2**, 14–16.

Rantsiou, K., Drosinos, E.H., Gialitaki, M., Urso, R., Krommer, J., Gasparik-Reichardt, J., Toth, S., Metaxopoulos, I., Comi, G., and Cocolin, L. (2005). Molecular characterization of *Lactobacillus* species isolated from naturally fermented sausages produced in Greece, Hungary and Italy. *Food Microbiology*. **22**, 19–28.

Rebecchi, A., Crivori, S., Sarra, P.G., and Cocconcelli, P.S. (1998). Physiological and molecular techniques for the study of bacterial community development in sausage fermentation. *Journal of Applied Microbiology*. **84**, 1043–1049.

Schillinger, U., Kaya, M., and Lucke, F.K. (1991). Behavior of listeria monocytogenes in meat and its control by a bacteriocin-producing strain of *Lactobacillus* sake. *Journal of Applied Microbiology*. **70**, 473–478.

She, X.J. (2005). *Studies on the Mechanism of Mature and Improvement of Processing in Jinhua*. Wuxi, Jiangsu, China: Jiangnan Univ: 59 p.

Somboon, T., Sanae, O., and Kazuo, K. (1998). Lactic acid bacteria found in fermented fish in Thailand. *The Journal of General and Applied Microbiology*. **44**, 193–200.

Sunesen, L.O. and Stahnke, L.H. (2003). Mould starter cultures for dry sausages-selection, application and effects. *Meat Science.* **65**, 935–948.

Toldrá, F., Aristou, M.C., and Flores, M. (2009). Relevance of nitrate and nitrite in dry-cured ham and their effects on aroma development. *Grasas Y Aceites.* **60**, 291–296.

Toldrá, F. and Flores, M. (1998). The role of muscle proteases and lipases in flavor development during the processing of dry-cured ham. *Critical Reviews in Food Science and Nutrition.* **38**, 331–352.

Wang, P.P. (2008). Iberian ham. *Meat Research (China).* **8**, 1.

Wang, W. (2006). Study on Processing Technology Optimization and Physical and Chemical Indices of Sichuan-Type Fermented Sausage. [MSc thesis]. Chongqing, China: Southwest Univ. 48 p.

Wang, X.H., Ma, P., Jiang, D.F., Peng, Q., and Yang, H.Y. (2006a). The natural microflora of Xuanwei ham and the no-mouldy ham production. *Journal of Food Engineering.* **77**, 103–111.

Wang, Z.J., Han, Y.B., and Yao, X.L. (2006b). *Food Technology.* Beijing, China: Metrology Publishing House.

Wei, F.S., Xu, X.L., Zhou, G.H., Zhao, G.M., Li, C.B., Zhang, Y.J., Chen, L.Z., and Qi, J. (2009). Irradiated Chinese Rugao ham: Changes in volatile N-nitrosamine, biogenic amine and residual nitrite during ripening and post-ripening. *Meat Science.* **81**, 451–455.

Wei, L.J., Guo, H., and Che, F.R. (2001). Application of starter culture and development of fermented sausage processing techniques. *Journal of Shenyang Agricultural University (China).* **32**, 394–398.

Wu, Y.C., Kimura, B., and Fujii, T. (2000). Comparison of three culture methods for the differentiation of micrococcus and staphylococcus in fermented squid skiokara. *Fisheries Science.* **66**, 142–146.

Xiao, R. and Xu, K.L. (1998). Characteristics of traditional processing for Xuanwei ham. *Science and Technology of Food Industry (China).* **5**, 49–51.

Yang, H.C. (2006). The traditional famous high quality meat products in China: Xuanwei ham. *Meat Research (China).* **6**, 20–21.

Zeng, T. (2006). *Study on the Knead-Salting Processing and Affects for the Flavor and Quality of Jinhua Ham.* Nanjing, Jiangsu, China: Nanjing Agricultural Univ: 60 p.

Zhang, W.G., Xiao, S., Samaraweera, H., Lee, E.J., and Ahn, D.U. (2010). Improving functional value of meat products. *Meat Science.* **86**, 15–31.

Zhang, X.G. (2007). A long history of Chinese ham culture. *Meat Research (China).* **12**, 1–2.

Zhang, X.L. (2008). *Research on Lipoxygenase and Flavor in Rugao Ham.* Nanjing, Jiangsu, China: Nanjing Agricultural Univ: 76 p.

Zhang, Y.J. and Chen, Y.L. (2002). Research on mature mechanism of traditional dry cured ham. *Meat Industry (China).* **11**, 16–20.

Zhao, G.M. (2004). *Studies on the Effects of Muscle Proteolytic Enzymes in the Processing of Jinhua Ham.* Nanjing, Jiangsu, China: Nanjing Agricultural Univ: 199 p.

Zhao, G.M., Liu, Y.X., Zhang, C.H., Huang, S.J., Zhou, L.M., and Gao, X.P. (2007). Effects of raw ham discrepancies on curing and product quality of Jinhua hams. *Food Science (China)*. **28**, 44–47.

Zhao, G.M., Zhou, G.H., Xu, X.L., Peng, Z.Q., Huan, Y.J., Jing, Z.M., and Chen, M.W. (2005). Studies on time-related changes of dipeptidyl peptidase during processing of Jinhua ham using response surface methodology. *Meat Science*. **69**, 165–174.

Zhao, Z.L. (2008). *Characterization and Application of Micrococci and Staphylococco Isolated from Fermented Sausages as Starter Cultures*. Jinan, Shangdong, China: Shandong Institute of Light Industry: 57 p.

Zhou, G.H. and Zhao, G.M. (2007). Biochemical changes during processing of traditional Jinhua ham. *Meat Science*. **77**, 114–120.

Zhu, J.H. (2005). *Comparison of Proteolytic and Lipolytic Products, and Volatile Flavor Compounds among Different Quality Grade of Jinhua Hams*. Nanjing, Jiangsu, China: Nanjing Agricultural Univ: 62 p.

Zu, S.W. (2004). The dry cured ham in foreign countries. *Meat Industry (China)*. **1**, 16–19.

Zu, S.W. and Zhang, C.R. (2004). Research on dry cured ham. *Science and Technology of Food Industry (China)*. **9**, 89–90.

6

SOLID-STATE FERMENTED SOYBEAN

YONGQIANG CHENG AND BEIZHONG HAN

College of Food Science & Nutritional Engineering,
China Agricultural University
Beijing, PR China

Contents

6.1 Douchi

6.1.1 *Introduction*

As a Chinese traditional fermented soy product, *douchi* has been studied in recent years. There are several types of douchi being consumed in China nowadays. The production conditions and nutritional value are important research fields for research. The history, production methods, nutritional value, and especially recent progress on the functionalities of douchi are summarized in this chapter. The trend of study on this product is also discussed.

6.1.1.1 History   Douchi, fermented from soybean, is an important condiment from ancient China. It was called *youshu* in ancient times, a term meaning boiled soybean fermented in an airtight environment. It is estimated that in China, douchi production has a history of more than 2000 years. Based on the evidence discovered in Han Tomb No. 1 at Ma-wangdui, there is no doubt that by the late Western Han Dynasty (BC 206–AD 25), douchi had become a major commodity in the economy of the realm (Needham, 2000). There is much literature from that time describing the scenes of douchi production.

During the period of the Sui (AD 581–618) and Tang (AD 618–907) dynasties, two types of douchi had been produced, i.e., Xian (salty) and Dan (bland). The Dan douchi was used as a part of Chinese traditional herbal preparations, and Xian douchi was used as a condiment with the addition of different spices, such as pepper powder. Depending on the water contents, douchi could also be divided into dry and wet (water) types. The dry douchi is mainly produced in the southwest part of China such as the Sichuan Province, and the middle area of China, such as the Hunan Province, giving an oily and glossy appearance to the beans, whereas water douchi is mainly produced in the north area of China, and it is soft with adhesive texture.

6.1.2 *Production Method and Areas*

In China today, most of the douchi are produced by natural fermentation. Generally, although there are different types of douchi

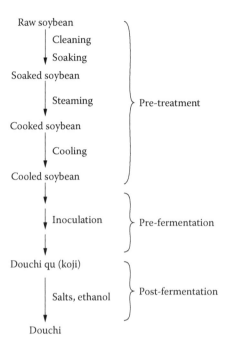

Figure 6.1   The schematic diagram for production of douchi.

in different areas in China, the production methods are very similar with only slight changes. The common production method of douchi is shown in Figure 6.1.

Soybeans are washed, and soaked for 3–4 hours at room temperature. After draining, the soybeans are steamed for 4 hours to allow about 52% water content for the soybean, and then cooled to 28°C. The soybeans mixed with 5% inoculum of fungus are incubated for 4 days under a controlled temperature (28°C–30°C) to make *koji*. After mixing the dressing, douchi is fermented and aged by storing in a closed bottle or sunning for more than half a year (at present, some factories have succeeded in shortening this period). For some varieties of douchi, spices are added together with salt into koji.

In different production areas of douchi, the prevalent microorganisms are different, resulting in douchi of three types: *Aspergillus*, *Mucor*, and bacteria. There are slight differences among these three types of douchi.

1. *Aspergillus* type: Most of the douchi produced in Hunan (middle China) and Guangdong Province (the south area of China) are this type. If using pure culture, the *Aspergillus orazye* 3.042 is commonly used. This *Aspergillus* strain is the most popular *Aspergillus* used as a pure starter for soy sauce, douchi, *sufu*, and soy paste production in China at present, screened by Chinese researchers in the 1960s. It is known that *Aspergillus* is a mild thermostable microorganism and for this reason, the *Aspergillus*-type douchi is seldom affected by the seasons.

2. *Mucor* type: *Actimomucor elegans* is the most commonly used *Mucor* for the production of douchi. This kind of douchi is often produced in the middle areas of China, such as Hunan Province. For the natural fermentation of douchi in these areas, the dominant microorganism is *Mucor,* and the low temperature is preferred for the koji production. In recent years, some Chinese researchers screened a pure culture of *Mucor* and improved the temperature stability of the mucor as well as shortened the fermentation period.

3. Bacteria type (water douchi): Water douchi is mainly produced in the north area of China (such as Shandong province) and southwest China (such as Yunnan Province). The production of water douchi is similar to *natto*, the Japanese fermented soy product. Although the dominant microorganisms in water douchi are bacteria, the pure culture of the bacteria has not been reported yet.

In recent years, with the development of the food industry in China, as an important condiment for Chinese dishes douchi also drew the attention of researchers. In particular, the nutritional and health benefits of douchi have been reported by many researchers.

6.1.2.1 Nutritional and Health Benefits   The main materials for douchi production are soybeans (*Glycine max* [L] Merrill) and/or black soybeans (*Glycine max* var; soybeans with black skin). It is well known that soybeans are rich in protein, fat, carbohydrate, and have a very high nutrition value based on recent studies. In fact, the black soybean has a higher protein content (about 45%–48% [d.b.], while for

Table 6.1 Components of Black Soybeans from Different Production Areas (g/100 g, d.b.)

PRODUCTION AREA	GUANGDONG	GUANGXI	HENAN	JIANGXI	VIETNAM
Total acids	1.69	1.72	1.91	1.82	1.90
Total amino acids	28.02	27.30	25.64	26.03	25.23
Amino acid nitrogen	0.74	0.70	0.65	0.62	0.63

soybeans it is about 35%–42%) than soybeans and has anthocyanin in the skin. The black soybean has a higher ash content (about 4.5%) than soybeans. There are slight differences in black soybeans from different areas as shown in Table 6.1.

After fermentation, it is considered that nutrient values are enriched, compared with the raw materials. Up to now, much physiological functionality has been found in douchi, which includes antioxidative activity (Wang et al., 2007b, 2008a), antihypertensive effects, fibrinolytic activity (Peng et al., 2003), antidiabetic properties, and acetylcholine inhibitory activity.

6.1.2.1.1 Antioxidative Activity The traditional fermented soybean foods such as sufu, *miso*, natto, and *tempeh*, have been found to exhibit remarkably strong antioxidative activity. Recent studies also revealed the antioxidative activity of douchi during its processing.

The 1,1-diphenyl-2-picrylhydrazyl radical (DPPH) and 2, 2′-azinobis (3-ethylbenzo)-thiazoline-6-sulfonic acid (ABTS) scavenging activity of douchi was shown in Figure 6.2.

The antioxidative properties during douchi prefermentation processing increased significantly ($p < 0.05$). This might have been due to the fact that antioxidative compounds were released during the fermentation.

6.1.2.1.2 Antihypertensive Effect For douchi, after fermentation, part of the protein in the raw materials is degraded into a peptide, which is considered to be one of the substances having antihypertensive effects. Zhang et al. (2006) studied the angiotensin-I converting enzyme (ACE)-inhibitory activity of douchi *qu* (koji) fermented by pure *Aspergillus egyptiacus*. Douchi fermented for 48 hours, and 72 hours were compared with douchi being secondary-fermented for 15 days. The results showed that ACE-inhibitory activities were

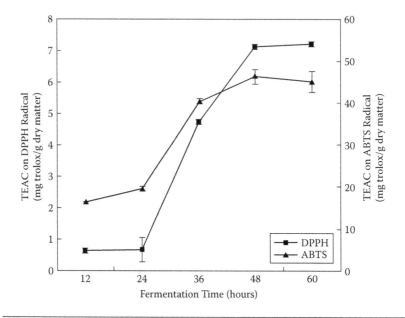

Figure 6.2   Scavenging activity of douchi extracts by DPPH method and ABTS method during the prefermentation. Results are the mean ± S.D. of three determinations.

improved following the fermentation. ACE-inhibitory activities of 48 hrs with primary fermented douchi qu did not change dramatically after preincubation with ACE, but increased greatly after preincubation with gastrointestinal proteases. The results suggest they were pro-drug-type or a mixture of pro-drug-type and inhibitor-type inhibitors (Figure 6.3). The ACE inhibitors in 48-hour fermented douchi qu were fractionated into four major peaks by gel filtration chromatography on Sephadex G-25. Peak 2, which had the highest activity, had only one peptide, composed of phenylalanine, isoleucine, and glycine with a ratio of 1:2:5.

6.1.2.1.3 Antidiabetic Properties   As one of the *in vitro* evaluation methods for screening of the potential components effective for blood glucose level control, the α-glucosidase inhibitory activity is commonly employed. The anti-α-glucosidase activity of aqueous douchi extract was reported (Chen et al., 2007). Thirty-one douchi samples collected from different parts of China exerted various degrees of inhibitory activity against rat intestinal α-glucosidase (Table 6.2). Among them, three samples, sourcing from Hunan, Sichuan and Jiangxi Provinces, respectively, showed significantly higher anti-α-glucosidase activities

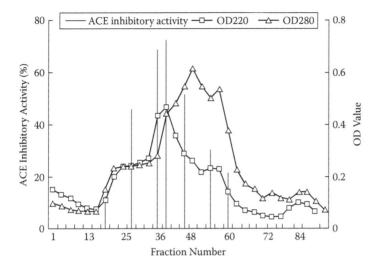

Figure 6.3   Elution profile of ACE inhibitor digested with gastrointestinal proteases. (Zhang, X.F. et al., 2006. *China Brewing.* **6**, 17–20, in Chinese.)

than other samples ($p < 0.05$). Moreover, three fungal strains, namely *Aspergillus oryzae*, *Actinomucor elegans*, and *Rhizopus arrhizus* were then used to prepare douchi in the laboratory. The α-glucosidase inhibitory activities of all soybeans increased slightly and no apparent differences were found in anti-α-glucosidase activity among the soybeans at the end of prefermentation. For maturation, different salt levels (5.0%, 7.5%, 10.0%, and 12.5%) were then added to the douchi qu resulted from prefermentation. The anti-α-glucosidase activity of douchi qu fermented with *A. oryzae* were higher than those of *A. elegans* and *R. arrhizus* and the highest anti-α-glucosidase activity was observed in douchi qu fermented with *A. orzyzae* at 5.0% and 7.5% salt levels. The results indicated that *A. oryzae* could utilize cooked black soybean to generate certain α-glucosidase inhibitor more effectively than *A. elegans* and *R. arrhizus*.

It is also reported that the water-extracted douchi exerted a strong inhibitory activity against rat intestinal α-glucosidase in foodstuffs. In borderline and developed diabetic subjects, 0.3 g of douchi-extract (DE) significantly inhibited postprandial blood glucose levels of rats (Fujita et al., 2001a). For safety evaluation, nine healthy subjects were given 1 g of DE before every meal (3 g/day) for 12 weeks. None of the subjects indicated changes in hematological and relevant biochemical parameters, body weight or BMI, suggesting the safety of DE. In a

Table 6.2   Brand Names, Origins, Type of Soybean Used, and Anti-α-Glucosidase Activities of Commercial Douchi Samples

SAMPLE NO.	BRANDS	ORIGINS	TYPES OF SOYBEAN	ANTI-α-GLUCOSIDASE ACTIVITIES[a]
1	Y.J.Q	Guangdong (south)	Black	8.474
2	S.S	Guangdong (south)	Black	7.526
3	X.Q	Guangdong (south)	Black	5.131
4	H.S.Q	Guangdong (south)	Black	5.705
5	Q.G	Hunan (middle)	Black	13.063
6	T.P.Q	Hunan (middle)	Black	7.467
7	L.Y.P.X	Hunan (middle)	Black	6.04
8	Q.W	Hunan (middle)	Black	4.739
9	R.M	Sichuan (southwest)	Black	1.812
10	A.J	Sichuan (southwest)	Black	8.930
11	Y.C	Sichuan (southwest)	Black	4.300
12	Y.C.J.X	Sichuan (southwest)	Black	5.300
13	C.Q.W.Z	Sichuan (southwest)	Yellow	2.856
14	E.W	Sichuan (southwest)	Black	5.746
15	W.N.H	Sichuan (southwest)	Black	13.863
16	J.X.Z.Z	Jiangxi (middle)	Yellow	0.790
17	D.X.Y	Jiangxi (middle)	Black	3.200
18	J.X	Jiangxi (middle)	Black	8.060
19	M.W	Jiangxi (middle)	Yellow	12.230
20	Q.X	Guizhou (southwest)	Yellow	2.376
21	G.Y	Guizhou (southwest)	Yellow	8.212
22	L.Y	Shandong (north)	Yellow	6.651
23	X.Y	Beijing (north)	Yellow	2.203
24	Y.M	Guangxi (south)	Black	5.240
25	N.N	Guangxi (south)	Black	3.060
26	Z.J	Hebei (north)	Yellow	6.238
27	J.L.D.J.1	Jilin (northeast)	Yellow	4.780
28	J.L.D.J.2	Jilin (northeast)	Yellow	1.182
29	TJ	Tianjin (north)	Yellow	4.135
30	K.M	Yunnan (southwest)	Yellow	3.476
31	X.I.D.J	Xinjiang (west)	Yellow	2.400

Source: Chen, Y. et al, 2007. *Food Chemistry.* **103**(4), 1091–1096.

[a] Anti-α-glucosidase activities of douchi samples were computed as the slope values from the curves of absorbance versus the concentration of aqueous douchi extract. The higher the slope value, the stronger the anti-α-glucosidase activity of the aqueous douchi extract.

noncomparative experiment, 18 type-2 diabetic patients ingested 0.3 g of DE before every meal (0.9 g/day) for 6 months (mo). Results indicated that blood glucose (mean: 9.31 ± 0.71 mmol/L) and HbA1c (mean: $10.24 \pm 0.58\%$) levels gradually decreased, and significant effects were elicited on the blood glucose levels (8.61 ± 0.66 mmol/L; $p < 0.01$) after 6 months and HbA1c after 3 ($9.13 \pm 0.43\%$; $p < 0.05$) and 6 months ($8.96 \pm 0.30\%$; $p < 0.05$) postingestion of DE. Indexes for serum lipids and total cholesterol level revealed moderate decreases with a slight increase in the high-density lipoprotein (HDL) level after DE ingestion. However, triglyceride (TG) levels significantly decreased at 3 ($p < 0.05$) and 6 mo ($p < 0.01$) postingestion of DE. At the same time, other biochemical parameters were not affected in any of the patients, and no one complained of any side-effects or abdominal distension. It is suggested that the extract of douchi that exhibited α-glucosidase inhibitory activity had an antihyperglycemic effect and may prove useful for improving glycemic control in patients suffering from noninsulin-dependent diabetic mellitus (Fujita et al., 2001b).

6.1.2.1.4 Antiacetylcholinesterase Activity Alzheimer's disease (AD) is an irreversible, progressive brain disease that slowly destroys memory and thinking skills and, eventually, the ability to carry out the simplest tasks of daily living. In most people with AD, symptoms first appear after age 60. AD is the most common cause of dementia among older people, but it is not a normal part of aging. A component that has acetylcholinesterase (AChE) inhibitory activity is considered to be potentially beneficial for preventing AD. After studying the douchi produced in the laboratory, Zou (Zou, 2006) found that at the initial 24 hours of the prefermentation, the IC_{50} of AChE inhibitory activity of douchi extract was 0.436 mg/ml, and decreased with the proceeding of fermentation. Liu et al (Liu et al., 2009) investigated the concentration and distribution of isoflavones in 19 representative commercial douchi products (Table 6.3) and measured their AChE inhibitory activity. Isoflavone aglycones are the predominant isoflavone forms in Chinese commercial douchi samples. The total content of isoflavones in douchi extracts were from 24 to 1471 µg/g (dry matter) (Figure 6.4). Among different samples, *Aspergillus*-type douchi had more isoflavone aglycones content than that of *Mucor*-type

Table 6.3 Brand Names, Origins, Type of Soybean Used, and Types of Commercial Douchi Samples

SAMPLE CODE	BRANDS	ORIGINS	TYPE OF SOYBEAN	TYPE OF DOUCHI
1	C.W	Sichuan (southwest)	Yellow	*Mucor*-type
2	T.C	Sichuan (southwest)	Black	*Mucor*-type
3	R.M	Chongqing (southwest)	Yellow	*Mucor*-type
4	Y.C	Chongqing (southwest)	Yellow	*Mucor*-type
5	Q.M	Chongqing (southwest)	Yellow	*Mucor*-type
6	A.J	Chongqing (southwest)	Yellow	*Mucor*-type
7	Y.F	Guangdong (south)	Black	*Aspergillus*-type
8	X.Q	Guangdong (south)	Black	*Aspergillus*-type
9	Y.J.Q	Guangdong (south)	Black	*Aspergillus*-type
10	Y.M	Guangxi (south)	Black	NC*
11	Q.X	Guizhou (southwest)	Yellow	*Bacillus*-type
12	K.M	Yunnan (southwest)	Yellow	*Bacillus*-type
13	T.M.S	Hunan (middle)	Black	*Aspergillus*-type
14	T.P.Q	Hunan (middle)	Black	*Aspergillus*-type
15	L.Y.P	Hunan (middle)	Black	*Aspergillus*-type
16	Q.W	Hunan (middle)	Black	*Aspergillus*-type
17	D.X.Y	Jiangxi (middle)	Black	*Aspergillus*-type
18	K.W.K	Shanghai (east)	Black	*Aspergillus*-type
19	Y.B	Jilin (northeast)	Yellow	NC[a]

Source: Chen, Y. et al, 2007. *Food Chemistry.* **103**(4), 1091–1096.
[a] Not confirmed.

and *Bacillus*-type douchi. Samples also indicated various extents of AChE inhibitory activity. The IC_{50} value of AChE inhibitory activities ranged from 0.040 to 2.319 mg/mL (Figure 6.4). *Aspergillus*-type douchi extracts exhibited significantly higher AChE inhibitory activity than the other two types (*Mucor* and *Bacillus*). Some brands of douchi that have low contents of isoflavone aglycones showed much lower inhibitory activity. However, high inhibitory activities did not mean high isoflavone content; maybe some other substances contributed to the inhibition, which should be further studied. However, in a different study, a soy diet that was rich in isoflavones was able to reverse the increase of AChE in the hippocampus of ovarectomized rats (Lee et al., 2004). These contradictive results need further study to identify active components (functional factor) in douchi.

It has been also reported that AChE activity in the soy isoflavones diet fed animals was significantly inhibited in the cortex, basal

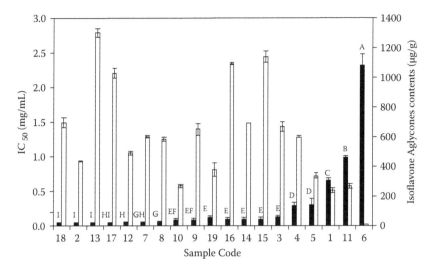

Figure 6.4 Isoflavoneaglycones contents and the AChE inhibitory activities of various douchi extracts. Values are means of 3 replications. Vertical bars indicate standard errors. Sample codes 1 to 19 are the same samples as in Table 6.4. Samples are rearranged in order of IC50 values. IC50: concentration of crude extract required to inhibit enzyme activity by 50%. Means for IC50 without a common letter are significantly different at $p < 0.05$. Note: ☐: Isoflavone aglycones contents, ■: IC50.

forebrain, and hippocampus compared with animals fed the control diet (Pal and Tandon, 1998; Monteiro et al., 2007). In a more recent study, soy isoflavones treatment could result in a significant decrease in acetylcholinesterase (AChE) activity and increase in the contents of some amino acid neurotransmitters such as glutamic and aspartic acids in the frontal cerebral cortex and hippocampus of mice (Liu et al., 2007).

6.1.3 Product Characteristics

Production conditions, such as processing, NaCl supplementation, have an influence on the chemical and physical properties of douchi. It is reported (Wang et al., 2007c) that 61% of the isoflavones in raw soybeans were lost when the NaCl content was 10%, which is mainly attributed to prefermentation (43%) and postfermentation (18%) during the douchi processing. While a pretreatment did not generate major differences in total isoflavone content during douchi processing, isoflavone composition was altered. The levels of aglycones increased, while the corresponding levels of α-glucosides, malonylglucoside, and

Table 6.4   Changes in Hunter L-Values, a-Values, and b-Values and Total Color Difference (ΔE) During Processing[a]

STEP	L-VALUES	a-VALUES	b-VALUES	ΔE
Soaked soybeans	54.40 ± 2.13A	N–4.03 ± 0.28H	14.39 ± 2.30A	—[b]
Steamed soybeans	46.31 ± 0.71B	0.35 ± 0.18EF	5.95 ± 1.30D	12.48
12 h	43.28 ± 0.24C	0.20 0 ± 01FG	9.54 ± 0.06B	12.85
24 h	41.74 ± 0.72D	0.40 ± 0.08E	5.90 ± 0.24D	15.87
36 h	40.92 ± 0.40DE	0.62 ± 0.06D	10.27 ± 0.10B	14.85
48 h	40.14 ± 0.24E	0.12 ± 0.08G	7.02 ± 0.20DC	16.58
60 h	37.76 ± 0.14F	1.6 ± 0.01C	6.52 ± 0.01DC	19.25
1wN10[c]	35.82 ± 0.52G	3.79 ± 0.09A	10.13 ± 0.06B	20.06
2wN10[c]	29.04 ± 0.20G	3.20 ± 0.07B	5.99 ± 0.07D	27.67
3wN10[c]	29.24 ± 0.24H	3.30 ± 0.04B	5.92 ± 0.08D	27.54

[a] Values represent the mean ± standard deviation; n = 9. Values in a column with different superscripts were significantly different ($p < 0.05$).
[b] A reference sample (soaked soybeans), cannot be calculated.
[c] 1w = 1 week, N10 = 10% NaCl content in douchi.

acetylglucoside decreased. Further, isoflavones in the form of aglycones exceeded 90% of total content following postfermentation. Changes in isoflavone isomer distribution were found to be related to α-glucosidase activity during fermentation, which was affected by NaCl supplementation.

The changes in Hunter L-values, a-values, b-values, and total color difference (ΔE) during the douchi processing are presented in Table 6.4. During the entire processing procedure from soaked soybeans to douchi, Hunter L-values decreased gradually, a-values increased, and b-values decreased but without a gradual tendency. All ΔE values were greater than 12, suggesting a great difference in the color using soaked samples as a control. The results suggested that the processing steps examined in this study, including soaking, cooking, and fermentation, could cause significant differences in Hunter L-values, a-values, b-values, and ΔE values ($p < 0.05$). Cooking and fermentation made the steamed soybeans darker, and decreased the L-values. Some new components were produced during the fermentation by complex biochemical reactions, similar to other soy products (Wang et al., 2006a).

Figure 6.5 shows the changes in firmness of douchi during postfermentation at various NaCl contents. The firmness of douchi decreased during postfermentation. The firmness of douchi with 10% salt

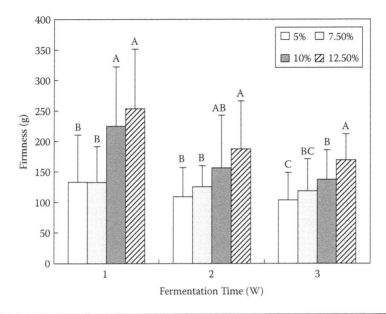

Figure 6.5   Changes in the firmness of douchi during douchi postfermentation at various NaCl content. Results are mean ± S.D. of 30 determinations. Bars with different letters were significantly different ($p < 0.05$).

concentration decreased rapidly from 225 g to 138 g within 3 weeks. The level of the firmness increased with the increase of NaCl content at the same fermentation time ($p < 0.05$) between 5% and 12.5% in salt content.

The pH value increased rapidly from 6.88 to 8.43 during prefermentation (Figure 6.6). It has been proved that the fungi produces protease and hydrolyzed protein during douchi fermentation. Most fermentation of soybean protein-rich materials with fungi caused a characteristic increase in pH value.

The pH value decreased gradually during postfermentation (Figure 6.7). The pH value decreased with the decrease of NaCl content at the same fermentation time. The pH values were 6.60 and 6.89 at 5% and 12.5% NaCl contents at the third week, respectively.

6.1.3.1 Characteristics of Douchi Production   Stated as above, there are many differences for the production of douchi compared with some other fermented soybean products, such as natto and tempeh. Table 6.5 summarized the differences of these three types of products.

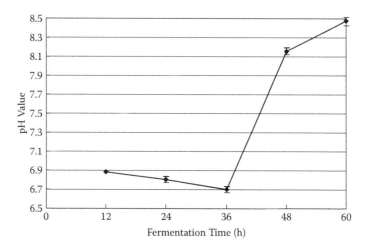

Figure 6.6   Changes in pH value during douchi prefermentation. Results are mean ± S.D. of three determinations.

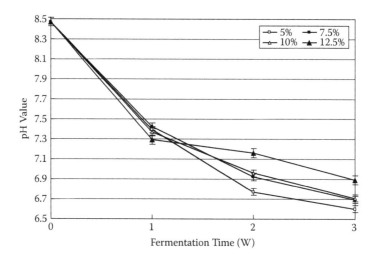

Figure 6.7   Changes in pH value during douchi postfermentation at various NaCl content. Results are mean ± S.D. of three determinations.

6.1.3.2 Potential Trends of the Future: A Study   Because it has been established as a traditional Chinese fermented soy product, studies on douchi's production and nutrition values have been conducted only in the past 2 decades. There are still many issues—such as the pure culture fermentation of production, thermo-stable starter suitable for production in hot seasons, automation of the production line, mechanism of the functionalities, even the way of popularization of

Table 6.5 Differences among the Production of Natto, Tempeh, and Douchi

ITEM	NATTO	TEMPEH	DOUCHI
Materials	Soybean, black soybean	Soybean, soybean-cereals	Soybean, black soybean
Pretreatment	Cleaning, soaking, draining,	Cleaning, soaking with acid, dehulling	Cleaning, soaking, draining,
Soaking	6–18 h in room temperature	12–15 h in room temperature, adding acid	3–10 h in room temperature
Steaming	1.5 kg/cm^2, 30–40 min, water content about 64%	100°C, 30 min	*Mucor* type: 1 h, water content 45%; *Aspergillus* type: 45 min, water content 55–56%; Bacteria: 30–40 min
Microorganism	*Bacillus Natto*	*Rhizopus*	*Mucor*, *Aspergillus*, Bacteria (*Bacillus subtilis*)
Inoculation temperature	37°C	38°C	Room temperature
Fermentation way	Pure starter	Pure starter or natural fermentation	Natural fermentation
Fermentation conditions	40°C, 18–24 h	37°C, 24 h	*Mucor*: 28–35°C, 15–21days, 20°C aging for more than 6 months; *Aspergillus*: 28°C–32°C, 5–7 days, 30°C–35°C aging 40 days; before aging, mixed with salt and/or spices
Post-treatment	Cool	Slice, frying, toasting	Drying, sterilizing

it to the world, etc.—that should be further studied. With the economic development of China, there will be much greater progress on the fermentation technologies for douchi production and improvement of nutrition values.

6.2 Sufu

6.2.1 Introduction

Sufu—*fu-ru* written in hieroglyphics—is a traditional Chinese fermented soybean curd and a highly flavored, soft creamy cheese-type product that can be used in the same way as cheese (Su, 1986; Steinkraus, 1996).

Sufu is the name for the product that first appeared in the literature (Wang and Hesseltine, 1970). Literally, sufu (fu-ru) means "molded milk" and tosufu (*dou-fu-ru*) means "molded soymilk." Because of the numerous dialects used in China and the difficulties of phonetic translation from Chinese into English, sufu has appeared in literature under many different names. The following synonyms for sufu in the literature have been found: *sufu, tosufu, fu-ru, dou-fu-ru, tou-fu-ru, toe-fu-ru, jiang-dou-fu, fu-yu,* and *foo-yue.* Sufu is also known as *tofuyo, nyu-fu,* or *fu-nyu* in Japan (Yasuda and Kobayashi, 1987), *chao* in Vietnam, *ta-huri* in the Philippines, *taokaoan* in Indonesia, and *tao-hu-yi* in Thailand (Beuchat, 1995). These names confuse Western people as well as the Chinese. Officially, sufu should be named *furu* (or *doufuru*) in Chinese.

Manufacture of tofu (soybean curd) began during the era of the Han Dynasty. The *Ben-Cao-Gang-Mu* (*Chinese Materia Medica*), compiled by Li Shi-zhen in 1597, indicated that tofu was invented by Liu An (179 BC to 122 BC), king of Weinan (Shi and Ren, 1993; Steinkraus, 1996). However, it is not known when sufu production began. Due to the long history and incomplete written records, no attempt was made to search for its origin. The first historical record mentioned that the sufu process was carried out in the Wei Dynasty (220–265 AD; Hong, 1985; Wang and Du, 1998). It became popular in the Ming Dynasty (1368–1644), and there are many books describing sufu processing technologies (Zhang and Shi, 1993).

Sufu products are manufactured both commercially and domestically, and the annual production is estimated over 300,000 metric tons in China. Sufu is consumed as an appetizer and a side dish, for example, with breakfast rice or steamed bread. Sufu adds zest to the bland taste of the rice and flour diet. Since it is made from soybeans and is an easily digested and nutritious protein food, Chinese people consider it a health food. Sufu is a highly flavored, creamy cheese-like product, so it would be expected to be suitable for use in western countries as a healthy, noncholesterol food from plant origin. Table 6.6 summarizes some major chemical constituents of sufu.

6.2.2 Classification of Sufu

There are many different types of sufu, which are produced by various processes in different localities in China (Li, 1997; Wang and Du, 1998).

Table 6.6 Proximate Composition of Commercial Sufu

COMPONENT	CONTENT[a]
Moisture (g)	58–70
Crude protein (g)	12–17
Crude lipid (g)	8–12
Crude fiber (g)	0.2–1.5
Carbohydrate (g)	6–12
Ash (g)	4–9
Calcium (mg)	100–230
Phosphorus (mg)	150–300
Iron (mg)	7–16
Thiamin (V_{B1}) (mg)	0.04–0.09
Riboflavin (V_{B2}) (mg)	0.13–0.36
Niacin (mg)	0.5–1.2
V_{B12} (µg)	1.7–22
Food energy (KJ)	460–750

Source: Wang, R.Z. and Du, X.X. (1998). *The Production of Sufu in China.* China Light Industry Press, Beijing, pp. 1–4; Su, Y.C. (1986). Sufu. In *Legume-Based Fermented Foods*, eds. N.R. Reddy, M.D. Pierson, and D.K. Salunkhe, CRC Press, Boca Raton, Florida, pp. 69–83.

[a] per 100 g sufu fresh weight.

Four types of sufu can be distinguished according to the processing technologies.

Mold-Fermented Sufu: Four steps are normally involved in making this type of sufu: (1) Preparing tofu, (2) Preparing *pehtze* (*pizi*) with a pure culture mold fermentation, (3) Salting, (4) Ripening.

Naturally Fermented Sufu: Four steps are also normally involved in making this type of sufu; (1) Preparing tofu, (2) Preparing pehtze (pizi) with natural fermentation, (3) Salting, (4) Ripening. (see: Traditional process with natural fermentation).

Bacteria-Fermented Sufu: Five steps are normally involved in making this type of sufu; (1) Preparing tofu, (2) Presalting, (3) Preparing pehtze (pizi) with a pure culture bacterial fermentation, (4) salting, and (5) ripening. During the presalting, the tofu adsorbs the salt till the salt content of tofu reaches about 6.5%, which takes about 2 days. Pehtze is prepared by using pure cultured *Bacillus* spp. or *Micrococcus* spp. at 30°C–38°C for about one week. In order to keep the shape of the final product, pehtze is dried at 50°C–60°C for 12 hours before salting. The ripening time normally takes less than 3 months. This sufu is made in some areas such as Kedong (Heilongjiang) and Wuhan (Hubei).

Enzymatically Ripened Sufu: Three steps are normally involved in making this type sufu: (1) Preparing tofu, (2) salting, and (3) ripening. Because there is no fermentation before ripening, some koji is added in the dressing mixture for enzymatic ripening. The ripening time takes 6–10 months. This product of sufu is produced only in a few areas of China, such as Taiyuan (Shanxi) and Shaoxin (Zhejiang).

According to the color and flavor, sufu can be classified into four types, which are mainly based on the different ingredients of dressing mixtures in the ripening stage.

Red Sufu: The dressing mixture of red sufu mainly consists of salt, *angkak* (red *kojic* rice), an alcoholic beverage, sugar, flour (or soybean) paste, and some spices. The outside color of the sufu is from red to purple, and the interior color is from light yellow to orange. Because red sufu possesses an attractive color and strong flavor, it is the most popular product all over China. Angkak *(anka,* red kojic rice or red qu) is a product produced by solid-substrate fermentation of cooked rice with various strains of *Monascus* spp., such as *M. purpureus.* It has a specific aroma and purple red color and has been used as a natural colorant in red sufu and some other traditional food.

White Sufu: White sufu has similar ingredients as red sufu in the dressing mixture but without angkak. It has an even light yellow color inside and outside. White sufu is a popular product in the south of China because it is less salty than red sufu.

Gray Sufu: The dressing mixture of gray sufu contains the soy whey left over from making tofu, salt and some spices. Gray sufu is ripened with a special dressing mixture, which could be dominated by both bacteria and mold enzymes and results in a product with a strong, offensive odor. The preparation of this type of sufu is top secret in the industry and is slowly becoming a lost art (Wang and Fang, 1986).

Other Types: Other sufu types are made by adding various ingredients to the dressing, including vegetables, rice, bacon, and even higher concentrations of alcohol. For instance, a dressing mixture containing high levels of ethanol results in a product having a marked alcoholic bouquet. This product is called *zui-fang* (or *tsui-fang*), which means "drunk sufu."

Sufu can also be classified according to size, such as "big sufu" and "small sufu," and shape, "square sufu," and "chess (round) sufu" (Huang, 1991).

6.2.3 Manufacturing Process

Most sufu products are produced by a similar principle, which involves four main steps: (1) Preparation of tofu, (2) preparation of pehtze, (3) salting, and (4) ripening. The schematic diagram for the production of sufu is shown in Figure 6.8 (Han et al., 2001b).

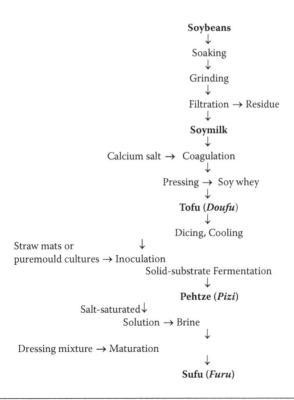

Figure 6.8   The schematic diagram for production of sufu.

6.2.3.1 Preparation of Tofu   The production of tofu is highly mechanized. The yield and quality of commercial tofu are affected by soybean composition (Murphy et al., 1997), soymilk characteristics (Lim et al., 1990), coagulant, and other factors (Molzuddin et al., 1999). Preparation of tofu used for sufu mainly follows the process technologies for commercial tofu except for slight differences in some steps. First, soybeans of selected quality are washed and ground with added water to a milky slurry in a steel mill. The slurry is then heated and filtered through cloth to separate soymilk. Coagulation is done at 70°C–80°C by addition of calcium sulphate and magnesium sulphate. Generally, 20% more coagulant is used to produce tofu for sufu than for regular tofu production. Coagulant added is 2.5%–3.5% of the dry starting weight of soybeans. Moreover, after the calcium salts are mixed with soymilk, the mixture needs to be agitated vigorously in order to get a homogenous coagulum, and then it is set aside for 10–15 min to complete the coagulation. The precipitate is pressed to remove excess water (soy whey) with cheesecloth in a mechanized press. Finally, a soft, cake-like tofu results, which can then be cut into desired sizes (normally rectangular pieces, approximately 3.2 × 3.2 × 1.6 cm). Moisture and pH of tofu are 70%–79% and 6–7, respectively, varying with the type of sufu to be produced (Li, 1998; Wang and Du, 1998).

6.2.3.2 Preparation of Pehtze   Pehtze, fresh bean curd overgrown with mycelium of molds, is produced by means of solid-substrate fermentation after inoculation with pure culture molds. The process of making pehtze was considered a natural phenomenon until 1920, when a microorganism believed to be responsible for sufu fermentation was isolated and identified (Su, 1986). Wai (Wai, 1929) studied the microorganisms found in sufu and isolated a pure culture strain of *Mucor* from the fermenting pehtze samples. The pehtze obtained using these fungal species were equally good in color and flavor.

Microorganisms: The fungal genera involved (*Actinomucor, Mucor,* and *Rhizopus*) all belong to the Mucoraceae. The mold used in fermentation of sufu is critical and has to possess certain characteristics. First of all, the mold must have enzyme systems with high proteolytic and lipolytic activities since it grows on tofu that is a protein and lipid-rich and carbohydrate-poor medium. Secondly, the mold must have white or, at most, slightly yellowish white mycelium to ensure that the sufu has an attractive appearance. Thirdly, the texture of the mycelial mat should be dense and tenacious so that a film formed on the surface of the pehtze will act as an overcasing to protect the final product of sufu from deformation. Finally, the mold growth should not develop any off-odor, astringent taste, or mycotoxins and the mold should resist undesired bacterial contamination during the fermentation.

Han et al. (2004a) reported the identity and phylogenetic relationships of mold starter cultures used for the preparation of pehtze. Starter cultures used in commercial pehtze fermentation were obtained from factories located in several provinces of China and Vietnam, isolated from their pehtze, and some were obtained from culture collections. They were identified as *Actinomucor repens (elegans)*, *A. taiwanensis, Mucor circinelloides, M. hiemalis, M. racemosus, Mucor sufu,* and *Rhizopusmicrosporus* var. *microsporus.*.

Inoculation: A pure culture inoculum of mold can be prepared, starting from agar slant culture, by liquid or solid-substrate culture in roux bottles. The medium used for solid-substrate culture consists of bran and water (1:1.2–1.4) and that for liquid-substrate culture contains soy whey (a by-product of tofu preparation) with added maltose (2%–3%) and peptone (1.5%–2.0%). After sterilization by autoclaving, the medium is incubated at 25°C–28°C for 3 days. The spore suspension ($\sim 10^5$ CFU/ml) is harvested and inoculated on the surfaces of the tofu with manually operated sprayers, comparable to those used for spraying plants.

Incubation: The inoculated tofu is placed, evenly spaced, in wooden or plastic trays, the bottoms of which are made of bamboo, wooden, or plastic strips. The loaded trays are piled up in an incubation room, where a controlled temperature (about 25°C), a relative humidity (about 90%), and good aeration are needed for optimum growth of the mycelia. The thin white mycelia are developed in 8–12 hours and a thick mycelial mat is formed after 36–40 hours of incubation. Then the room temperature is decreased by aeration to prevent overgrowth of mold, until a slightly yellowish white color appears, at which point formation of fresh pehtze is complete. The total cultivation time is about 48 hours, which is much less than in the traditional way (5–15 days). Before pehtze is transferred to the salt treatment, the mold mycelial mat should be flattened by hand, in the same way as done in the traditional way.

Chou et al. (1988) reported that the incubation temperature, humidity, and cultivation time greatly affected the growth of and enzyme production by *A. taiwanensis* growing on tofu. Optimum conditions observed for growth of *A. taiwanensis* were 25°C–30°C at 97% relative humidity when tofu of 65% moisture content was inoculated. Under these conditions, a maximum production of protease, lipase, α-amylase, and α-galactosidase was achieved.

Growth and production of protease, lipase, α-amylase, glutaminase, and α-galactosidase by *A. elegans* and *R. oligosporus* as influenced by temperature and relative humidity (RH) during the sufupehtze preparation were studied (Han et al., 2003a). Optimum conditions for biomass formation rate of *A. elegans* and *R. oligosporus* were 25°C at RH 95%–97%, and 35°C at RH 95%–97%, respectively. Maximum yields of protease (108 U/g pehtze), lipase (172 U/g) and glutaminase (176 U/g) by *A. elegans* were achieved after 48 hours at 25°C and RH 95%–97%, and highest yields of α-amylase (279 U/g pehtze) and α-galactosidase (227 U/g) at 30°C and RH 95%–97% after 48 hours and 60 hours of incubation. Highest protease (104 U/g pehtze), and lipase (187 U/g) activities of *R. oligosporus* were observed after 48 hours at 35°C and RH 95%–97%, while maximum α-amylase (288 U/g pehtze) and glutaminase (187 U/g) activities were obtained after 36 hours at 35°C and RH 95%–97%. Maximum α-galactosidase activity (226 U/g) by *R. oligosporus* was found after 36 hours at 30°C and RH 95%–97%.

6.2.3.3 Salting   Freshly prepared pehtze has a bland taste. Like in other fermented foods such as cheese (Messens et al., 1999) and miso (Chiou et al., 1999), salt has multiple roles in sufu. During the salting period, the pehtze absorbs the salt and free water until the salt content of pehtze reaches an equilibrium level. The absorbed salt will later impart a salty taste to the sufu, but it will control the microbial growth and enzyme activity in sufu as well. In addition, salt influences physical and biochemical changes in the product retards the growth of undesirable microbial contaminants.

The added salt also plays another important role in releasing the mycelia-bound proteases. Wang and Hesseltine (Wang and Hesseltine,

1970) mentioned that the fungal proteases were not extracellular and that they are loosely bound to the mycelium, possibly by ionic linkage. The release of proteinase from mycelium of *M. hiemalis* grown in a soybean medium was studied by Wang (Wang, 1967). Only a small fraction of the proteinase produced by the organism appeared in the culture filtrate, whereas the bulk of the enzyme was bound to the mycelial surface. Inclusion of NaCl or other ionizable salts in the growth medium, however, resulted in the liberation of the loosely bound enzyme from the mycelium as it was formed. On the other hand, the mold grows just on the surfaces of the tofu pieces, and the mycelium hardly penetrates into the tofu. The salting could enable the enzymes to diffuse into the tofu for substrate degradation.

Pehtze can be salted in many ways, for example:

1. The pehtze is transferred to containers, each having a volume of 10–20 liters and each layer of pehtze being sprinkled with a layer of salt in accordance with a traditional method. This method not only takes longer, but also the pieces of pehtze have a widely varying salt concentration.

2. The pehtze is transferred to vessels, each layer of pehtze being sprinkled with a layer of salt, and then the vessels are filled with a saturated salt solution. After 4–5 days at room temperature, the salt content of the pehtze can reach over 12% and the moisture content decreased by 10%–15%. Final levels of moisture content may vary in the range 50%–65% (w/w).

3. Pehtze is immersed in an alcoholic saline solution consisting of about 12% NaCl and 10% ethanol (distilled liquor or rice wine is used). Pehtze immersed in this solution can be sold as such, which normally is used for white sufu production in some areas. Actually, this method combines the salting and the ripening together in one step.

Wai (Wai, 1964) attempted to preserve and ripen pehtze after dipping it in the following solutions: (1) 0.1% benzoic acid + 0.05% ethyl acetate; (2) 2% acetic acid + 0.05% ethyl acetate; (3) 1% propionic acid + 0.05% ethyl acetate; (4) 0.01% benzaldehyde + 0.05% ethyl acetate; (5) 1% orange peel oil + 0.05% ethyl acetate; (6) 1% acetic acid + 2% NaCl and Chinese sorghum wine (liquor) to make up a 5% ethyl alcohol content; (7) Chinese sorghum wine (liquor) to make up a 5%

ethyl alcohol content + 2% NaCl, 0.01% anise, and 1% red pepper; (8) Chinese sorghum wine (liquor) to make up a 10% ethyl alcohol content + 0.01% benzaldehyde + 2% NaCl; (9) Chinese sorghum wine (liquor) to make up a 10% ethyl alcohol content + 1% acetic acid + 4% NaCl; (10) Chinese sorghum wine (liquor) to make up a 10% ethyl alcohol content + 5% NaCl ; and (11) 10% ethyl alcohol + 5% NaCl. The pehtze were immersed in the above solutions and preserved in closed glass bottles. Results showed that solutions (8), (9), (10), and (11) were appropriate for the preservation of pehtze both at room temperature (20°C–35°C) and at 10°C in a refrigerator for more than 6 months. Furthermore, a layer of sesame oil added to the solution made the procedure more effective.

6.2.3.4 Ripening   The flavor and aroma of sufu develop during the ripening step. During this period, the enzymes produced by the mold act upon their respective substrates, and it is likely that hydrolysis of protein and lipid provide the principal compounds of the mild, characteristic flavor of sufu. The soybean proteins are hydrolyzed by the proteinases into peptides and amino acids, and soybean lipids are also hydrolyzed to some extent into fatty acids.

Salted pehtze is ripened in various jars or bottles, ranging in size from approximately 0.25–10 liters, containing a dressing mixture that varies with the type of sufu. The dressing mixture that is most commonly used for red sufu includes salt (final salt content 10%–12%), angkak (red kojic rice) 2%, flour (or bean) paste 3%–5%, alcohol content 8%–12%, sugar 5%–10%, and some spices. Additional essence is also added to the dressing mixture to supply a special flavor. For instance, rose essence is added to the dressing mixture for rose sufu preparation. The addition of hot pepper to a dressing mixture would make hot sufu. Therefore, the flavor and aroma of sufu, in addition to its own characteristics, can be easily improved or modified by the ingredients of the dressing mixture.

Ripening requires extravagant time and space, which all are increasingly expensive throughout the world. Although nowadays the ripening time is shorter than the 6 months that the traditional process took, modern processes still take about 2–3 months. Reduction of ripening times can be achieved by using smaller cubes of tofu, lowering the salt content from ~14% to <10%, lowering the alcohol content from ~10% to ~6%, keeping the ripening temperature at a higher and more constant level, and using smaller jars. The high concentrations of salt are considered to retard the hydrolysis of protein and lipid. In addition, human consumption of sufu is also limited because of its saltiness. Unfortunately, lowering salt content could cause other problems, such as shorter shelf life and a reduction of safety.

In an attempt to overcome the problem caused by lowering salt content, a coating of whole blocks of pehtze with paraffin (m.p. 60°C) was studied (Wai, 1964). The pehtze was first mixed with salt (7% of pehtze weight) and then coated with a layer of melted paraffin. Upon solidification of the paraffin, the product was stored in a glass container. After one month at room temperature, the paraffin layer was removed and the contents were subjected to sensory evaluation. The resultant sufu was found to be satisfactory. In order to accelerate the ripening of sufu, stem bromelain was used as a coagulant of soymilk and a hydrolytic enzyme (Fuke and Matsuoka, 1984). Increases in ripening rate and enhanced flavor were obtained by the addition of stem bromelain.

Han (2003) studied microbial changes and protein and lipid degradation in sufu production, with special attention to the effect of different salt contents. Microbial spoilage was observed in sufu with 5% salt content after 20 days of ripening as evidenced by souring and off-flavor. Sufu with 8% salt was similar to sufu with 11% salt and had a stable and inactive microflora. However, in the 8% salt, sufu protein and lipid were degraded to a higher extent. Conventional red sufu has a salt content of over 11%, and takes more than 3 months for ripening. Based on the relevant distinctive features, sufu with 8% salt content would be qualified to be termed as a "finished product" after only 45 days of ripening.

In an industrial setting, pilot-scale trials were carried out with sufu having 7% salt content. Judged by an expert panel, sufu with 7% salt content at the age of 45–180 d fulfilled all the criteria of conventional sufu, and was highly appreciated by all panelists (Han, 2003). The only problem was that due to its ongoing enzymatic degradation, 8 months after the manufacturing date, the sufu became too soft to be taken out of the container using a chopstick. Pasteurization of sufu at the end of ripening could be an option to inactivate the enzymes involved. Further evaluation of the feasibility and acceptability of pasteurized sufu would be required.

6.2.4 Microbial and Biochemistry Changes during Production

6.2.4.1 Microbial Changes   Although a pure culture is used in pehtze fermentation, the process of sufu manufacture itself is carried out under nonsterile conditions, so, in addition to the mold starter, other microorganisms may be expected to be present in sufu. The quantitative evolution of microflora of sufu was studied throughout the production, with special attention to the effect of different salt contents during the ripening (Han, 2003). Total counts of mesophilic aerobic bacteria (TMAB), bacterial endospores (B. endospores), *Bacillus cereus*, lactic acid bacteria (LAB), Enterobacteriaceae, and fungi increased from around 10^4, 10^4, <10, 10^4, <10, and 10^3 cfu/g (in tofu) to around 10^8, 10^8, 10^3, 10^7, 10^7, and 10^7cfu/g (in pehtze), respectively. LAB, Enterobacteriaceae and fungi decreased 1–2 log cfu/g, and TMAB, B. endospores and *B. cereus* decreased <1 log cfu/g after the salting of pehtze. TMAB and B. endospores in sufu with 8% and 11% salt content decreased to around 10^6 cfu/g during the ripening, and about 90% of B. endospores were identified as *Bacillus subtilis*. *B. cereus* remained at levels of around 10^2–10^3cfu/g during the ripening, and 87.5% of presumptive *B. cereus* were confirmed as such. LAB in sufu with 8% and 11% salt content decreased gradually from 10^5–10^6 cfu/g to <10^2 cfu/g. LAB in sufu with 5% salt content increased to 10^9 cfu/g during the ripening. These were identified as most probably *Lactobacillus curvatus*, and caused sour spoilage of sufu by pH decrease (from 6–7 to 4). Enterobacteriaceae and fungi decreased to the nondetectable level after 20 days and 30 days of ripening, respectively, in all samples. There was no obvious difference in changes of detected microflora (except for LAB) between different types of sufu (i.e., red and white) in the study. Figure 6.9 shows microflora changes during the production of a red sufu with 9% salt content (Han et al., 2002).

The relationship between microbiological quality and brine composition of sufu was studied (Pao, 1994). High levels of aerobic plate

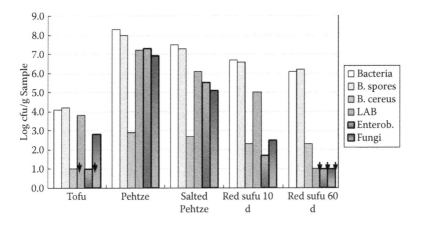

Figure 6.9 Microflora of tofu, pehtze, salted pehtze, red sufu 10 d, red sufu 60 d. (From Han, B.Z. et al. 2002. Changes in the microbial profile and chemical composition during the production of red sufu—a fermented soybean food. Paper presented at *International Conference on Soybean Technology*, Beijing, China; Han, B.Z. 2003. Characterization and Product Innovation of Sufu, a Chinese Fermented Soybean Food: Wageningen Universiteit, Wageningen University.)

counts (>10^5 cfu/g) and spore counts (>10^4 cfu/g) were observed in 80% of all samples tested. In addition, 60% of all samples had lactic acid bacterial counts greater than 10 cfu/g. Halophiles were found in most samples, and high levels of moderate halophiles (>10^5 cfu/g) were found in 60% of all samples. Predominant halophiles were identified as *Tetragenococcus halophila* (previously *Pediococcus halophilus*). The role of *T. halophila* in sufu ripening was analyzed by sensory evaluation. In general, the addition of *T. halophila* at early stages of the fermentation helped the overall sensory quality development of sufu. This result showed that the presence of *T. halophila* in sufu fermentation appear to be desirable.

6.2.4.2 Biochemical Changes   During the fungal fermentation process a substantial breakdown occurs of the soybean storage proteins, lipids, and carbohydrates due to the production of fungal proteases, lipases, and carbohydrases (De Reu, 1995). Proteolysis is the principal and most complex biochemical event that occurs during ripening of sufu. The soybean proteins are hydrolyzed by the proteinases into peptides and amino acids. The pleasant and palatable taste is considered to be related to the content of free amino acids, mainly glutamic acid, in Oriental fermented food (Chou et al., 1993). Purification and some properties of glutaminase (catalyses hydrolysis of glutamine to gluta-mate) from *A. taiwanensis* were studied by Lu et al. (Lu et al., 1996). Glutaminase was stable at a temperature up to 35°C and at pH values of 6.0 ~ 8.0. In the presence of 10% NaCl, the enzyme activity was inhibited 50%. The alcohol also retards the degradation of soybean proteins (Chou and Hwan, 1994). A higher protein solubility and a higher content of peptides and amino acids were observed in the sufu ripened in a brine solution without ethanol.

Han et al. (2003b) reported the SDS-PAGE profiles for salted pehtze and subsequent red and white sufu with different salt con-tents at various ripening stages. The major protein subunits can be clearly identified in salted pehtze and in sufu containing 14% salt after 10 days' ripening. Nevertheless, after 10 days, most protein subunits had been degraded to a large extent in sufu containing 8% and 11% salt. All protein subunits had disappeared in sufu containing 8% salt at the age of 60 days. Some protein subunits were still visible in sufu with 14% salt content at the same age.

The increase of the ratio of free amino nitrogen to total nitrogen (FAN/TN) during sufu ripening gives a further indication of the hydrolysis of protein. The FAN/TN in white sufu with 8% salt con-tent increased from 0.16 at 10 days to 0.44 at 80 days' ripening. The ratio of free amino acid to crude protein (FAA/CP) is also an indi-cator of protein degradation. The FAA/CP in white sufu of a lower salt content (8%) exceeded 0.25 after 80 days ripening. Comparing FAN/TN with FAA/CP in sufu, FAN/TN was much higher than FAA/CP (Han et al., 2003b). As an example, taking into account the absence of protein bands in SDS-PAGE, data for white sufu with 8% salt after 80 days indicate that approximately 27% of N occurs as monomeric amino acids, and remaining N as peptides and amino acid degradation products.

Soybean lipids form the second largest substance in sufu after protein, and they are also hydrolyzed to some extent into fatty acids during the sufu ripening. Crude lipids slightly decreased during the sufu ripening, regardless of red or white sufu with various salt content (Han et al., 2003b). Fungal lipases can decompose crude lipid into free fatty acids and di- and monoglycerides. Chou and Hwan (Chou and Hwan, 1994) observed that lipid content fluctuated during the sufu ripening was prepared with either *A. taiwanenis* or *A. elegans*. The free fatty acid content increased then decreased during the ripening period, which probably results from the formation of fatty acid esters from free fatty acids and alcohol present in the dressing mixtures. The added alcohol reacts with the fatty acids chemically or enzymatically to form esters, providing the pleasant odor of the product. Regardless of the starter mold used for sufu, linoleic acid (18:2) was the highest followed by oleic acid (18:1), palmitic acid (16:0), linoleic acid (18:3), and stearic acid (18:0).

It is commonly assumed that the flavor, aroma, and texture of sufu that mainly develop during the ripening are determined by the enzymes produced by the mold in pehtze preparation. However, significant levels of bacteria (mainly *Bacillus* spp.) were also present in pehtze and sufu product (Han, 2003). It has been shown that *Bacillus* spp. show considerably higher and different proteolytic activity compared to, for example, *Rhizopus* spp., resulting in the degradation of virtually all major soy protein subunits (Kiers et al., 2000). So *Bacillus* spp. could also contribute to the formation of hydrolytic enzymes that benefit the sufu ripening. The function of microbes other than fungal starter cultures should be investigated in the future.

Hardness and smoothness were identified as important factors influencing consumer acceptability of tofu (Ji et al., 1999). These textural properties are also important factors influencing consumer acceptability of sufu. The Texture Profile Analysis (TPA) of red and white sufu with various salt contents was studied by Han et al. (2003b). Hardness, elasticity, and adhesiveness of red sufu were slightly higher than that of white sufu at the same salt content; this is probably due to the different composition of dressing mixtures for red and white sufu. The salt levels in sufu greatly affected the textural changes. Higher salt contents resulted in higher values for hardness and elasticity, with lower adhesiveness. Hardness of sufu with 8% salt decreased rapidly from ~600 g to ~100 g within 45 d, followed by a slow decrease. Even at the lowest (8%) salt content, elasticity values did not decrease below 0.5. The TPA was widely used to evaluate the textural properties of tofu (Hou et al., 1997; Ji et al., 1999), and they could also be used as parameters for judging the extent of ripening of sufu during its production. In the absence of comparative standards and based on limited subjective evaluations by sufu manufacturing experts, Han et al. (2003b) proposed that sufu may be considered ready for packing and distribution when hardness and elasticity have decreased to around 100 g and 0.55, respectively.

6.2.5 Chemical Composition and Nutritional Quality of Sufu

From a nutritional point of view, sufu has a higher content of protein-nitrogen than other oriental soybean foods, such as miso and natto (Su, 1986). Nutritionally, soybean milk, tofu, and sufu have the same importance to people of Asia as cow milk and cheese do to the people of the Western Hemisphere. Asians prefer the salt-coagulated bean curd, not only because it has the desired texture, but also because it serves as an important source of calcium (Wang and Hesseltine, 1970; Zhao, 1997).

The chemical and nutritional compositions of sufu are listed in Table 6.6 (Su, 1986; Li, 1998; Wang and Du, 1998). In spite of their differences in color and flavor, most types of sufu have a similar proximate composition.

Chemical parameters of 23 commercial samples were analyzed (Han et al., 2001a), which included moisture, pH, free amino N, NaCl, ethanol, sucrose, glucose, and fructose. Concentrations of NaCl, ethanol, glucose, and fructose varied from 6.2%, 0.5%, 0%, and 0% to 14.8%, 6.3%, 6.2%, and 4.8%, respectively. In addition, large variation in the volume of NaCl, ethanol, glucose, fructose, and water among samples from the same brand indicate that most companies do not closely control the composition of their products.

Table 6.7 shows textural properties and chemical parameters of tofu, pehtze, salted pehtze, and red and white sufu with 8% salt (Han, 2003; Han et al., 2003b).

Since soybean foods serve as a source of protein in the Chinese diet, their amino acid content and pattern are important from a nutritional point of view. In addition, amino acids contribute to the taste of foodstuffs (Nishimura and Kato, 1988) and influence the consumers' acceptance of sufu. The amino acid content of sufu is presented in Table 6.8 (Su, 1986; Liu and Chou, 1994; Wang, 1995; Wang and Du, 1998; Han et al., 2004b). Glutamic acid and aspartic acid were the most abundant amino acids found in red sufu and gray sufu. The ratio of (glutamic acid + aspartic acid): (total amino acid content) was around 30%, which provides sufu with a delicious taste. The cystine

Table 6.7 Textural Properties and Chemical Parameters of Tofu, Pehtze, Salted Pehtze, and Red/White Sufu With 8% Salt

SAMPLE	HARDNESS (g)	ELASTICITY (−)	ADHESIVENESS (J)	PH	FAN (mm/g)[a]	MOISTURE (g kg⁻¹)[a]	C PROTEIN (g kg⁻¹)[b]	FAA (g kg⁻¹)[b]	C LIPIDS (g kg⁻¹)[b]	FFA (g kg⁻¹)[c]
Tofu	560 ± 63[d]	1 ± 0	0.1 ± 0	6.9 ± 0.21	0.04 ± 0.002	743 ± 42	631 ± 45	1.3 ± 0.1	301 ± 18	37 ± 1.6
Pehtze	860 ± 55	1 ± 0	0.2 ± 0.03	7.0 ± 0.35	0.36 ± 0.02	703 ± 37	628 ± 38	18 ± 1.1	308 ± 21	64 ± 2.9
Salted pehtze	1380 ± 103	1 ± 0	3.8 ± 0.21	7.0 ± 0.27	0.26 ± 0.01	563 ± 41	375 ± 26	11 ± 0.9	184 ± 12	59 ± 3.3
Red sufu 8% 10 d[e]	580 ± 75	0.85 ± 0.08	2.6 ± 0.41	6.7 ± 0.25	0.23 ± 0.02	649 ± 46	403 ± 31	30 ± 2.3	223 ± 17	93 ± 5.5
Red sufu 8% 80 d	70 ± 14	0.50 ± 0.02	14.6 ± 1.13	5.9 ± 0.18	0.49 ± 0.03	644 ± 38	361 ± 22	94 ± 4.9	190 ± 20	139 ± 9.1
White sufu 8% 10 d	560 ± 81	0.87 ± 0.09	22.0 ± 50	6.8 ± 0.20	0.24 ± 0.01	716 ± 41	458 ± 29	36 ± 2.7	238 ± 21	97 ± 7.2
White sufu 8% 80 d	70 ± 15	0.50 ± 0.02	13.0 ± 1.8	5.9 ± 0.25	0.61 ± 0.04	701 ± 39	402 ± 30	110 ± 6.4	210 ± 24	149 ± 11

Note: FAN = Free Amino Nitrogen; C Protein: Crude Protein; FAA = Free Amino Acid; C Lipids: Crude Lipids; FFA = Free Fatty Acids.
[a] Fresh weight basis; [b] dry matter basis; [c] of crude lipids; [d] averages ± standard deviations; [e] sufu at 10 (or 80) days of ripening.

Table 6.8   Profiles of Total Amino Acid (TAA), Free Amino Acid (FAA) in Different Sufu

AMINO ACID	TAA			FAA		
	SUFU[a] (mg/g PROTEIN)	RED SUFU[c] (mg/g SUFU)	GRAY SUFU[b] (mg/g SUFU)	WHITE SUFU[c] (mg/g SUFU DM)	RED SUFU[c] (mg/g SUFU DM)	GRAY SUFU[c] (mg/g SUFU DM)
Asp	51	32.0	6.6	7.9	5.2	0.7
Thr	20	11.0	2.3	4.7	2.4	0.1
Ser	23	12.9	2.7	0.2	2.6	0.2
Glu	6	54.1	20.8	18.5	14.4	0.1
Gly	44	12.6	4.2	3.7	2.2	0.0
Ala	100	15.8	7.0	7.9	4.9	8.9
Val	53	15.8	5.8	6.5	3.5	6.8
Met	7	1.8	1.4	1.7	1.0	0.0
Ile	48	17.6	5.8	7.3	3.7	7.1
Leu	88	27.7	9.5	10.3	6.1	9.6
Tyr	22	11.8	2.5	0.9	4.1	3.7
Phe	46	19.7	5.9	7.4	4.8	6.5
Lys	70	16.0	2.9	7.5	4.0	0.2
His	14	6.3	1.8	1.8	1.0	0.1
Arg	21	12.3	2.7	0.0	0.2	0.0
Pro	24	16.6	2.9	5.2	2.9	0.0
Trp	6	3.8	0.5			
Cys	4	6.9	2.0			

[a]   Su, (1986): commercial sample nonspecific.
[b]   Wang, (1995) and Wang and Du, (1998).
[c]   Han et al., (2003d): sufu with 11% salt content at 80 days of ripening.

and methionine contents of gray sufu may be lower than those of red sufu because of their degradation or conversion to other sulfur compounds during ripening, which may contribute to the offensive odor of gray sufu.

Total amino acid (TAA) and free amino acid (FAA) profiles were determined during consecutive stages of sufu manufacture, for example, tofu, pehzte (fungal fermented tofu), salted pehzte, and in white, red, and gray sufu, ripening in dressing mixtures of different salt content (8, 11, and 14% w/w; Han et al., 2004b). TAA in tofu, pehzte, and salted pehzte totaled 547, 551, and 351 mg/g dry matter, respectively. FAA increased from total 1.3 to 15.6 mg/g dm in pehzte after fermentation of tofu by *A. elegans* and to 11.9 mg/g dm in salted pehzte. During ripening up to 80 days, total FAA in red sufu increased from 28 to 88 mg/g (8% salt), 28 to 63 mg/g (11% salt) or 26 to 42 mg/g

(14% salt). In white sufu the levels of FAA were higher and the effect of salt was stimulating rather than inhibiting the formation of FAA. Levels of FAA in white sufu increased in 80 days from 33 to 104 mg/g (8% salt), 27 to 92 mg/g (11% salt), and 19 to 73 mg/g (14% salt). While FAA increased during ripening of red and white sufu, the ratio of each amino acid remained essentially constant, and glutamic acid, leucine, aspartic acid, alanine, phenylalanine, and lysine were found in large quantities. However, changes of free amino acid in gray sufu evolved in a totally different pattern compared with red and white sufu. All hydrophobic amino acids (except for methionine) increased to a large extent from 6.8 to 33.7 mg/g (5.0-fold), which dominated the proportion of total free amino acid. The dramatic decrease in some amino acids (glutamic acid, glycine, and proline) during ripening suggests that they were consumed or transformed somehow at a greater rate than they were formed by proteolytic activity. Gray sufu is not yet described in published literature. Obviously its different amino acid profile during ripening, as well as its strong, offensive odor and gray (or sometimes blue) color formed during 5–15 days of ripening, warrant further study.

Volatile compounds formed from amino acids by decarboxylation, deamination, transamination, and other transformations can make substantial contributions to sufu flavor. This is a major reason for the many different volatile compounds encountered in sufu. The common volatile compounds detected in red/white types of sufu are shown in the Table 6.9 (Chung, 1999, 2000).

Chung (Chung, 1999) found a total of 111 compounds from three commercial brands of white sufu. Brands A, B, and C contained 90, 90, and 82 compounds, respectively. Among the 63 compounds

Table 6.9 Common Volatile Compounds Detected in Red/White Types of Sufu

Alcohols	Ethanol, 2-butanol, 1-propanol, 2-methylpropanol, 2-pentanol, 1-butanol, 1-pentanol, 2-methoxyphenol, benzenemethanol, benzeneethanol, 4-methylphenol, 4-ethylphenol, 4-ethyl-2methoxyphenol
Esters	Ethyl butanoate, ethyl 2-ethylbutanoate, ethyl pentanoate, ethyl hexanoate, ethyl lactate, ethyl octanoate, ethyl 3-hydroxybutanoate, ethyl 3-phenylproprionate
Miscellaneous	Phenol, n-hexanal, benzaldehyde, 2-pentylfuran, naphthalene, pyridine, 2-methylpyrazine, 2,5-dimethylpyrazine, 2,6-dimethylpyrazine, 2-heptanone, 3-octanone, 3-hydroxy-2-butanone, 3-hydroxy-2-methyl-4H-pyran-4-one, 3-(methylthio)propanol

Source: Chung et al., 1999, 2000.

common to all three brands, a majority of them belonged to alcohols (19) and esters (15). The rest of the common compounds included pyrazines (7), ketones (7), pyridines (5), aldehydes (4), miscellaneous compounds (3), sulfur-containing compounds (2), and furan (1). Twelve common compounds, including six esters, four alcohols, one ketone, and one miscellaneous compound, had a dry weight of >1000 μg/kg of sample. The quantity of ethanol detected might produce large numbers of ethyl esters that contribute to the desirable fruity and floral notes for the white sufu products.

Chung (Chung, 2000) also analyzed volatile flavor components in red sufu. A combined total of 91 compounds of volatiles were found, and 89 were identified from the three red sufu samples. The numbers of components identified in samples (A, B, and C) were 87, 85, and 81, respectively. They belonged to 10 classes of compounds including acid (1), aldehydes (7), esters (22), miscellaneous compounds (8), other nitrogen-containing compounds (8), pyrazines (4), other oxygen-containing compounds (5), alcohol (24), ketones (8), and sulfur-containing compounds (3). There were 78% compounds found in all three samples, but their quantitative levels varied. In white sufu 70% of them were identified (Chung, 1999). Ethyl 2-methylpropanoate, 2,3-butanedione, ethyl butanoate, ethyl 2-methylbutanoate, 3-(methylthio) propanal, benzeneacetaldehyde, and ethyl 3-phenlpropionate were found to be important common components contributing to the characteristic aroma of red sufu.

The complex flavor of sufu was reported by Hwan and Chou (Hwan and Chou, 1999). A total of 61 volatile compounds were identified including 22 esters, 18 alcohols, 7 ketones, 3 aldehydes, 2 pyrazines, 2 phenols, and 7 other compounds from the 75-day aged sufu products. Ripening in the presence of ethanol resulted in higher levels of volatiles. Ho et al. (1989) compared the volatile flavor compounds of red sufu and white sufu. Red sufu contains much larger amounts of alcohols, esters, and acids, which may be due to the fermentation of angkak by *Monascus* spp. The esters give red sufu its characteristic fruity aroma. White sufu contains a large quantity of anethol, which seems to be the major contributor to its flavor.

Soybeans contain high levels of oligosaccharides, notably α-galactosides of sucrose, such as raffinose and stachyose. These may have prebiotic properties, but excessive levels may also result in intestinal gas production (flatulence). These oligosaccharides are mainly removed by soaking and cooking of soybeans (Ruiz-Terán and Owens, 1999) and are probably further degraded by molds since *A. elegans* and *R. oligosporus* can produce α-galactosidase in pehtze preparation (Han et al., 2003a).

Isoflavones are a type of phytoestrogen, plant-based compounds with weak estrogenic activity. Asian countries have seen a low incidence of several chronic diseases mentioned previously that plague Western countries. Possibly, the ingestion of soybean isoflavones contributes to this. It has been reported that the level of some isoflavones (i.e., genistein and daidzein) increased by about 20 times after 50 days of sufu ripening, which thus offers an easily absorbable source of isoflavones (Zhang et al., 2002).

6.2.5.1 Microbiological Safety and Quality The microbiological safety and quality of commercial sufu were investigated (Han et al., 2001a). Twenty-three samples of three different types of sufu were obtained, mainly in China and some in the Netherlands. Microbiological analyses were done for total count of mesophilic aerobic bacteria (TMAB), bacterial endospores, total count of halotolerant bacteria at 10% (THB10) and at 17.5% NaCl (THB17.5), lactic acid bacteria (LAB), fungi, Enterobacteriaceae, and the following pathogens: *Bacillus cereus, Clostridium perfringens, Staphylococcus aureus*, and *Listeria monocytogenes*. High levels (>10^5 cfu/g) of TMAB and bacterial endospores were found in most samples, and 85% of TMAB was identified as Gram-positive. THB10 and THB17.5 were detected in all samples, and the number of halotolerant bacteria was inversely proportional to the NaCl concentration in the media. Most of the isolates of THB10 grew after having been inoculated onto plate count agar (PCA) without NaCl, indicating that these were not obligate halophilic organisms. On the other hand, most isolates transferred from THB17.5 failed to grow on PCA without NaCl, so these could be obligate halophiles. Considerable levels (10^5 and 10^7 cfu/g) of LAB were detected in two samples of white sufu, and isolates of LAB were identified as most probably *Lb. casei*.

In all samples analyzed, fungi were absent (< 10 cfu/g). This appears to differ from the data of Pao (Pao, 1994) who reported 10^2–10^6 cfu/g of fungi in sufu. Shi and Fung (Shi and Fung, 2000) mentioned that no molds were detected in sufu samples after 1, 2, and 3 months of ripening. Most likely, the fungi, particularly the mold starters, do not survive after the pehtze preparation, owing to the combination of salt and ethanol in the dressing mixtures applied for the ripening of sufu.

One-third of the samples contained less then 10^3 cfu/g *B. cereus*, but three samples had over 10^5 cfu/g indicating potential hazard to consumers, which should be of concern for the sufu industry (Han et al., 2001a). A quarter of the samples contained low numbers of *C. perfringens* (< 10^3 cfu/g), except a special sample (~10^5 cfu/g). Most Enterobacteriaceae do not tolerate elevated salt levels. Whereas they were found at high levels (~10^7 cfu/g) in pehtze (Han, 2003), they do not survive the salting and ripening steps. According to our expectation, no Enterobacteriaceae were detectable in any of the samples. Similar results were observed in an earlier study (Pao, 1994), in which coliforms were not found in the investigated samples. Although Enterobacteriaceae are unlikely survivors of the final stages of the sufu process, their ability to grow and possibly produce endotoxins during the early pehtze-making stage might constitute a hazard. Consequently, toxicological studies are required to assess the potential hazard posed by such toxins during sufu production.

S. aureus is often associated with human skin. It occurs in commercial tofu (Van Kooij and De Boer, 1985), and it is likely to be present in sufu as this involves manual operations on tofu. Due to its salt tolerance, this pathogen expectedly survives in sufu. However, it was impossible to detect *S. aureus* in the samples tested due to the fact that there were no visible haloes surrounding the suspect colonies. The competitive microflora (mainly bacilli) prevented the formation of such haloes during the incubation of *S. aureus*, which was confirmed with artificially contaminated samples. Qualitative detection tests on staphylococcal enterotoxins in sufu samples showed that some of the gray sufu contained enterotoxin A. Other enterotoxins (B, C, D, and E) were below the level of detection. Although *L. monocytogenes* is a salt-tolerant pathogen being able to grow in 10% NaCl and to survive in the presence of 30% NaCl, it was not found in any of the samples analyzed by Han et al. (2001a).

Control of the foodborne pathogens *Escherichia coli* O157:H7, *Salmonella typhimurium*, *S. aureus*, and *L. monocytogenes* during sufu fermentation and ripening was evaluated by Shi and Fung (2000). Before fermentation, pathogens were inoculated onto tofu at 10^5 cfu/g or 10^3 cfu/g, and starter culture (*A. elegans*) was inoculated at 10^3 cfu/g. After 2 days of fermentation at 30°C, the four pathogens reached 10^7–10^9cfu/g, and the mold count reached 10^6–10^7cfu/g. After fermentation, sufusample (phetze) were ripened in a solution of 10% alcohol + 12% NaCl. After 1 month of ripening, the total bacterial count was 10^6–10^7cfu/g, but all foodborne pathogens and mold were reduced to nondetectable levels. From commercial sufu samples, 17 *Bacillus* spp. (56.7%), 2 *Enterococcus durans* (6.7%), 5 miscellaneous Gram-negative bacilli (16.7%), 5 *Corynebacterium aquaticum* (16.7%), and 1 *Shewanella putrefaciens* (3.3%) were obtained (Shi and Fung, 2000).

Angkak (red kojic rice), an important ingredient in the dressing mixture of red sufu as a natural colorant, is traditionally obtained by fungal solid-state fermentation of cooked rice, mainly with *Monascus* spp. (Nout and Aidoo, 2002). Angkak is also regarded to be a health-promoting food ingredient (Wang et al., 1997). It was reported that the angkak mold *Monascus ruber* (ATCC 96218) produced the mycotoxincitrinin at levels up to 6.8 mg per g glucose when grown in aerated shaking flask cultures in a chemically defined liquid growth medium (Hajjaj et al., 1999, 2000). We did a limited sampling of angkak produced traditionally on rice. Samples from different regions (Guangdong, Jiangsu, Hunan, Fujian, and Beijing) did not contain detectable levels (< 1 ppb) of citrinin (courtesy by K.E. Aidoo, Glasgow Caledonian University, UK).

Yen (Yen, 1986) reported that the average amine contents in 15 samples of commercial sufu from Taiwan, China were: cadaverine (0.039 mg/g), histamine (0.088 mg/g), beta-phenylethylamine (0.063 mg/g), putrescine (0.473 mg/g), tryptamine (0.150 mg/g), and tyramine (0.485 mg/g). Tyramine and putrescine were the major amines found, and these might have a potential harmful effect on human beings if levels are very high.

6.2.6 Conclusion

Consumption of soybean foods is increasing because of reported beneficial effects on nutrition and health (Liu, 2000). These effects include lowering of plasma cholesterol, prevention of cancer, diabetes, and obesity (Friedman and Brandon, 2001). But soybean protein by itself lacks desirable flavors. To overcome this deficiency, fermentation can either add desirable flavor or destroy unpleasant flavor. This is especially true in sufu and other soybean products made by fermentation.

In China, sufu is one of the most important traditional fermented soybean foods. It has been widely consumed as a relish by Chinese people for more than 1000 years, and further, it can be used in the same way as cheese. Throughout history, the Chinese seemed to follow their own ways in developing the product without foreign influences. Although some new methods (e.g., a pure culture) for preparing sufu have been developed, further studies are still needed to guarantee uniform high quality products. If innovations in taste, flavor, and product quality are made, sufu may become more widely popular all over the world.

References

Beuchat, L.R. (1995). Indigenous fermented foods. In *Biotechnology: Enzymes, Biomass, Food and Feed*, eds. G. Reed and T.W. Nagodawithana, VCH Press, Weinheim, New York, Basel, Cambridge and Tokyo, pp. 9, 523–525.

Chen, J., Cheng, Y.-Q., Yamaki, K., and Li, L.-T. (2007). Anti-α-glucosidase activity of Chinese traditionally fermented soybean (douchi). *Food Chemistry.* **103**(4), 1091–1096.

Chiou, R.Y.Y., Ferng, S., and Beuchat, L.R. (1999). Fermentation of low-salt miso as affected by supplementation with ethanol. *International Journal of Food Microbiology.* **48**(1), 11–20.

Chou, C., Yu, R., and Tsai, C. (1993). Production of glutaminase by *Actinomucor elegans, Actinomucor taiwanensis* and *Aspergillus oryzae. Journal-Chinese Agricultural Chemical Society.* **31**, 78–78.

Chou, C.C., Ho, F.M., and Tsai, C.S. (1988). Effects of temperature and relative humidity on the growth of and enzyme production by *Actinomucor taiwanensis* during sufu pehtze preparation. *Applied and Environmental Microbiology.* **54**(3), 688–692.

Chou, C.C. and Hwan, C.H. (1994). Effect of ethanol on the hydrolysis of protein and lipid during the ageing of a Chinese fermented soya bean curd: Sufu. *Journal of the Science of Food and Agriculture.* **66**(3), 393–398.

Chung, H.Y. (1999). Volatile components in fermented soybean (Glycine max) curds. *Journal of Agricultural and Food Chemistry.* **47**(7), 2690–2696.

Chung, H.Y. (2000). Volatile flavor components in red fermented soybean (Glycine max) curds. *Journal of Agricultural and Food Chemistry.* **48**(5), 1803–1809.

De Reu, J.C. (1995). Solid-Sate Fermentation of Soya Beans to Temp: Process Innovations and Product Characteristics. Wageningen Agricultural University, PhD-thesis, Wageningen, The Netherlands.

Friedman, M. and Brandon, D.L. (2001). Nutritional and health benefits of soy proteins. *Journal of Agricultural and Food Chemistry.* **49**(3), 1069–1086.

Fujita, H., Yamagami, T., and Ohshima, K. (2001a). Fermented soybean-derived water-soluble touchi extract inhibits alpha-glucosidase and is anti-glycemic in rats and humans after single oral treatments. *Journal of Nutrition.* **131**(4), 1211–1213.

Fujita, H., Yamagami, T., and Ohshima, K. (2001b). Long-term ingestion of a fermented soybean-derived Touchi-extract with alpha-glucosidase inhibitory activity is safe and effective in humans with borderline and mild type-2 diabetes. *Journal of Nutrition.* **131**(8), 2105–2108.

Fuke, Y. and Matsuoka, H. (1984). Preparation of fermented soybean curd using stem Bromelain. *Journal of Food Science.* **49**(1), 312–313.

Hajjaj, H., Blanc, P., Groussac, E., Uribelarrea, J.L., Goma, G., and Loubiere, P. (2000). Kinetic analysis of red pigment and citrinin production by *Monascus ruber* as a function of organic acid accumulation. *Enzyme and Microbial Technology.* **27**(8), 619–625.

Hajjaj, H., Blanc, P.J., Groussac, E., Goma, G., Uribelarrea, J.L., and Loubiere, P. (1999). Improvement of red pigment/citrinin production ratio as a function of environmental conditions by *Monascus ruber. Biotechnology and Bioengineering.* **64**(4), 497–501.

Han, B.Z. (2003). Characterization and Product Innovation of Sufu, a Chinese Fermented Soybean Food: Wageningen Universiteit (Wageningen University).

Han, B.Z., Beumer, R.R., Rombouts, F.M., and Robert Nout, M.J.R. (2001a). Microbiological safety and quality of commercial sufu: A Chinese fermented soybean food. *Food Control.* **12**(8), 541–547.

Han, B.Z., J.H., W., and Y.L., Z. (2002). Changes in the microbial profile and chemical composition during the production of red sufu: A fermented soybean food. Paper presented at International Conference on Soybean Technology, Beijing, China.

Han, B.Z., Kuijpers, A.F.A., Thanh, N.V., and Nout, M.J.R. (2004a). Mucoraceous moulds involved in the commercial fermentation of Sufu Pehtze. *Antonie Van Leeuwenhoek International Journal of General and Molecular Microbiology.* **85**(3), 253–257.

Han, B.Z., Ma, Y., Rombouts, F.M., and Nout, M.J.R. (2003a). Effects of temperature and relative humidity on growth and enzyme production by *Actinomucor elegans* and *Rhizopus oligosporus* during sufu pehtze preparation. *Food Chemistry.* **81**(1), 27–34.

Han, B.Z., Rombouts, F.M., and Nout, M.J.R. (2001b). A Chinese fermented soybean food. *International Journal of Food Microbiology.* **65**(1–2), 1–10.

Han, B.Z., Rombouts, F.M., and Nout, M.J.R. (2004b). Amino acid profiles of sufu, a Chinese fermented soybean food. *Journal of Food Composition and Analysis.* **17**(6), 689–698.

Han, B.Z., Wang, J.H., Rombouts, F.M., and Nout, M.J.R. (2003b). Effect of NaCl on textural changes and protein and lipid degradation during the ripening stage of sufu, a Chinese fermented soybean food. *Journal of the Science of Food and Agriculture.* **83**(9), 899–904.

Ho, C.T., Zhang, Y., Shi, H., and Tang, J. (1989). Flavor chemistry of Chinese foods. *Food Reviews International.* **5**(3), 253–287.

Hong, G.Z. (1985). The history of sufu. *China Brewing.* **1**, 44–45 (in Chinese).

Hou, H.J., Chang, K.C., and Shih, M.C. (1997). Yield and textural properties of soft tofu as affected by coagulation method. *Journal of Food Science.* **62**(4), 824–827.

Huang, Z. (1991). *Handbook of Chinese Seasonings Process Technology.* Beijing, China: China Standards Press.

Hwan, C.H. and Chou, C.C. (1999). Volatile components of the Chinese fermented soya bean curd as affected by the addition of ethanol in ageing solution. *Journal of the Science of Food and Agriculture.* **79**(2), 243–248.

Ji, M.P., Cai, T.D., and Chang, K.C. (1999). Tofu yield and textural properties from three soybean cultivars as affected by ratios of 7S and 11S proteins. *Journal of Food Science.* **64**(5), 763–767.

Kiers, J.L., Van Laeken, A.E.A., Rombouts, F.M. and Nout, M.J.R. (2000). *In vitro* digestibility of *Bacillus* fermented soya bean. *International Journal of Food Microbiology.* **60**(2–3), 163–169.

Lee, Y.B., Lee, H.J., Won, M.H., Hwang, I.K., Kang, T.C., Lee, J.Y., Nam, S.Y., Kim, K.S., Kim, E., Cheon, S.H., and Sohn, H.S. (2004). Soy isoflavones improve spatial delayed matching-to-place performance and reduce cholinergic neuron loss in elderly male rats. *Journal of Nutrition.* **134**(7), 1827–1831.

Li, Y.J. (1997). Sufu: A health soybean food. *China Brewing.* **4**, 1–4 (in Chinese).

Li, Y.J. (1998). The modified process of sufu production. *China Brewing.* **4**, 6–11 (in Chinese).

Lim, B.T., DeMan, J.M., DeMan, L. and Buzzell, R.I. (1990). Yield and quality of tofu as affected by soybean and soymilk characteristics: Calcium sulfate coagulant. *Journal of Food Science.* **55**(4), 1088–1092.

Liu, K. (2000). Expanding soybean food utilization. *Food Technology.* **54**(7), 46–58.

Liu, Y.H. and Chou, C.C. (1994). Contents of various types of proteins and water soluble peptides in Sufu during aging and the amino acid composition of tasty oligopeptides. *Journal of the Chinese Agricultural Chemical Society.* **32**(3), 276–283.

Liu, Y.Q., Wang, L.J., Cheng, Y.Q., Saito, M., Yamaki, K., Qiao, Z.H., and Li, L.T. (2009). Isoflavone content and anti-acetylcholinesterase activity in commercial douchi (a traditional chinese salt-fermented soybean food). *Jarq-Japan Agricultural Research Quarterly.* **43**(4), 301–307.

Liu, Y.Q., Xin, T.R., Lu, X.Y., Ji, Q., Jin, Y., and Yang, H.D. (2007). Memory performance of hypercholesterolemic mice in response to treatment with soy isoflavones. *Neuroscience Research.* **57**(4), 544–549.

Lu, J.M., Yu, R.C., and Chou, C.C. (1996). Purification and some properties of glutaminase from *Actinomucor taiwanensis*, starter of sufu. *Journal of the Science of Food and Agriculture.* **70**(4), 509–514.

Messens, W., Dewettinck, K., and Huyghebaert, A. (1999). Transport of sodium chloride and water in Gouda cheese as affected by high-pressure brining. *International Dairy Journal.* **9**(8), 569–576.

Molzuddin, S., Johnson, L.D., and Wilson, L.A. (1999). Rapid method for determining optimum coagulant concentration in tofu manufacture. *Journal of Food Science.* **64**(4), 684–687.

Monteiro, S.C., Mattos, C.B., Scherer, E.B.S., and Wyse, A.T.S. (2007). Supplementation with vitamins E plus C or soy isoflavones in ovariectomized rats: Effect on the activities of Na+,K+-ATPase and cholinesterases. *Metabolic Brain Disease.* **22**(16), 156–171.

Murphy, P.A., Chen, H.P., Hauck, C.C., and Wilson, L.A. (1997). Soybean protein composition and tofu quality. *Food Technology.* **51**(3), 86–110.

Needham, J. (2000). *Science and Civilization in China, Volume 6, Biology and Biological Technology, Part V: Fermentations and Food Science*, by H.T. Huang, Cambridge University Press.

Nishimura, T. and Kato, H. (1988). Taste of free amino acids and peptides. *Food Reviews International.* **4**(2), 175–194.

Nout, M.J.R. and Aidoo, K.E. (2002). Asian fungal fermented food.

Pal, P. and Tandon, V. (1998). Anthelmintic efficacy of *Flemingia vestita* (Fabaceae): Genistein-induced alterations in the esterase activity in the cestode, *Raillietina echinobothrida*. *Journal of Biosciences.* **23**(1), 25–31.

Pao, S.C. (1994). Halophilic Organisms in Sufu, Chinese Cheese, The Ohio State University, PhD thesis.

Peng, Y., Huang, Q., Zhang, R.H., and Zhang, Y.Z. (2003). Purification and characterization of a fibrinolytic enzyme produced by *Bacillus amyloliquefaciens* DC-4 screened from douchi, a traditional Chinese soybean food. *Comparative Biochemistry and Physiology B: Biochemistry and Molecular Biology.* **134**(1), 45–52.

Ruiz-Terán, F. and Owens, J.D. (1999). Fate of oligosaccharides during production of soya bean tempe. *Journal of the Science of Food and Agriculture.* **79**(2), 249–252.

Shi, X. and Fung, D.Y.C. (2000). Control of foodborne pathogens during sufu fermentation and aging. *Critical Reviews in Food Science and Nutrition.* **40**(5), 399–425.

Shi, Y.G. and Ren, L. (1993). *The Manufacturing Process of Soybean Products.* Beijing, China: China Light Industry Press.

Steinkraus, K.H. (1996). Chinese Sufu. In *Handbook of Indigenous Fermented Foods*. New York, Basel, Hong Kong: Marcel Dekker. pp. 633–641.

Su, Y.C. (1986). Sufu. In *Legume-Based Fermented Foods*, eds. N. R. Reddy, M. D. Pierson, and D. K. Salunkhe. Boca Raton, Florida: CRC Press, pp. 69–83.

Van Kooij, J.A. and De Boer, E. (1985). A survey of the microbiological quality of commercial tofu in the Netherlands. *International Journal of Food Microbiology.* **2**(6), 349–354.

Wai, N. (1929). A new species of mono-mucor, mucor sufu, on Chinese soybean cheese. *Science.* **70**, 307–308.

Wai, N. (1964). Soybean cheese. *Bulletin of the Institute of Zoology, Academia Sinica.* **18**, 75–94.

Wang, D., Wang, L.J., Zhu, F.X., Zhu, J.Y., Chen, X.D., Zou, L., Saito, M., and Li, L.T. (2008). *In vitro* and *in vivo* studies on the antioxidant activities of the aqueous extracts of douchi (a traditional Chinese salt-fermented soybean food). *Food Chemistry.* **107**(4), 1421–1428.

Wang, H.L. (1967). Release of proteinase from mycelium of *Mucor hiemalis*. *Journal of Bacteriology.* **93**(6), 1794–1799.

Wang, H.L. and Fang, S.F. (1986). History of Chinese fermented foods. *Mycological memoir.*

Wang, H.L. and Hesseltine, C.W. (1970). Sufu and lao-chao. *Journal of Agricultural and Food Chemistry.* **18**(4), 572–575.

Wang, J., Lu, Z., Chi, J., Wang, W., Su, M., Kou, W., Yu, P., Yu, L., Chen, L., Zhu, J.S. and Chang, J. (1997). Multicenter clinical trial of the serum lipid-lowering effects of a Monascus purpureus (red yeast) rice preparation from traditional Chinese medicine. *Current Therapeutic Research - Clinical and Experimental.* **58**(12), 964–978.

Wang, L.J. (2006). Studies on the fermentation mechanism and volatile components of *Aspergillus*-type douchi. Ph.D. thesis, China Agricultural University, Beijing, China [In Chinese].

Wang, L.J., Li, D., Zou, L., Chen, X.D., Cheng, Y.Q., Yamaki, K., and Li, L.T. (2007a). Antioxidative activity of douchi (a Chinese traditional salt-fermented soybean food) extracts during its processing. *International Journal of Food Properties.* **10**(2), 385–396.

Wang, L.J., Yin, L.J., Li, D., Zou, L., Saito, M., Tatsumi, E. and Li, L.T. (2007b). Influences of processing and NaCl supplementation on isoflavone contents and composition during douchi manufacturing. *Food Chemistry.* **101**(3), 1247–1253.

Wang, R.Z. (1995). Sufu quality and ripening (postfermentation) control. *Journal of China Brewing Industry.* **2**, 31–35.

Wang, R.Z. and Du, X.X. (1998). *The Production of Sufu in China.* Beijing: China Light Industry Press, pp. 1–4.

Yasuda, M. and Kobayashi, A. (1987). Preparation and characterization of Tofuyo (fermented soybean curd). Trends in food biotechnology. *Proceedings of the 7th World Congress of Food Science and Technology,* Singapore.

Yen, G.C. (1986). Studies on biogenic amines in foods. I. Determination of biogenic amines in fermented soybean foods by HPLC. *Journal of the Chinese Agricultural Chemical Society.* **24**(2), 211–227.

Zhang, G.Y. and Shi, Y.G. (1993). The history of sufu production. *Journal of Food and Fermentation Industry.* **6**, 72–74.

Zhang, J.H., Tatsumi, E., Ding, C.H., and Li, L.T. (2006). Angiotensin I-converting enzyme inhibitory peptides in douchi, a Chinese traditional fermented soybean product. *Food Chemistry.* **98**(3), 551–557.

Zhang, X.F., Li, L.T., Yin, L.J., Li, Z.G., and Tatsumi, E. (2002). Study on the changes of soybean isoflavone during the sufu making. *China Brewing.* **6**, 17–20 (in Chinese).

Zhao, D.A. (1997). Legend of sufu. *Journal of China Brewing Industry.* **5**, 33–34.

Zou, L. (2006). The *In Vitro* Oxidative and Acetylcholine Esterase Inhibitory Activities of Douchi. Master thesis of China Agricultural University. Beijing, China [in Chinese].

7

FERMENTED VEGETABLES

FANG FANG

*Key Laboratory of Industrial Biotechnology, Ministry of Education,
and School of Biotechnology, Jiangnan University, China*

Contents

7.1 Introduction

Vegetables are plants cultivated for food. Generally, they are less sweet than fruits and often require processing to increase their edibility (Little et al., 1973). Fermentation, as described in previous chapters, is a "slow decomposition process of organic substances induced

199

by microorganisms, or by complex enzymes of plant or animal origin" (Walker, 1988). This technology is one of the oldest biotechnologies used in food processing. The exact origins of preservation of vegetables by fermentation are unknown, whereas the birth of fermented vegetables is thought to be in China, where a fermented mixture of vegetables was given to laborers during the construction of the Great Wall around 300 BC (Pederson, 1979). Fresh vegetables are easy to contaminate due to the high water activity (a_w) and nutrient content. Vegetable fermentation initially comes from food preservation, which results in products with a longer shelf life than fresh vegetables. Hence, fermented vegetables guarantee longer quality maintenance and availability of vegetables during the off-season. Although a number of other vegetable preservation methods are available today, fermentation is still taken as a highly appropriate technology in developing countries and remote areas because of the high efficiency and low energy costs for food manufacture. Besides being preserved, fermented vegetables are known to have remained palatable and have no or less antinutrients (Muchoki et al., 2010). This is another reason why it remains an important process in both developed and developing countries. Apart from the improvement of food security and nutrition, vegetable fermentation is culturally and economically important, which benefits people through the following aspects: (1) increasing income and employment, (2) improving cultural and social well being, and (3) health benefits (Battcock and Azam-Ali, 1998).

7.2 Fermented Vegetables around the World

7.2.1 Introduction

Fermented vegetables are manufactured and consumed around the world. The most popular fermented vegetables, such as sauerkraut made from cabbage and pickles made from cucumber (Battcock and Azam-Ali, 1998), are favorites in both Asia and Europe. *Kimchi* is a solid-fermented vegetable product manufactured in Korea, of which 2.1 million tons is consumed by local people or the world market annually. There are more than 200 kimchi manufacturers in Korea (Yi et al., 2001). Nearly 70% of kimchi are Chinese cabbage kimchi, which is the favorite and the best-known type. In many Nepali

Table 7.1 Fermented Vegetables from Around the World

TYPE	NAME	VEGETABLES	ORIGIN	REFERENCES
Kimchi		Cabbage, cucumber	Korea	Kim and Chun, 2005
Sauerkraut		Cabbage		
Pickles	Pickle	Cucumber	Africa, Asia, Latin America	Battcock and Azam-Ali, 1998
	Achar tandal	Cauliflower	India	Battcock and Azam-Ali, 1998
	Hom-dong		Thailand	Battcock and Azam-Ali, 1998
	Nukazuke	Cucumbers, carrots, turnips	Japan	Battcock and Azam-Ali, 1998
	Nukamiso-zuke	Vegetables	Japan	Battcock and Azam-Ali, 1998
	Pak-gard-dong	*Brassica juncea*	Tailand	Boon-Long, 1986
	Torshi felfel	Sweet peppers	West Asia and Africa	Battcock and Azam-Ali, 1998
	Torshi betingen	Aubergines	West Asia	Battcock and Azam-Ali, 1998
	Olives	Olives		
Others	Gundruk	Dried vegetables	Nepal	Tamang et al., 2007

communities and remote areas, a fermented and dried vegetable product called *gundruk* is an important source of minerals particularly during the off-season when the diet consists primarily of starchy tubers and maize, which tend to be low in minerals. Gundruk is normally served as a side dish or used as an appetizer in a bland, starchy diet. The annual production of gundruk in Nepal is estimated at 2000 tons. Table 7.1 shows examples of represented fermented vegetable products around the world.

7.2.2 Sauerkraut

Sauerkraut is German for "sour cabbage." This type of fermented food is thought to have originated in the north of China and was introduced in Europe 1000 years ago. Eastern Europeans, in particular, consume a large amount of sauerkraut. Jews adopted sauerkraut as part of their cuisine and are thought to have introduced it in the northern countries of Western Europe and the United States (Kitchenproject). Generally, sauerkraut is made by cutting fresh cabbage into strips, mixing with a certain amount of salt, and packing it into an airtight container. The fermentation vessel is kept at about 23°C for 3 days, and then left in cooler temperatures for 8 weeks. Sometimes, other vegetables and spices are added for making sauerkraut.

7.2.3 Pickles

According to what archeologists and anthropologists believe, the ancient Mesopotamians were making pickled products in 2400 BC. Another record stated that cucumbers brought from India helped begin a tradition of pickling in the Tigris Valley 4000 years ago. Today, pickles are one of the most popular side dishes in the world. In the United States, 2500 metric tons of pickles are consumed annually (Terebelski and Ralph, 2003). Generally, there are two types of pickles: brined (fermented) pickles and quick (unfermented) pickles. Pickles mentioned in this chapter refer to vegetables manufactured by solid fermentation. Besides cucumbers, many vegetables such as radishes, carrots, turnips and leafy vegetables can be fermented by pickling. Pickled red onion (*hom-dong*) is made in Thailand, aubergines (*torshi betingen*) are pickled in West Asia, and *sunki*, is an unsalted, fermented vegetable traditional to the Kiso area of Japan (Endo et al., 2008). *Torshi felfel*, fermented sweet peppers are produced in West Asia and Africa. Cauliflower stalks are fermented to produce *Achar tandal* in India. *Pak-Gard-Dong* is a fermented mustard leaf (*Brassica juncea*) product made in Thailand (Boon-Long, 1986).

7.2.4 Kimchi

Kimchi is a Korean traditional fermented side dish, made of vegetables with varied seasonings (Park et al., 2012). The oldest record of producing kimchi can be found from 2600 to 3000 years ago. The common seasonings for preparation of kimchi include brine, scallions, spices, ginger, chopped radish, garlic, saeujeot, and *aekheot* (fish sauce). There are a number of vegetables suitable for making kimchi, such as the napa cabbage, cucumber, radish, and green onion. Kimchi has gained popularity around the world because it has a very pleasing flavor and is a quite healthy addition to the diet.

7.2.5 Lafun

Lafun is a fermented cassava food product produced in West Africa (Padonou et al., 2010). The fibrous powdery form product is produced in a traditional way. Fresh cassava roots are cut into chunks, washed, and steeped (naturally fermented) for 3–4 days or until the roots become soft. Then the fermented roots are peeled, broken up into small pieces, and sun dried. Finally, the dried pieces are milled into powder.

7.3 Production Methods

Typical production processes of making kimchi and pickles are shown in Figures 7.1 and 7.2. Fermented vegetables, whether being

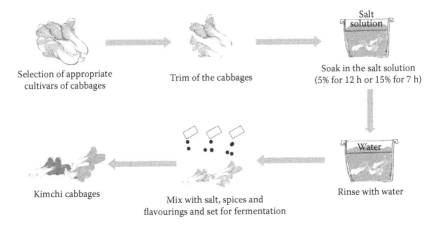

Selection of appropriate cultivars of cabbages

Trim of the cabbages

Soak in the salt solution (5% for 12 h or 15% for 7 h)

Kimchi cabbages

Mix with salt, spices and flavourings and set for fermentation

Rinse with water

Figure 7.1

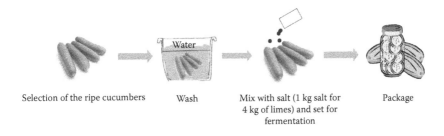

Selection of the ripe cucumbers

Wash

Mix with salt (1 kg salt for 4 kg of limes) and set for fermentation

Package

Figure 7.2

homemade or manufactured in well-equipped factories, are produced by either steeping in brine or solid fermentation. Here, the latter process is explained. Vegetable fermentation is normally performed salting or a low concentration of oxygen process. Pickling is a common method for the production of fermented vegetables. Pickling in a solid form is called *dry-salted pickling*. The typical process for preparation of dry salted pickles fermented vegetables includes the selection of vegetables, washing in clean water, mixing with salt (1 kg salt for 15–20 kg vegetables), and fermentation. Salt used in this process has to be pure to avoid quality problems. For instances, contamination of iron can cause blackening of the vegetables, magnesium gives a bitter taste, and carbonates can result in pickles with a soft texture (Battcock and Azam-Ali, 1998). Unlike in the making of liquidized pickles, salt is the only ingredient for producing dry-salted pickled products and without using vinegar (acetic acid), spices, and sugar.

The salting and low concentration of oxygen during solid vegetable fermentation are harsh conditions for some bacteria to grow in, allowing lactic acid bacteria to be the dominant species in this specific niche. Lactic acid bacteria are microaerophilic, and can grow in the presence of reduced amounts of atmospheric oxygen. Lactic acid bacteria can tolerate relatively high salt concentrations. The salt tolerance gives them an advantage over other species and allows the lactic acid bacteria to begin metabolism, which produces acid that further inhibits the growth of undesirable microorganisms. Thus, to control spoilage and to inactivate pathogens in vegetables, vegetables are intentionally fermented in a solid form with high osmotic pressure and a low concentration of oxygen.

7.4 Fermentation Engineering

Contrary to other fermentation processes, vegetable fermentation is mostly uncontrolled. For spontaneous fermentations (i.e., homemade fermented vegetables or vegetables fermented in a traditional way), microorganisms used in solid fermentation are neither pure strains nor a constant inoculum. They are a mixture of microorganisms originally coming from vegetables, the environment, and containers used for the fermentation. The composition of the microbiota and its development are critical factors affecting fermentation and final product quality.

Microbiologists have characterized *Lactobacillus* as the major genus in fermented vegetables. However, lactic acid bacteria are a small part of the autochthonous microbiota of raw vegetables. Spontaneous fermentations normally result from the competitive activities of a variety of autochthonous and contaminating microorganisms. Those best adapted to the fermentation conditions will eventually dominate. Initiation of a spontaneous process takes a relatively long time and may have a high risk for failure. Failure of the fermentation process can result in spoilage or the survival of pathogens, thus generating unexpected health risks for consumers. The discovery of the presence of lactic acid bacteria in fermented vegetables inspired the idea of using starter cultures for vegetable fermentation. Companies in Southwest China, especially the Sichuan province where most of the Chinese pickles are produced, are carrying on vegetable fermentation using starter *Lactobacillus* strains. The use of starter cultures would lead to a rapid acidification process and inhibit the growth of spoilage and pathogens, and to a short fermentation process with desirable consistent product quality (Leal-Sánchez et al., 2003).

7.4.1 Bioreactions in Vegetable Fermentation

Bioreactions in vegetable fermentation normally involve changes in both nutrients and other physiochemical properties of vegetables. Microorganisms present in the fermentation process contribute to food functionality through their enzyme activities and the release of metabolites. Changes in vegetable nutrients include increasing free amino acids and vitamins, improvement in protein digestibility, and development of desirable flavors and colors. In addition, bioactive compounds such as γ–amino butyric acid, conjugated linoleic acid may be synthesized during fermentation (Gobbetti et al., 2010).

7.4.2 Bacterial Community in Fermented Vegetables

A fermented vegetable, as a specific niche, is actually a diverse bacterial community. Table 7.2 demonstrates the dominance of lactic acid bacteria in vegetable fermentation. It was established that bacteria are the common dominance (360 and 50 times greater than archaea and yeast) in the whole microbial communities of various types of

Table 7.2   Identified Microorganisms in Fermented Vegetables

PRODUCT	VEGETABLE	MICROBIAL COMMUNITY	REFERENCES
Sauerkraut	Cabbage	*Leuconostoc citreum*	Plengvidhya et al., 2007
		Leuconostoc argentinum	
		Lactobacillus paraplantarum	
		Lactobacillus coryniformis	
		Weissella sp	
		Leuconostoc fallax	
		Lactobacillus brevis	
		Pediococcus pentosaceus	
Pickles	Eggplant	*L. plantarum*	Sanchez et al., 2004
		Lactobacillus brevis	
		Lactobacillus pentosus	
		L. fermentum	
Kimchi	Chinese cabbage	*Weissella*	Kim and Chun, 2005
		Leuconostoc	
		Lactobacillus	
Lafun	Cassave	*Bacillus cereus*	Wilfrid Padonou et al., 2009
		Klebsiella pneumonia	
		Pantoea agglomerans	
		L. fermentum	
		L. plantarum	
		Weissella confuse	
		Saccharomyces cerevisiae	
		Pichia scutulata	
		Kluyveromyces marxianus	
		Hanseniaspora guilliermondi	
		Pichia rhodanensis	
		Pichia kudriavzevii	
		Candida glabrata	
		Candida tropicalis	
		Trichosporon asahii	

kimchi (Park et al., 2009). In the bacterial community of kimchi, lactic acid bacteria have been proved to be the major representing bacteria by a culture-independent method, which includes 46% of *Weissella*, 39% of *Leuconostoc*, and 15% of *Lactobacillus* species. Among them, *W. koreensis, Lb. sakei,* and *Leuconostoc spp.* were found to be the major players in mature kimchi in five tested samples (Kim and Chun, 2005). Other lactic acid bacteria, *Lb. delbruecikii, Lb. fermentum,* and *Lb. plantarum*, were also detected as the dominance during prepara-tion of *sunki*, a fermented vegetable product in Japan (Endo et al.,

2008). *Leuconostoc mesenteroides*can tolerate fairly high concentrations of salt and sugar (50%), and is associated with sauerkraut and pickle fermentations. *Lb. plantarum* and *Lactobacillus pentosus* are regarded as the main species leading olive fermentation, being often used as a starter in guided olive fermentations. *Nukadoko*, a naturally fermented rice bran mash used for pickling vegetables in Japan, exhibits a *Lactobacillus*-dominated microbiota (Sakamoto et al., 2011).

7.4.3 Control of Fermentation

Control of vegetable fermentation is the process to ensure that only the desired microorganisms grow on the substrate. This process targets the inhibition of growth of other microorganisms that may be either pathogens or causing spoilage during fermentation process. Many food spoilage organisms cannot survive in either acidic or alcoholic environments. Thus, the development of either of these end products can prevent vegetables from spoilage. Oxidation reduction is another way to limit the growth of certain microorganisms. Both the type of organisms present and the reaction conditions determine the nature of the fermentation and the final products. Microbial reactions can be controlled to produce expected results by manipulating the external reaction conditions. There are a few ways of altering the fermentation environment to stimulate the growth of desirable organisms.

7.4.3.1 Low Concentration of Oxygen Microorganisms can be broadly classified into two groups—aerobes and anaerobes. Oxygen is essential for aerobes to carry out metabolic activities, whereas anaerobes grow in the absence of atmospheric oxygen. Facultative anaerobes can adapt to the prevailing conditions and grow in either the presence or absence of oxygen. Microaerophilic organisms grow in the presence of reduced amounts of oxygen. Therefore, controlling the microbial community of a food can be achieved by controlling the availability of free oxygen. Solid fermentation of vegetables is an anaerobic process. Vegetables are deprived of air, and the containers are tightly sealed during fermentation. Fungi do not grow well in anaerobic conditions; therefore, they are not important in terms of food spoilage, neither are they dominant microorganisms in fermented vegetables in terms of low oxygen availability. Many lactic acid bacteria are microaerophilic or facultative anaerobes. Therefore, reduction of atmosphere oxygen is an efficient way for controlling fermentation of vegetables by lactic acid bacteria.

7.4.3.2 Temperature   According to tradition, vegetables are normally fermented at room temperature. The relatively low temperatures (about 22°C) favor the growth of *Lactobacillus* strains. In some cases, temperature control is an important factor in the fermentation process. For instance, a temperature between 18°C–22°C is most desirable for initiating sauerkraut fermentation. Variations of just a few degrees from the recommend temperature alter the activity of the microbial process and may affect the quality of the product.

7.4.3.3 pH   Generally, the pH for solid vegetable fermentation is not controlled except for special purposes. However, the pH value during solid fermentation will decrease due to the production of lactic acid by lactic acid bacteria. The control of pH is normally required to achieve a good shelf life. It has been shown that red beets, green beans, cucumbers, and green tomatoes fermented by lactobacilli with controlled pH at 3.8 or below were microbiologically stable (Fleming et al., 1983). The pH-controlled fermentation process offers microbiological stability during 12-months's storage at 24°C.

7.4.3.4 Inhibitors   Inhibitors in vegetable fermentation are those bacteriocins or bacteriocin-like substances synthesized by *Lactobacillus* strains. These antimicrobial substances produced by lactobacilli are found to be more active against Gram-positive bacteria than to Gram-negative bacteria. These inhibitors can be used to preserve kimchi by inhibiting lactobacilli from over-ripening kimchi, rather than to inhibit spoilage microorganisms during the fermentation process (Choi and Park, 2000).

7.4.4 Metabolism

The microbial metabolism in fermented vegetables are determined mainly by the lactic acid bacteria, which will be specified in this chapter. The lactic acid bacteria are the most diverse bacteria group. Strains in this genus belong to two main groups, the homofermenters and the

Table 7.3 Major Lactic Acid Bacteria in Fermented Vegetables

HOMOFERMENTER	FACULTATIVE HOMOFERMENTER	OBLIGATE HETEROFERMENTER
Enterococcus faecium	*Lactobacillus bavaricus*	*Lactobacillus brevis*
Enterococcus faecalis	*Lactobacillus casei*	*Lactobacillus buchneri*
Lactobacillus acidophilus	*Lactobacillus coryniformis*	*Lactobacillus cellobiosus*
Lactobacillus lactis	*Lactobacillus curvatus*	*Lactobacillus confusus*
Lactobacillus delbrueckii	*Lactobacillus plantarum*	*Lactobacillus coprophilus*
Lactobacillus leichmannii	*Lactobacillus sake*	*Lactobacillus fermentatum*
Lactobacillus salivarius		*Lactobacillus sanfrancisco*
Streptococcus bovis		*Leuconostoc dextranicum*
Streptococcus thermophilus		*Leuconostoc mesenteroides*
Pediococcus acidilactici		*Leuconostoc paramesenteroides*
Pedicoccus damnosus		
Pediococcus pentocacus		

Source: Beuchat, L.R. (1995). *Food Technology.* **49**(1), 97–99.

heterofermenters. The final products of strains from both groups contain lactic acid though two different pathways of lactic acid production. Homolactic and heterolactic fermentations by lactic acid bacteria are demonstrated below. Homofermenters produce mainly lactic acid, via the glycolytic (Embden-Meyerhof) pathway. Heterofermenters produce lactic acid plus appreciable amount of ethanol, acetate and carbon dioxide, via the 6-phosphoglucanate/phosphoketolase pathway (Axelsson, 1998). Homolactic fermentation converts one mole of glucose into two moles of lactic acid, whereas the heterolactic fermentation converts one mole of glucose to one mole each of lactic acid, ethanol, and carbon dioxide. The fermentation panels of lactic acid bacteria in fermented vegetables are detailed in Table 7.3. Homolactic fermentation is shown as,

$$\underset{\text{glucose}}{C_6H_{12}O_6} \rightarrow \underset{\text{lactic acid}}{2\ CH_3CHOCOOH},$$

while heterolactic fermentation is

$$\underset{\text{glucose}}{C_6H_{12}O_6} \rightarrow \underset{\text{lactic acid}}{2\ CH_3CHOHCOOH} + \underset{\text{ethanol}}{C_2H_5OH} + \underset{\text{carbon dioxide}}{CO_2}.$$

Except for processes using lactic acid bacteria starter strains, vegetable fermentation is carried on with a complex bacterial community. Acids other than lactic acids (oxalic, citric, malic, succinic, lactic, formic, acetic, propionic, and butyric acids) were also detected in lactic acid fermented vegetables (Andersson and Hedlund, 1983). Acetic acid fermentation, mainly trigged by *Acetobacter*, which includes conversion of glucose into ethyl alcohol by yeast and the transformation of ethanol into acetic acid by *Acetobacter*, will be described in Chapter 8 of this book.

Phenolic compounds are components in vegetables that are directly related to food flavor, astringency, and color when degraded by microorganisms. Phenolic compounds are considered nutritionally undesirable due to the affection on utilization of vitamins and minerals. These substrates can inhibit digestive enzymes, thus reducing the nutritional values of foods. Recent studies show that phenolic compounds have nutritional and antioxidant properties. The presence of phenolic compounds in the diet may benefit host health because of their chemopreventive activities against carinogenesis and mutagenesis (Rodríguez et al., 2009). Lactic acid bacteria, especially *Lb. plantarum*, can produce tannase (Osawa et al., 2000), phenolic acid decarboxylase (Rodríguez et al., 2008), and benzyl alcohol dehydrogenase (Landete et al., 2008) to degrade some phenolic compounds providing sensory elements in foods. However, most of the metabolic pathways of biosynthesis or degradation of phenolic compounds in lactic acid bacteria remain unknown.

Biogenic amines are nitrogenous compounds present in food, especially fermented food (e.g., sauerkraut 110–300 mg kg^{-1}; Křížek and Pelikánová, 1998), which induce toxicological risks and health troubles. Biogenic amines are formed by decarboxylation of amino acids or by amination and transamination of aldehydes and ketones. The formation of biogenic amines is thought to occur under these conditions: availability of free amino acid, conditions that allow bacterial growth, decarboxylase synthesis, and decarboxylase activity (Bodmer et al., 1999). Bacteria of many genera such as *Bacillus*, *Citrobacter*, *Pseudomonas*, and *Lactobacillus*, *Pediococcus*, and *Streptococcus* can decarboxylate one or more amino acids (Santos, 1996). Cadaverine, histamine, putrescine, spermidine, and tyramine were detected in the lactic acid-bacteria fermented vegetables, ranging from 1 to 15 mg kg^{-1} (Simon-Sarkadi et al., 1993).

7.5 Function of Fermented Vegetables

Fermentation can improve food safety and nutrition by removal of antinutritional factors such as naturally occurring toxins and antinutritional compounds. The most common antinutrients in vegetables include nitrites, phenols, cyanogenic glycosides, glucosinolates, oxalates, and saponins (Teutonico and Knorr, 1984). These antinutrients or toxins can be removed or detoxified by microorganisms during fermentation. In addition, fermentation enhances the nutritional value of a food product through increased vitamin levels and improved digestibility and may have a health benefit to consumables.

Fermented vegetables also benefit consumers through providing natural antimicrobial substance and functional or bioactive elements. Kimchi, the Korean fermented vegetable, shows growth inhibitory effects against pathogens such as *Bacillus*, *Listeria*, and *Staphylococci* (Kim et al., 2008). Probiotic bacteria (Bifidobacteria or lactobacilli)-fermented broccoli can inhibit some pathogenic bacteria such as *Streptococcus mutans*, *Porphylomonas gingivalis*, and *Candida albicans* (Suido and Miyao, 2008). An attempt of producing β-carotene (by *Rhodotorula glutinis*) in fermented radish brine was also achieved (Malisorn and Suntornsuk, 2008).

7.6 Safety Issues of Fermented Vegetables

The presence of lactic acid bacteria, the bacteriocin producers, in vegetable fermentation can promote the microbial stability of both fermented and nonfermented vegetable food products (Settanni and Corsetti, 2008). Therefore, fermented vegetables produced with a good quality control process are generally considered safe. However, contamination of pathogens or the presence of biogenic amines and other toxins attracts concerns on safety issues of fermented vegetables. *Bacillus cereus*, a bacterium that causes food poisoning is detected in traditional fermented Nigerian fermented vegetables (Oguntoyinbo and Oni, 2004). Other pathogens such as *Escherichia coli* O157:H7, *Listeria monocytogenes*, *Clostridium botulinum*, *Aeromonas hydrophila*, *Samonella* spp., and *Campylobacter jejuni* were found to be frequently associated with fresh vegetables (Schuenzel and Harrison, 2002). Thus, the presence of spoilage microorgamisms consequently, reduced the safety and quality of fermented vegetables. Biogenic amines are toxic when consumed at high concentrations that were detected in *Doenjang* (a Korean fermented soybean paste) (Shukla et al., 2010). Japanese pickled radish was also found to contain N-nitroso-N-methylurea (NMU) (<0.3–72 ng), a carcinogen, mutagen, and teratogen (Sen et al., 2001). Fortunately, safety of fermented vegetables can be improved with proper controls of the fermentation process and pre- or posttreatments. Biogenic amines formed during fermentation can be reduced by gamma irradiation (Kim et al., 2005). Prewashing with sodium chlorite can reduce pathogenic bacteria in fermented vegetables (Inatsu et al., 2005).

7.7 Perspective

Solid vegetable fermentation is both a culturally and economically important food processing technology. This process confers nutrients, sensory flavors, and long storage life to vegetables. Using characterized strains as the starter culture for vegetable fermentation will probably benefit us through the following aspects: probiotic fermentation of vegetables can be achieved (Gupta and Abu-Ghannam, 2011); a non-biogenic amine producer can be selected as a starter in vegetable fermentation; a selected starter (e.g., kimchi prepared with *Leuc. citreum*) can inhibit growth or survival of foodborne pathogens (Chang and Chang, 2011) during fermentation; and production can be achieved at low NaCl concentration with expected shortened fermentation time (Beganovic et al., 2011).

References

Andersson, R. and Hedlund, B. (1983). Hplc analysis of organic acids in lactic acid fermented vegetables. *Zeitschrift fur Lebensmittel-Untersuchung und -Forschung.* **176**(6), 440–443.

Axelsson, L. (1998). *Lactic Acid Bacteria: Classification and Physiology.* New York: Marcel Dekker.

Battcock, M. and Azam-Ali, S. (1998). *Fermented Fruits and Vegetables: A Global Perspective.* Rome: Food and Agriculture Organization of the United Nations Rome.

Beganovic, J., Pavunc, A.L., Gjuracic, K., Spoljarec, M., Suskovic, J., and Kos, B. (2011). Improved sauerkraut production with probiotic strain *Lactobacillus plantarum* L4 and *Leuconostoc mesenteroides* LMG 7954. *Journal of Food Science.* **76**(2), M124–129.

Beuchat, L.R. (1995). Application of biotechnology to fermented foods. *Food Technology.* **49**(1), 97–99.

Bodmer, S., Imark, C., and Kneubühl, M. (1999). Biogenic amines in foods: Histamine and food processing. *Inflammation Research.* **48**(6), 296–300.

Boon-Long, N. (1986). *Traditional Technologies of Thailand: Traditional Fermented Food Products.* Mysore, India: Central Food Technological Research Institute.

Chang, J.Y. and Chang, H.C. (2011). Growth inhibition of foodborne pathogens by kimchi prepared with bacteriocin-producing starter culture. *Journal of Food Science.* **76**(1), M72–78.

Choi, M.H. and Park, Y.H. (2000). Selective control of lactobacilli in kimchi with nisin. *Letters in Applied Microbiology.* **30**(3), 173–177.

Endo, A., Mizuno, H., and Okada, S. (2008). Monitoring the bacterial community during fermentation of sunki, an unsalted, fermented vegetable traditional to the Kiso area of Japan. *Letters in Applied Microbiology.* **47**(3), 221–226.

Fleming, H.P., Mcfeeters, R.F., Thompson, R.L., and Sanders, D.C. (1983). Storage stability of vegetables fermented with pH control. *Journal of Food Science.* **48**(3), 975–981.

Gobbetti, M., Cagno, R.D., and De Angelis, M. (2010). Functional microorganisms for functional food quality. *Critical Reviews in Food Science and Nutrition.* **50**(8), 716–727.

Gupta, S. and Abu-Ghannam, N. (2011). Probiotic fermentation of plant based products: Possibilities and opportunities. *Critical Reviews in Food Science and Nutrition.* **52**(2), 183–199.

Inatsu, Y., Maeda, Y., Bari, M.L., Kawasaki, S., and Kawamoto, S. (2005). Prewashing with acidified sodium chlorite reduces pathogenic bacteria in lightly fermented Chinese cabbage. *Journal of Food Protection.* **68**(5), 999–1004.

Kim, J.-H., Kim, D.-H., Ahn, H.-J., Park, H.-J., and Byun, M.-W. (2005). Reduction of the biogenic amine contents in low salt-fermented soybean paste by gamma irradiation. *Food Control.* **16**(1), 43–49.

Kim, M. and Chun, J. (2005). Bacterial community structure in kimchi, a Korean fermented vegetable food, as revealed by 16s rrna gene analysis. *International Journal of Food Microbiology.* **103**(1), 91–96.

Kim, Y.S., Zheng, Z.B., and Shin, D.H. (2008). Growth inhibitory effects of kimchi (Korean traditional fermented vegetable product) against *Bacillus cereus*, *Listeria monocytogenes*, and *Staphylococcus aureus*. *Journal of Food Protection.* **71**(2), 325–332.

Kitchenproject. The History of Sauerkraut. Retrieved 2012-02-09, 2012.

Křížek, M. and Pelikánová, T. (1998). Determination of seven biogenic amines in foods by micellar electrokinetic capillary chromatography. *Journal of Chromatography A.* **815**(2), 243–250.

Landete, J.M.A., Rodríguez, H.C., De Las Rivas, B., and Muñoz, R. (2008). Characterization of a benzyl alcohol dehydrogenase from *Lactobacillus plantarum* WCFS1. *Journal of Agricultural and Food Chemistry.* **56**(12), 4497–4503.

Leal-Sánchez, M.V., Ruiz-Barba, J.L., Sánchez, A.H., Rejano, L., Jiménez-Díaz, R., and Garrido, A. (2003). Fermentation profile and optimization of green olive fermentation using *Lactobacillus plantarum* LPCO10 as a starter culture. *Food Microbiology.* **20**(4), 421–430.

Little, W., Fowler, H.W., and Coulson J. (1973). *The Shorter Oxford English Dictionary.* Oxford University Press.

Malisorn, C. and Suntornsuk, W. (2008). Optimization of β-carotene production by *Rhodotorula glutinis* DM28 in fermented radish brine. *Bioresource Technology.* **99**(7), 2281–2287.

Muchoki, C.N., Lamuka, P.O., and Imungi, J.K. (2010). Reduction of nitrates, oxalates and phenols in fermented solar-dried stored cowpea leaf vegetables. *African Journal of Food, Agriculture, Nutrition and Development.* **10**(11), 4398–4412.

Oguntoyinbo, F.A. and Oni, O.M. (2004). Incidence and characterization of *Bacillus cereus* isolated from traditional fermented meals in Nigeria. *Journal of Food Protection.* **67**(12), 2805–2808.

Osawa, R., Kuroiso, K., Goto, S., and Shimizu, A. (2000). Isolation of tannin-degrading lactobacilli from humans and fermented foods. *Applied and Environmental Microbiology.* **66**(7), 3093–3097.

Padonou, S.W., Nielsen, D.S., Akissoe, N.H., Hounhouigan, J.D., Nago, M.C., and Jakobsen, M. (2010). Development of starter culture for improved processing of Lafun, an African fermented cassava food product. *Journal of Applied Microbiology.* **109**(4), 1402–1410.

Park, E.J., Chang, H.W., Kim, K.H., Nam, Y.D., Roh, S.W., and Bae, J.W. (2009). Application of quantitative real-time pcr for enumeration of total bacterial, archaeal, and yeast populations in kimchi. *The Journal of Microbiology.* **47**(6), 682–685.

Park, E.J., Chun, J., Cha, C.J., Park, W.S., Jeon, C.O., and Bae, J.W. (2012). Bacterial community analysis during fermentation of ten representative kinds of kimchi with barcoded pyrosequencing. *Food Microbiology.* **30**(1), 197–204.

Pederson, C.S. (1979). *Microbiology of Food Fermentations.* Westport: AVI Publishing Co.

Plengvidhya, V., Breidt, F., Jr., Lu, Z., and Fleming, H.P. (2007). DNA fingerprinting of lactic acid bacteria in sauerkraut fermentations. *Applied Environmental Microbiology.* **73**(23), 7697–7702.

Rodríguez, H., Curiel, J.A., Landete, J.M., De Las Rivas, B., De Felipe, F.L., Gómez-Cordovés, C., Mancheño, J.M., and Muñoz, R. (2009). Food phenolics and lactic acid bacteria. *International Journal of Food Microbiology.* **132**(2–3), 79–90.

Rodríguez, H., Landete, J.M., Curiel, J.A., De Las Rivas, B., Mancheño, J.M., and Muñoz, R. (2008). Characterization of the p-coumaric acid decarboxylase from *Lactobacillus plantarum* CECT 748t. *Journal of Agricultural and Food Chemistry.* **56**(9), 3068–3072.

Sakamoto, N., Tanaka, S., Sonomoto, K., and Nakayama, J. (2011). 16s rRNA pyrosequencing-based investigation of the bacterial community in nukadoko, a pickling bed of fermented rice bran. *International Journal of Food Microbiology.* **144**(3), 352–359.

Sanchez, I., Seseña, S., and Palop, L.L. (2004). Polyphasic study of the genetic diversity of lactobacilli associated with "almagro" eggplants spontaneous fermentation, based on combined numerical analysis of randomly amplified polymorphic DNA and pulsed-field gel electrophoresis patterns. *Journal of Applied Microbiology.* **97**(2), 446–458.

Santos, S.M.H. (1996). Biogenic amines: Their importance in foods. *International Journal of Food Microbiology.* **29**(2–3), 213–231.

Schuenzel, K.M. and Harrison, M.A. (2002). Microbial antagonists of foodborne pathogens on fresh, minimally processed vegetables. *Journal of Food Protection.* **65**(12), 1909–1915.

Sen, N.P., Seaman, S.W., Baddoo, P.A., Burgess, C., and Weber, D. (2001). Formation of N-nitroso-N-methylurea in various samples of smoked/dried fish, fish sauce, seafoods, and ethnic fermented/pickled vegetables following incubation with nitrite under acidic conditions. *Journal of Agricultural and Food Chemistry.* **49**(4), 2096–2103.

Settanni, L. and Corsetti, A. (2008). Application of bacteriocins in vegetable food biopreservation. *International Journal of Food Microbiology.* **121**(2), 123–138.

Shukla, S., Park, H.-K., Kim, J.-K., and Kim, M. (2010). Determination of biogenic amines in Korean traditional fermented soybean paste (doenjang). *Food and Chemical Toxicology.* **48**(5), 1191–1195.

Simon-Sarkadi, L., Holzapfel, W.H., and Halasz, A. (1993). Biogenic amine content and microbial contamination of leafy vegetables during storage at 5°C. *Journal of Food Biochemistry.* **17**(6), 407–418.

Suido, H. and Miyao, M. (2008). *Bifidobacterium longum*-fermented broccoli supernatant inhibited the growth of *candida albicans* and some pathogenic bacteria *in vitro. Biocontrol Science.* **13**(2), 41–48.

Tamang, J., Thapa, N., Rai, B., Thapa, S., Yonzan, H., Dewan, S., Tamang, B., Sharma, R., Rai, A., Chettri, R., Mukhopadhyay, B., and Pal, B. (2007). Food consumption in sikkim with special reference to traditional fermented foods and beverages: A micro-level survey. *Journal of Hill Research.* Supplementary Issue, **20**(1), 1–37.

Terebelski, D. and Ralph, N. (2003). Pickle history timeline. New York Food Museum.

Teutonico, R. and Knorr, D. (1984). Plant tissue culture: Food applications and the potential reduction of nutritional stress factors. *Food Technology Journal.* **38**(2), 120–126.

Walker, P.M.B. (1988). *Chambers Science and Technology Dictionary*. Chambers, Cambridge University Press.

Wilfrid Padonou, S., Nielsen, D.S., Hounhouigan, J.D., Thorsen, L., Nago, M.C., and Jakobsen, M. (2009). The microbiota of lafun, an African traditional cassava food product. *International Journal of Food Microbiology*. **133**(1–2), 22–30.

Yi, H., Kim, J., Hyung, H., Lee, S., and Lee, C.H. (2001). Cleaner production option in a food (*kimchi*) industry. *Journal of Cleaner Production*. **9**, 35–41.

8

SOLID-STATE FERMENTED CONDIMENTS AND PIGMENTS

GANRONG XU AND BOBO ZHANG

Key Laboratory of Industrial Biotechnology, Ministry of Education, School of Biotechnology, Jiangnan University, Wuxi, China

Contents

8.1 Vinegar

8.1.1 Brief Introduction to Vinegar Fermentation

Vinegar is the product of a two-stage fermentation process. Primary fermentation involves converting sugars and starchy materials into alcohol. Vinegar is produced when bacteria convert the alcohol into acetic acid.

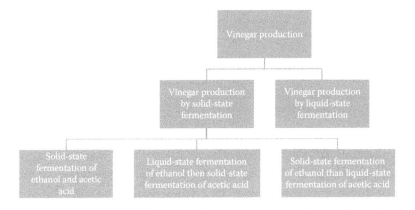

Figure 8.1   Vinegar production processes.

The chemical reactions related to vinegar production are summarized as follows:

1. $(C_6H_{10}O_5)_n + nH_2O \rightarrow nC_6H_{12}O_6$
2. $C_6H_{12}O_6 \rightarrow 2\ C_2H_5OH + 2\ CO_2$
3. $C_2H_5OH + O_2 \rightarrow CH_3COOH + H_2O$

Vinegar production involves ethanol fermentation and acetic acid fermentation. No matter whether ethanol fermentation or vinegar fermentation is being used, they can all be run in a solid state or liquid state. Therefore, there are several types of vinegar production processes as given in Figure 8.1.

In the process of ethanol and acetic acid production by solid-state fermentation, the former process is a traditional way like the fermentation of Chinese rice wine (see also Chapter 9), whereas the rice wine can be used as raw material for the latter process. The raw materials for vinegar production by solid-state fermentation in Asia are rice, wheat, millet, sorghum, corn, sweet potato, and wheat bran (Chen et al, 2009). Wheat bran, rice hull, and sorghum hull are used as a filling agent in solid-state fermentation, which makes the substrate of solid-state fermentation loose and facilitates oxygen transfer.

Traditional *qu*—a fermented product with growth of molds containing several hydrolysis enzymes—is used as a saccharifying agent and fermenting agent for ethanol fermentation (Yue et al., 2007). Various kinds of qu can be used for the ethanol fermentation, including *da qu* (wheat and bean as raw materials), wheat qu (only wheat as raw

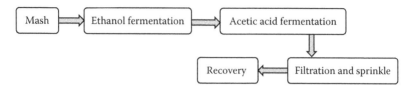

Figure 8.2 The flowchart of vinegar production in solid-state fermentation.

material), rice qu (rice as raw material), and *fu qu* (wheat bran as main raw material). All the qu is made by traditional solid-state cultivation.

Large-scale production of vinegar uses the combination of two stages: liquid-state fermentation for ethanol production and solid-state fermentation for acetic acid production (Couto and Sanromán, 2006).

Liquid-state fermentation of ethanol takes place in big fermenters. The saccharifying agents used in the ethanol production stage include various commercial enzyme preparations, crude enzymes, and an enzyme complex produced by *Aspergillus niger* or *Aspergillus oryzae*. Nowadays, commercial enzymes are widely used for ethanol production. Active yeast starter is used as the fermenting agent. The detailed techniques used in rice wine fermentation are introduced in Chapter 9. Here, only the solid-state fermentation for acetic acid production is discussed in detail. The solid-state fermentation of acetic acid is carried out in large-scale aeration tanks with a screw agitator.

In any case, the process of solid-state fermentation of vinegar is almost the same for the traditional or modern process and can be summarized in Figure 8.2.

Manual or mechanical operations for stirring and mixing the fermenting mash are necessary to supply oxygen for the acetic acid fermentation, which is one characteristic of the traditional solid-state fermentation processes.

Acetic acid fermentation process can be divided into three stages: microbial growth, acetic acid formation, and esterification.

1. *Microbial growth stage:* The fermentation mash (ethanol concentration 6%, moisture 60%) is inoculated with the seed culture of *Acetobacter* and then maintained at 38°C–44°C. Under appropriate conditions, *Acetobacter* grows rapidly, and the microbial growth stage lasts about 13 days after inoculation. Vinegar fermentation by *Acetobacter* is conducted aerobically. In solid-state fermentation, dissolved oxygen is guaranteed not only by

adding a large amount of porous filling agent in the fermentation mash, but also by aeration and mixing the fermenting substrates intermittently. In a traditional operation, the aeration is achieved by mixing and exposing the mash to the air.

2. *Acetic acid formation stage:* Acetic acid fermentation lasts about 20 days. During this stage, the ethanol in the mash is oxidized to acetic acid by the bacteria. The oxygen supply is reduced gradually due to the reduced mixing operation. The bacteria then enter the death stage. The ethanol concentration in the mash deceases gradually while the acetic acid concentration increases accordingly. When the acetic acid concentration no longer increases, the temperature begins to decrease, and then the mash should be sealed tightly to prevent further oxidation. By depletion of oxygen, the acetic acid in the mash cannot be further oxidized into carbon dioxide and water.

3. *Esterification stage:* During this period, it is important to avoid the acetic acid being further fermented to carbon dioxide. At the same time, the reaction of esterification takes place. Various aromatic substances are formed during this stage.

8.1.2 Microorganisms Involved in Acetic Acid Fermentation and Cultivation Methods

8.1.2.1 Microorganisms Most of the bacteria used in vinegar fermentation belong to *Acetobacter* spp. (Sievers et al., 1992; Tesfaye et al., 2002). The appropriate temperature for their growth is above 30°C. Most of these strains can grow at 30°C–40°C. These strains oxidize ethanol to acetic acid. If the conditions are appropriate, the acetic acid can be further oxidized to carbon dioxide and water, which should be avoided in vinegar production.

The strain *A. orleanense* is a strain used for production of grape vinegar. This strain is not an ideal acetic acid producer, but can tolerate a high concentration of acetic acid.

The strain *A. rancens* can grow on the medium surface to form wrinkled film. The film can grow along with the wall of container. The acetic acid concentration can reach to 6%–9% by using this strain.

The strain *A. aceti* 1.01, with the advantage of a high conversion rate from ethanol to acetic acid, is a popular strain used by vinegar manufacturers on the mainland of China. The conversion rate can reach to 93%–95%.

8.1.2.2 Cultivation Methods for Acetobacter *spp.*   The methods of cultivation of seed culture for vinegar fermentation are classified into solid-state cultivation and liquid-state cultivation.

Traditionally, microorganisms from air, raw materials, qu were the main sources of fermenting agents. High-quality vinegar fermentation mash is used as seed culture for the next batch. Now pure cultures of *Acetobacter* spp. are commonly used in vinegar production by modern manufacturing enterprises.

1. *Liquid-state cultivation of the pure cultures.* Culture medium is composed of: yeast extract 1% and glucose 0.3%. 100 mL culture medium is transferred into 500 mL conical flask and sterilized at 121°C for 30 min. After cooling, 95% ethanol is added to the medium to the final concentration at 4%. A slant culture of *Acetobacter* spp. is inoculated to the medium. The culture is usually incubated statically or with shaking at 30°C for 5–7 days.
2. *Solid-state cultivation of the pure cultures.* The solid-state reactor for seed culture cultivation is a vessel with a perforated bottom. The fermenting mash is placed on the perforated bottom. Liquid can flow out from the outlet in the bottom and is recycled to the surface of the fermenting mash.

The flask seed culture is added to mature fermented mash with an ethanol concentration of 4%–5%. The inoculum volume of seed culture is 2%–3% and then the vessel is mixed well and covered with a lid. Bacteria begins to grow in the mash, then the temperature of the mash rises to about 38°C after 2 days of the inoculation. Then the vinegar liquid is recycled from the outlet to the surface of the vinegar mash. When the acetic acid concentration in the mash reaches up to 4%, the seed culture is checked by microscopic examination. If there is no contamination by other contaminants, the seed culture can be used as an inoculum.

8.1.3 Typical Vinegar Production Processes in Solid-State Fermentation

Traditional vinegar production techniques in China were described in a book entitled *qiminyaoshu*. The traditional techniques are characterized by using qu (saccharifying and fermenting agents) made by solid-state cultivation, and ethanol and vinegar fermentation, both conducted in a solid state. The vinegar production techniques used in northern China are different from southern China in many aspects, including the raw materials, saccharifying and fermenting agents. Several typical vinegar production techniques used in northern China and in southern China are introduced below.

8.1.3.1 Shanxi Aged Vinegar

8.1.3.1.1 Brief Introduction Shanxi Province is located in northwest China. The techniques employed in Shanxi aged vinegar production are in accordance with the traditional techniques in *qiminyaoshu*. The saccharifying and fermenting agent used in ethanol fermentation is da qu (the same qu used in Chinese spirits production). The ethanol fermentation time is very long. The vinegar fermentation temperature is relatively higher than usual. After vinegar fermentation, half of the fermented mash is smoked gradually, and the other half of the fermented mash is trickled through the smoked vinegar mash. The new vinegar should undergo the maturation stage during the period from summer to winter. During the summer, the vinegar is matured in the environment with a high temperature, and during the winter season, the vinegar is matured in the environment with a very low temperature, so that the water in the vinegar is frozen to ice, then the ice is dredged up from the vinegar. So the acetic acid concentration in Shanxi aged vinegar is usually above 9%, the highest among ordinary vinegar. The color of Shanxi aged vinegar is dark brown, and its flavor is very rich.

8.1.3.1.2 Raw Materials The main ingredients for Shanxi aged vinegar production are sorghum, 100 kg; da qu, 62.5 kg; wheat bran, 73 kg; rice bran, 73 kg; salt, 5 kg; various spices (pepper, aniseed, cinnamon, clove), 0.05 kg; and total water, 340 kg.

8.1.3.1.3 Pretreatment of Raw Materials Sorghum is coarsely milled. 50% water is added (based on the sorghum's weight) into the sorghum and mixed well. After the water is absorbed thoroughly, the sorghum is cooked and then transferred to jars. Boiled water is added into the cooked sorghum and further absorbed by the sorghum. Finally, the cooked sorghum is cooled to 25°C–26°C.

8.1.3.1.4 Da qu Making Da qu, the saccharifying and fermenting starter culture that is actually used as an enzyme preparation and microbial seed, is made by wheat and peas in the ratio of 7 to 3. They are coarsely milled and mixed well. The mixture is moistened with water and pressed in a mold with a mold-press machine to bricks with the dimension of length × width × height = 10–12 cm × 8–10 cm × 4–6 cm. The da qu bricks are transferred into an incubation room for cultivation under controlled conditions. During the cultivation, the mold multiply on the surface of the da qu bricks and yield liquefying amylase, saccharifying amylase, acid proteases, lipase, and esterase. Temperature control of da qu bricks in the incubation room is the key operation during cultivation. The maximal temperature is usually controlled below 50°C. The da qu is cultivated in the incubation room for about 1 month and matured for several months before use.

8.1.3.1.5 Process Flow for Shanxi Aged Vinegar

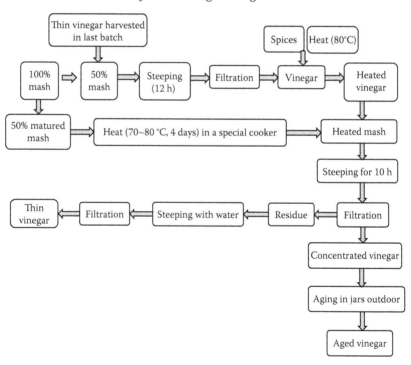

8.1.3.1.6 Ethanol Fermentation The cooked and cooled sorghum is mixed with da qu powder and cold water with the ratios of 150:62.5:65, and the moisture content of the mixture is adjusted to about 55%. The mixture is transferred into jars for fermentation. The initial temperature of the mash is controlled between 20°C and 24°C. In the first fermentation stage lasting about 4 days, the fermentation temperature increases gradually. To avoid the temperature rise beyond the limit, the fermenting mash is stirred twice every day in the first period. After fermentation for 3 days, the temperature may rise to 30°C. The maximal temperature reaches about 36°C in the fourth day, after which the temperature decreases gradually. In order to maintain the fermentation temperature at an appropriate range, the jars are covered with plastic film and insulation mats. In the second fermentation stage lasting 15 days or more, the temperature is controlled between 18°C and 20°C. During the fermentation, the starch is hydrolyzed to reducing sugar, and then the reducing sugar is converted to ethanol.

8.1.3.1.7 Acetic Acid Fermentation The *fermentation mash preparation* is 13 kg fermented mash containing ethanol at the concentration about 15% mixed with 6 kg wheat bran and 7 kg rice bran. The mixture is transferred to fermentation jars. The moisture of the mash is adjusted to 60%~64%, while the ethanol concentration of the mash is adjusted to between 4.5%–5%.

Seed culture for inoculation: There is no special cultivation for seed culture in a traditional operation. The fermented mash from a previous batch with the temperature between 38°C and 45°C is used as seed culture for the acetic acid fermentation. Inoculum volume is 10% based on the fermentation mash. The seed is mixed well with the fermentation mash and then fermentation begins.

During the whole fermentation period, the most important operation is to stir the fermenting mash and exchange the mash among different jars whose temperature is abnormal. Stirring the mash can supply the oxygen to the microbes in the mash and disperse unpleasant flavors in the mash.

Acetic acid fermentation in different jars may progress fast or slowly, resulting in the temperature difference in these jars. So part of the mash with a higher temperature is taken out and exchanged with the mash of lower temperature. In this way, the temperatures of fermenting mashes in all jars can be kept equal.

Further fermentation for maturation of the fermenting mash: The fermenting mashes in several jars are transferred to a bigger vat and pressed tightly for further fermentation. Spread some common salt onto the surface of the mash and seal the jars with plastic film. The mashes are further fermented for another 10 to 15 days.

8.1.3.1.8 Heating, Filtration, and Aging   After fermentation, half of the mash is heated by slow fire for 4 days in heating equipment, and the other half of the mash is mixed with thin vinegar from the last batch, steeped, then filtered. The filtered vinegar is mixed with the heated mash, steeped for 10 h, and finally filtered to obtain the concentrated vinegar which then is aged for a long time. The process flow is as follows:

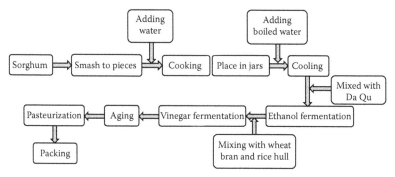

The operation of heating 50% matured mash by slow fire is a traditional process. The heating equipment is composed of a ceramic vat mounted over a pit where a charcoal fire is burning. As the thermoconductivity of the ceramic container is poor, the mash is heated slowly. During the period of heating, starch and protein in the mash are hydrolyzed slowly under weak acidic conditions to reduce sugar and amino acids. Then the Maillard reaction takes place in the heated mash to form a brown substance, resulting in the formation of the typical dark brown color of Shanxi aged vinegar. Usually, 100 kg of sorghum yields 400 kg of the final product, the Shanxi aged vinegar. The acetic acid concentration in the vinegar is 9%.

The vinegar is contained in ceramic jars, which are placed outdoors. After being stored for 1 year, the vinegar is filtered to remove the precipitate, then it is sold.

8.1.3.2 Zhenjiang Vinegar

8.1.3.2.1 Brief Introduction   Zhenjiang vinegar originated with the Hengshun Vinegar Company in the city of Zhenjiang, in the eastern coastal province of Jiangsu, China. The traditional production process and modern production process are currently applied side by side by the Hengshun Vinegar Company (Liu et al., 2004; Huang et al., 2008).

8.1.3.2.2 Traditional Process for Zhenjiang Vinegar Production In the traditional process, Zhenjiang vinegar is made from glutinous rice and wheat *koji*. The main process flow is described in Figure 8.3.

The typical traditional operations of Zhenjiang vinegar production are introduced in Figures 8.4–8.10, and brief descriptions follow.

Raw materials (for each batch of operation): glutinous rice, 500 kg; herbal qu, 2 kg; wheat koji, 30 kg; wheat bran, 850 kg; rice hull, 450 kg. Rice steeping: 0°C, 30 h or 5°C–10°C, 24 h; 15°C–20°C, 18 h; 25°C–30°C, 15 h.

Fermenting ethanol mash, 1500–1650 kg; 26°C–28°C (2 d) → 35°C (7 d); stir; recycle the liquid 3–4 times; ethanol, 10%–14%.

Vinegar mash preparation: ethanol mash/wheat bran/rice hull; inoculation with matured vinegar mash: 10%; fermentation at 45°C–46°C for about 15 days; stir once every day.

8.1.3.2.3 Large-Scale Production of Zhenjiang Vinegar The process of large-scale production of Zhenjiang vinegar is shown in Figure 8.11.

In the large-scale production process, ethanol fermentation is conducted by liquid-state fermentation. Wheat koji, the traditional saccharifying and fermenting agent, is substituted with various enzyme preparations. Only acetic acid fermentation is carried out in solid-state fermentation. The fermentation reactors are specially designed pools with a mechanical stirrer, without perforated bottoms and aeration systems.

8.1.3.2.4 Ethanol Fermentation The flowchart of ethanol fermentation in the large-scale production of Zhenjiang vinegar is shown in Figures 8.12–8.13.

Milled rice powder is mixed with 65°C–70°C water. The ratio of rice powder to water is 1:3–4. The mixture is stirred, and amylase is added into the mash. It is heated to 105°C ± 5°C, and the starch is liquefied. After liquefaction, the mash is cooled to 60°C and then glucoamylase is added into the mash. After saccharification for about 30 min, the mash is cooled to 26°C~28°C, and 0.1% active yeast starter

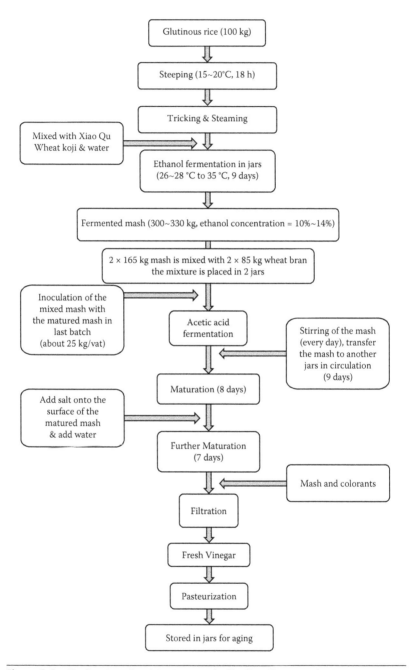

Figure 8.3   The flowchart of traditional process for Zhenjiang Vinegar production.

Figure 8.4   Cooked rice is placed in ceramic jars.

Figure 8.5   The cooked rice is inoculated with a yeast starter.

Figure 8.6   Ethanol fermentation.

and 3%–6% of wheat koji are added to the cooled mash. Ethanol fermentation is conducted in big fermenters of which the maximal volume is 300–400 M³. The maximal temperature is controlled below 34°C–36°C. Ethanol concentration will reach 10%–12% (v/v).

8.1.3.2.5 Acetic Acid Fermentation   Acetic acid fermentation is conducted in a fermenter, which is a concrete pool with a perforated bottom and a set of stirrers. The fermented mash containing ethanol is

Figure 8.7 Acetic acid fermentation.

Figure 8.8 Seal the fermenting mash tightly (further fermentation for 15 days).

mixed with wheat bran and rice hull. The mixture is called "ethanol mash." The ratio of the ethanol mash is: rice:wheat bran:rice hull:water = 1:1.7:(0.9–1.0):4–3 (see Figure 8.14).

Before loading the ethanol mash into the concrete pool, a part of wheat bran and rice hull is underlayed in the perforated bottom of the

Figure 8.9 Filtration of vinegar (a traditional method). For 1000 kg vinegar: dark reddish brown, 135 kg; salt, 20 kg; sucrose, 6 kg are needed.

Figure 8.10 Maturation of vinegar outdoors. Vinegar is stored in jars and placed outdoors for 6 months or more.

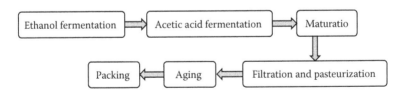

Figure 8.11 The flowchart of large-scale production of Zhenjiang vinegar.

concrete pool. Then the ethanol mash is put into the pool. The ethanol mash is inoculated with starter mash (the fermenting mash prepared in a previous batch is used as a starter mash). The inoculum volume of the starter mash is 10%.

Acetic acid fermentation is carried out aerobically and anaerobically by turns in the concrete pool due to the characteristics of the

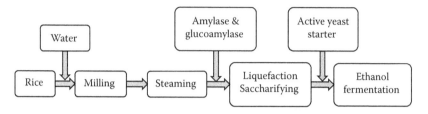

Figure 8.12   The flowchart of ethanol fermentation in large-scale production of Zhenjiang vinegar.

Figure 8.13   Ethanol fermentation in fermenters.

solid-state substrate. After inoculation, the starter mash placed in line or in heap in the pool is covered with rice hull. The starter mash is not mixed with the ethanol mash. The bacteria in a starter mash start to multiply in 2–3 days. When the starter mash under the rice hull begins to generate heat, it is an indication that the bacteria is growing well, then the starter mash is mixed thoroughly with the ethanol mash. This is called "vinegar mash."

8.1.3.2.6 Stirring   After fermentation for 2–3 days, the bacteria (mainly *Acetobacter*) grow well on the surface of the vinegar mash. It is known that the acetic acid fermentation is aerobic. Due to the absence of aeration systems and perforated bottoms in the fermenters, the oxygen supply must be achieved by stirring the vinegar mash intermittently. Manual and mechanical stirring operations are shown in Figure 8.15. Usually, stirring is performed once each day and

Figure 8.14 Large-scale production of Zhenjiang vinegar.

Figure 8.15 Fermenters for vinegar fermentation (stir by hand or stirrer).

downwards gradually. It takes 8–9 days for the bacteria to grow well in the vinegar mash in the bottoms of the fermenters.

As the stirring operation is conducted intermittently, the vinegar mash is actually fermented aerobically and anaerobically by turns in the concrete pools. The fermenting mashes in different heights in the fermentation pools are different in bacteria concentrations and other characters, such as the ethanol concentrations, acetic acid concentrations.

When acetic acid concentration in the vinegar mash reaches to 4.5%, the temperature of the mash has the tendency of decline and it is the time to transfer the mash to hermetically sealed vessels. The mash is pressed tightly and the vessels are sealed with plastic film. Otherwise, the acetic acid in the mash will be oxidized to CO_2 and H_2O. Usually, the vinegar mash is sealed for 7 days.

8.1.3.2.7 Aging of the Mash   After the mash is sealed for 7 days, the mash is dug up, mixed, returned to the pool, and sealed again. This operation can help release the unpleasant smells and reduce the temperature of the mash. The aging time is about 20–30 days.

8.1.3.2.8 Filtration and Aging of the Vinegar   Before vinegar filtration, a special colorant is prepared. The colorant is made of fried rice steeped in hot water. The matured vinegar mash is transferred to a filtration tank with a perforated bottom, and the colorant is pumped into the mash. Salt in proper amounts is mixed with the mash. The vinegar mash is steeped in diluted vinegar or water. After steeping for 3–18 hours (according to the liquid used for steeping), the filtration is carried out to collect the vinegar. The harvested vinegar is placed in tanks for precipitation for 3–5 days, and the clarified vinegar is heated to a boiling point for 20–30 min. The heated vinegar is placed into ceramic kegs for storage for 2–3 years.

8.1.3.3 Wheat Bran Vinegar

8.1.3.3.1 Brief Introduction   Wheat bran vinegar is a traditional type of vinegar popular in the Sichuan Province, in southwest China (Zhang and Li, 2009). In the traditional process for making wheat bran vinegar, three processes, including saccharification, ethanol fermentation, and acetic acid fermentation, take place concomitantly.

The characteristic of this vinegar is that glutinous rice is brewed to rice wine by using herbal qu as saccharifying and fermenting agents, and then the rice wine, which is used as a yeast starter, is mixed with a relatively large quantity of wheat bran. Usually, the wheat bran quantity in common vinegar production is 1.5- or 2-fold of the cereal, but wheat bran quantity in the vinegar produced in the Sichuan Province is 20-fold of the rice. That is the main reason why the product is named *wheat bran vinegar*. Wheat bran is rich in C5-sugar (pentose), which is the basic substance forming the Maillard reaction, so the wheat bran vinegar is very dark in color and aromatic in flavor.

8.1.3.3.2 Wheat Bran qu Preparation   Wheat bran qu is crude enzymes fermented by *Aspergillus niger* 3758 by using wheat bran as raw material.

8.1.3.3.3 Herbal qu Preparation   The herbal qu is made by adding some herb powder to the raw materials for qu making. Seed of water chestnut is milled to powder whose main substance is starch. A lot of Chinese medicines and herbal medicines, such as tangerine peel, liquorice, Chinese red pepper, *Atractylis,* and *lovage rhizome*, are sun cured and milled to powders. All the powders are mixed with water and shaped to brick weighed about 2 kg. The herbal qu is cultivated in an incubation room for 6–7 days, letting the mold grow on the surface of the qu. The herbal qu is used as a yeast starter for ethanol fermentation. As the herbal qu is a natural fermented product, the microbial flora in the herbal qu is unknown but rich in yeasts and *Acetobacter* populations.

8.1.3.3.4 Yeast and Acetobacter Starter Preparation   Glutinous rice 30 kg is steeped and then steamed. Next, 100 kg water and 0.3 kg herbal koji are added to the steamed rice, then 1.5 kg of laliao juice is added to the mixture. Cover the jar and fermentation begins. After fermentation for about 7 days, the yeast and *Acetobacter* starter can be used as seed cultures for inoculation into the raw materials, mainly based on wheat bran.

8.1.3.3.5 Acetic Acid Fermentation   The fermenter is a simple tank made of wood, with a length of 2.4 m, width of 1.25 m, and height of 0.7 m. The tank is without a cover (in some area, the substrates are placed simply onto the ground). About 650 kg wheat bran is placed in the tank. The starch content in the substrate bed is controlled within the range of 21%–23%. The yeast and *Acetobacter* starter mentioned before, and about 40 kg of water are added into the tank, mixed well, and heaped loosely. The moisture of substrate is 50%–52%. The bed is stirred once each day. After fermentation for 3 days, the top of substrate bed becomes hot, indicating that the acetic acid fermentation is progressing well. The substrate bed is gradually warmed up downwards. After 8 days of fermentation, the temperature of the substrate begins to decrease. By the time the fermentation has been continuing for about 14 days, the substrate bed is full of acetic acid flavor. Then, the substrate is transferred to ceramic jars for further fermentation.

The substrate should be tightly pressed in the jars, and salt is sprinkled onto the top of the substrate to prevent its putrefaction. The further fermentation in the jar lasts about 3 months, day and night, outdoors. For 650 kg wheat bran and 30 kg glutinous rice, about 1200 kg of vinegar product can be harvested.

In the innovated process for wheat bran vinegar production, the herbal qu is substituted with a pure culture of *Aspergillus niger*, yeast, and *Acetobacter*. Fermented substrate in the last batch is used as an inoculum for the next batch. In this way, the fermentation time is shortened. Glucoamylase is added in the wheat bran substrate for improving the utilization efficiency of starch. Starch content in the substrate is lowed to 16%–18%, and the moisture of the substrate is controlled to 60%–62%. A backflow of the vinegar process can shorten the fermentation cycle.

8.1.3.4 Fujian Hong qu Vinegar

8.1.3.4.1 Brief Introduction   *Hong qu* is red mold rice fermented by *Monascus*, and can be used in rice wine brewing as a saccharification and fermentation starter. The rice wine then is fermented to vinegar (Liu et al., 2008).

The distinct feature of the Fujian hong qu vinegar is that the rice wine brewing fermented by hong qu is carried out in solid-state fermentation, then vinegar fermentation is carried out in liquid-state fermentation for clarified rice wine and in solid-state fermentation for residual cake. The process flowchart is simplified as in Figure 8.16.

8.1.3.4.2 Rice Steeping   Glutinous rice at 285 kg is steeped in a steeping tank for 6–8 h in summer or 10–12 h in winter.

8.1.3.4.3 Steaming   The steeped rice is steamed for about 30 min. After steaming, the rice is cooled to 35°C in summer or 38°C in winter.

8.1.3.4.4 Inoculation with Red Mold Rice   The quantity of red mold rice inoculated is 25% of the glutinous rice. The cooled rice is mixed with the red mold rice and then put in a ceramic jar.

8.1.3.4.5 Fermentation   Red mold rice made by *Monascus* plays versatile roles in brewing of the red rice wine in which it is important

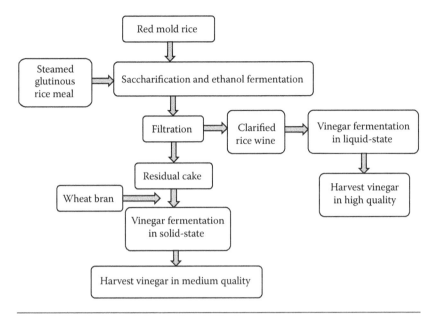

Figure 8.16 The flowchart of production of Fujian hong qu vinegar.

for liquefaction, saccharification, and ethanol fermentation. The red mold rice is also a part of the raw materials in the brewing. The whole fermentation process can be divided into two stages: the stage of liquefaction and saccharification and the stage of ethanol fermentation.

During the stage of liquefaction and saccharification, the inoculated rice meal (50 kg) is added with 60 kg cold boiled water, and mixed well in ceramic jars. The temperature is controlled below 38°C. After 24 hours, 40 kg cold boiled water is added into the mash again, and the ethanol fermentation begins. The mash is stirred once every day. After 5 days' fermentation, the mash is stirred once every other day. The ethanol concentration in the mash can reach 10% (v/v).

After about 2 months' fermentation, the mash is filtered, and then clarified red rice wine and residual cake are obtained. Both of them are used for vinegar fermentation in different ways.

Vinegar fermentation by using clarified red qu rice wine: The vinegar production is carried out by the fed-batch fermentation.

Vinegar fermented for 3 years is harvested 50% (by volume), and another 50% of the vinegar is left in the original jar. Then vinegar that was fermented for 2 years (50%) is transferred to the jar and another 50% of the 2 years' vinegar is mixed with 50% of the vinegar fermented for only 1 year. The vinegar fermented for 1 year (50%) is

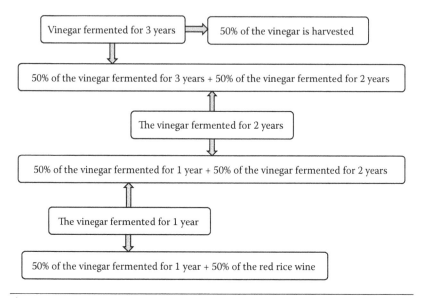

Figure 8.17   The flowchart of the fermentation process of Fujian hong qu vinegar.

mixed with red rice wine (50%) to compose the vinegar fermentation mash. The arrangement of the operation is shown as Figure 8.17.

8.1.3.4.6 Vinegar Fermentation by Using Residual Cake   After filtration of the red rice wine, the residual cake of the red rice wine mash is mixed with wheat bran, then the mixture fermentation is conducted with acetic acid fermentation in a semisolid state.

8.1.3.5 Novel Process for Vinegar Production
8.1.3.5.1 Brief Introduction   There are two important innovations in the new process: commercial enzyme preparations are substituted for the traditional qu, and the solid-state fermenters with aeration and backflow are substituted for the traditional ceramic jars or aeration tanks with a mash-turning machine. All the operations are conducted in the liquid state except acetic acid fermentation, which happens in the solid state.

8.1.3.5.2 Raw Materials and Pretreatment   Rice is used as the main raw material for ethanol fermentation, and the rice is steeped and ground into the thick slurry (70 mash, 18–20°Be′; Table 8.1)

8.1.3.5.3 Liquefaction   An α-amylase preparation is added to the thick slurry: amylase, 5 U/g rice; 90°C–100°C; 30 min.

Table 8.1   The Quantities of Materials Used in the New Process of Vinegar Production

RAW MATERIALS	QUANTITIES
Rice	1200 kg
Water used for liquefaction and saccharification	1300 kg
Water used for ethanol fermentation	3250 kg
Amylase and glucoamylase preparations	According to the enzyme activities
Yeast seed culture	500 kg
Seed starter for acetic acid fermentation	200 kg
Wheat bran	1400 kg
Rice hull	1650 kg
Salt	100 kg

8.1.3.5.4 Saccharification   Wheat bran qu (crude enzyme preparation, especially glucoamylase, made by *Aspergillus niger* or *Aspergillus usamii*) is added, cooked at 63°C for 3 h, then cooled at 27°C.

8.1.3.5.5 Ethanol Fermentation   Yeast seed culture is mixed with the saccharified mash, 33°C–37°C for 64 h. Ethanol concentration is 8.5%; acidity is 0.3%–0.4%.

8.1.3.5.6 Bioreactor with Aeration and Backflow for Acetic Acid Fermentation Before introducing the conditions for acetic acid fermentation, the solid-state fermenter system with aeration and backflow of vinegar is given in details below.

The fermenter is similar to the packed bed reactor, which is constructed from cement with a height of 2.4 m and diameter of 4 m. The total volume of the bioreactor is about 30 m³.

The bioreactor has a perforated bottom, which is 15–20 cm above the bottom of the bioreactor. The perforated bottom is made of a stainless steel or bamboo mat, which separate the bioreactor into two parts. Above the perforated bottom, the fermenting bed is placed, and under the perforated bottom, the vinegar liquid is collected and pumped back to the surface of the fermenting bed through spray pipes with small holes. The spray pipes can be revolved, so that the vinegar is sprayed uniformly onto the fermenting bed. Just below the perforated bottom, there are 12 aeration holes, which are annularly arranged around the bioreactor.

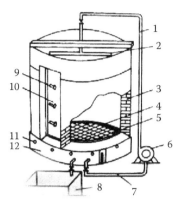

Figure 8.18   Bioreactor for vinegar fermentation: (1) Backflow pipe, (2) spray pipe, (3) cement tank wall, (4) perforated frame, (5) perforated bottom, (6) pump, (7) vinegar backflow pipe, (8) vinegar tank, (9) temperature gauge, (10) door, (11) aeration pipe, and (12) vinegar store.

Water-jacketing of the side walls of the bioreactor is not necessary, as the axial and radial temperature gradient in the bed could be prevented or minimized by circulation of the vinegar (Figure 8.18).

8.1.3.5.7 Acetic Acid Fermentation   The ethanol mash, wheat bran, rice hull, and *Acetobacter* seed culture are mixed well in a mixer, then they are transferred to the bioreactor. More *Acetobacter* seed cultures are added onto the surface of the mash. The initial fermentation temperature is controlled in the range of 35°C–38°C. After 24 hours, when the temperature may rise to above 40°C, the mash is stirred up and down. Then the vinegar collected under the perforated bottom is pumped back to the surface of the mash to regulate the temperature of the mash. Backflow of the vinegar to the fermenting mash is an effective method to prevent temperature rise. It is required to control the temperature not beyond 42°C–44°C. If the temperature rises beyond the limit, it is necessary to stop aeration through the holes. Backflow of the vinegar is routinely 6 times each day. Usually the fermentation time is about 20–25 days, and in summer, the fermentation time is about 35–40 days. If the mash's temperature is below the set point, the vinegar for backflow should be heated up.

After fermentation, the acetic acid concentration reaches to 65–67 g/L. In order to prevent the acetic acid becoming further oxidized to CO_2 and H_2O, salt is placed onto the surface of the mash and dissolved by the backflow of the vinegar. Finally, the vinegar is

Table 8.2   The Composition of Some Famous Vinegars in China

VINEGAR	TOTAL ACID CONC. (g/100 mL)	TOTAL ESTER (g/100 mL)	°BE′ (g/100 mL)	REDUCING SUGAR (g/100 mL)	AMINO ACID (g/100 mL)
Zhenjiang vinegar	6.40	4	10	2.5	0.25
Shanxi aged vinegar	9.5	3.8	18	4.5	0.3
Wheat bran vinegar	6.00		15.5	2.5	0.1
Fujian red Qu vinegar	8.00	1	5	2.18	0.22

harvested by filtration as the standard procedures. Usually, 1 kg of the rice can produce 8 kg of a vinegar product (see Table 8.2).

8.2 Sauce

8.2.1 Soy Sauce

8.2.1.1 Introduction
Soy sauce is a condiment produced by fermented soybeans (or soya bean meal) with *Aspergillus oryzae*, along with roasted grain or wheat flour or wheat bran, water, and salt. It is a traditional ingredient in East and Southeast Asian cuisines (Luh, 1995; Han et al., 2001). All varieties of soy sauce are salty condiments. Soy sauce originated in China 2500 years ago and later spread to East and Southeast Asia, and now widely used as a particularly important flavoring in Chinese, Japanese, Thai, and Korean cuisine (Bhumiratana et al., 1980; Röling et al., 1996; Peng et al., 2007).

The typical process of soy sauce production is shown in Figure 8.19, which shows the soy sauce production process divided into three major steps: qu making, soy sauce fermentation, and refining.

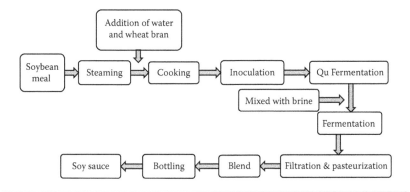

Figure 8.19   Soy sauce production process.

Qu are made by mixing cooked soybeans (soya bean meal) and grain with cultures of *Aspergillus oryzae* (Wicklow et al., 2007). Qu making is always carried out by solid-state fermentation. During the qu-making process, the *Aspergillus oryzae* multiply and produce varieties of hydrolysis enzymes, especially proteinase. So the matured qu is actually a mixture of hydrolysis enzymes and unhydrolyzed cereals.

Soy sauce fermentations start with matured qu mixed with salty water. The fermentations are actually further enzymatic hydrolysis of the residual proteins and starch in qu. Soy sauce fermentation processes are classified into the solid-state fermentation process and liquid-state fermentation process. The soybean proteins are converted into peptides and amino acids, and starch in grain is hydrolyzed to glucose in fermentation processes. Amino acids and glucose are easily polymerized into melanin during fermentation and exposed to high temperature.

Refining includes filtration or squeezing and pasteurization of the soy sauce.

8.2.1.2 Microorganisms   Microorganisms involved in soy sauce production are mainly *Aspergillus oryzae* and *Aspergillus soyae* (Yong and Wood, 1976). There are many varieties of *Aspergillus oryzae*, among which *Aspergillus oryzae* var. 3.042 is a frequently-used strain in China. This strain is characterized by its rapid growth and high activity of neutral proteinase.

Mixed culture of *Aspergillus oryzae* with other microorganisms was practiced to improve the quality of qu. The other microorganisms used can be the variants of *Aspergillus oryzae*, *Aspergillus niger*, and *Monascus* spp.

As the qu-making process is not a strictly aseptic operation, contamination by noninoculated microorganisms is unavoidable. Rapid growth of *Aspergillus oryzae* can help prevent risks of contamination by the other microorganisms.

Some *Lactobaccillus* and yeast strains, which are beneficial to improvement of soy sauce flavor and aroma, are usually inoculated in the fermentation stage.

8.2.1.3 qu Making   The qu-making processes are typical solid-state cultivation processes and usually divided into two stages: seed qu preparation and qu cultivation.

8.2.1.3.1 Seed qu Preparation   The objective of seed qu preparation is to multiply plenty of spores of *Aspergillus oryzae* that are used as inoculum for making a large quantity of qu.

The process of seed qu preparation is: slant culture → conical flask culture → shallow tray culture. A typical multistaging procedure generates a 2%–5% (w/w) inoculum for the next stages.

The mediums for each culture are shown as follows:

Slant culture medium: potato, 200 g; dextrose, 20 g; agar, 20 g; water 1000 mL.

Flask culture medium: wheat bran, 80%; soybean meal, 10%; wheat flour, 10%; solid substances:water = 1:1.1 (w/w). The thickness of culture medium is about 1 cm. The medium in the flasks is sterilized at 121°C for 60 min.

Shallow tray culture medium: wheat bran, 80%; soybean meal, 15%; wheat flour, 5%; solid substances:water = 1:1.1 (w/w). The medium is sterilized at 100°C for 60 min.

After cooled to 28°C, the culture media are inoculated with spores of *Aspergillus oryzae*. During the cultivation period, the qu cultures in the flasks are shaken twice to avoid the culture agglomeration. The temperature of the qu culture in the shallow tray is controlled in two stages with the first stage being 28°C–30°C and the second stage being 25°C–28°C. The maximal temperature is controlled below 35°C.

Cultivation of seed qu in the shallow tray lasts 72 hours in the cultivation room. The seed qu should have a large quantity of spores. The specifications of qualified seed qu are the following: spore number, 5×10^9 cfu/g (dry weight); percentage of germination of the spores, more than 90%.

Finally, the seed qu culture can be cultivated in a shallow tray bioreactor, which is shown in Figure 8.20. In the modern factory, tray bioreactors are used for seed qu preparation as shown in Figure 8.21.

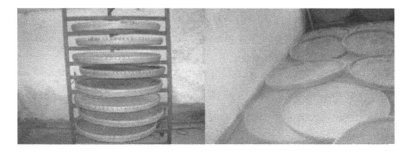

Figure 8.20   Seed qu incubated in a shallow tray.

Figure 8.21   Tray bioreactors used for seed qu preparation.

8.2.1.3.2 qu Cultivation   Qu is the source of amylolytic and proteolytic enzymes for the decomposition of starch and protein. Qu making is multipurpose:

—to produce various hydrolysis enzymes by cultures of *Aspergillus oryzae*;
—to partly hydrolyze the proteins and starch in the raw materials;
—to grow some useful microorganisms other than *Aspergillus oryzae* during qu cultivation when an open fermentation system is used.

8.2.1.3.2.1 Raw Materials for qu Making   The raw materials for soy sauce are soybeans, defatted soya bean cake (meal), wheat (or roasted wheat), wheat flour, and wheat bran. Traditionally, soybeans, wheat meal, and roasted wheat have been used as raw materials for qu making. After fermentation, the liquid (soy sauce) in the fermenting paste is squeezed or filtered while the residues of fermentation paste are further processed to *fermented bean paste*. Nowadays, defatted soybean cake (meal), wheat meal, and wheat bran are used as the raw materials for solid-state cultivation of qu.

There are two types of soybean cake, depending on the oil manufacturing process. The solid residue from a hot pressed oil process is called *hot pressed soybean cake*, in which the protein content is higher, and the proteins are heated and denatured during the oil-manufacture process.

Wheat bran was not used in traditional soy sauce production before, but it has been widely used as a raw material in mainland China since the 1960s. Wheat bran is obviously not an ideal raw material for traditional soy sauce production, but it is a good material for qu making as the aeration and moisture of the culture could be improved when an appropriate ratio of wheat bran is used in solid-state cultivation of qu, where the thickness of the fermenting bed of qu is increased to about 30 cm.

Soybean meal, wheat meal, and wheat bran for qu making varies in different companies. The ratios of mixtures in raw materials are listed as follows:

For example,

Soybean meal: wheat meal: wheat bran = 100:30–40:10–20
Soybean meal: wheat meal: wheat bran = 15:5:1
Soybean: roasted wheat = 100:100

The soybean is rich in protein while wheat is rich in starch. So the proportion of soybean meal and wheat meal will determine the final compositions of soy sauce. See Figure 8.22.

8.2.1.3.2.2 Conventional Cooking Methods for Raw Materials Conventional methods for treating the raw materials are cooking at high temperature and pressure. The temperature and heating time are the most important cooking conditions that significantly affect the utilization efficiency of the proteins and starch in the raw materials. The most common cooking vessel used is a 5 m³ cooker (Figure 8.23).

The mixture of raw materials is composed of the following: "Cold drawn" soybean meal, 1050 kg; wheat bran, 300 kg; wheat meal, 150 kg. First, soybean meal is put into the cooker, and steam is added while rotating the cooker. When the temperature of the soybean meal rises to about 110°C, the steaming process can be ended. After the

Figure 8.22 Soybean cake.

Figure 8.23 Rotating steaming cooker.

pressure in the cooker drops to zero, 800 kg water is added into the cooker gradually. The cooker is rotated again for about 20 min, after which wheat bran and wheat meal are added into the cooker and rotated for 10 min again. Meanwhile, steam is added into the cooker and the pressure in the cooker is increased to 50 kPa. Keep the pressure stable at 50 kPa for about 10 min. Stop steaming till the pressure drops to zero. Then the steam is added again to increase the temperature to 120°C. Keep the pressure and temperature stable for 3–4 min. When the steam in the cooker is exhausted, a hydraulic jet pump is turned on to cool the mixture in a vacuum.

The frequent change of pressure in the cooker is beneficial for breaking up the granules in the raw materials and releasing the nutrients. The protein denaturation is achieved fully. At the same time, the sterilization efficiency is increased significantly due to the frequent change of pressure in the cooker.

8.2.1.3.2.3 Extrusion Processing for Treating the Raw Materials An extrusion has the advantages of transportation, mixing, heating, and pressurizing at the same time in a simple extruder; the raw materials are facilitated to be dissolved and hydrolyzed under shearing and puffing actions. The matured soy sauce qu prepared with extruded raw materials exhibits higher hydrolytic enzyme activities than the traditional process. The soy sauce products fermented from this process has a higher protein utilization rate.

8.2.1.3.2.4 Inoculation of the Cooked Mixture with Seed Qu   After steam cooking, the mixture of soybean meal and wheat meal is cooled to about 28°C by vacuum cooling and forced air cooling and then inoculated with seed qu. The inoculum volume is 300 g seed qu/1000 kg mixture (wet). The seed qu should be distributed uniformly in the cooked raw materials by using a spiral chute conveyor and a special inoculator. In another way, the seed qu is mixed with a part of the cooked raw materials, and the mixture is cultivated at 28°C–35°C for several hours. Then the germinating mixture is blended with the remaining cooked raw materials.

The maximum height of a substrate bed in the perforated bottom is 30 cm. The inoculated substrates should be flattened-off in the bioreactors.

8.2.1.3.2.5 qu Cultivation   The quality of qu is very important in soy sauce production. The traditional qu-making process is performed in an open system. After steam cooking, the mixture of soybean and wheat cools to 27°C and is inoculated with the qu seed culture. The inoculated mixture is incubated in an aeration tank (or disc qu-making bioreactor) where temperature and humidity are carefully controlled and monitored. Here, the spores germinate, appearing initially as small white spots on the surface of the beans and wheat. During its growth, mold produces enzymes that convert proteins, fats, and starches to simpler and more easily fermentable substances.

With increasingly stringent requirement in large-scale production of soy sauce, several new qu-making processes have been developed to shorten the cycle of qu making and improve the activities of qu.

The inoculated mixture is incubated in a forced aeration with an intermittently-mixed bioreactor where temperature and humidity are carefully controlled and monitored.

Since most of the factories do not use depth filtration systems to treat the air, the qu-making process is performed in an open system. The key parameters during qu incubation processes are fermentation temperature, moisture of the substrate, and aeration rate. It is also very important to choose the right moment for the bed agitation.

About 18 h after inoculation, when white spots of the mold appear on the surface of the substrate bed and begin to become agglomeration, the qu substrate temperature will increase sharply. Aeration and stirring intermittently is the effective and quick measure to prevent the temperature rising beyond the set point. Stirring two or three times during the period of qu incubation can help to loosen the substrate bed.

Under controlled conditions of temperature and moisture, mold growth covers the entire mass and the fermenting qu becomes white and finally turns greenish. Qu incubation time is usually 48 h.

8.2.1.3.2.6 Temperature Control During the qu Cultivation Process
Temperature control is most important for qu making. The temperature control strategy is based on the requirement for both rapid growth of the microorganism and high enzyme production.

The appropriate temperature for germination of the spores and growth of the mycelium of *Aspergillus oryzae* ranges from 28°C–35°C. Spore germination is slow when the temperature is below 28°C, and contamination by bacteria could easily happen when temperature is above 35°C. The suitable temperature for the enzyme production by the microorganism ranges from 28°C–32°C. So the temperature control strategy during the qu-making process is set with two or three temperature ranges indicated by the temperature control curves:

High temperature period-intermediate temperate period-low temperature period: 32°C(16–18 h)–30°C(18–20 h)–25°C (6–8 h);

High temperature period–intermediate temperate period: 32°C–30°C; and

Intermediate temperate period–high temperature period–intermediate temperate period: 32°C–34°C–30°C.

8.2.1.3.2.7 Moisture Control during the Qu Cultivation Process   Initial moisture of the fermenting cultures is very important and usually set in the range of 47%–50%. Bacteria grow quickly when the initial moisture of the cultures is too high, and hydrolysis of proteins in the qu becomes difficult when the initial moisture of the culture is too low. During the first period of 0–20 h, the relative humidity in the incubation room should be kept above 95%. Normally, the relative humidity in incubation rooms should be within 85%–90% or at least more than 85%.

8.2.1.3.2.8 Qu Quality Control   A qu bed should be loose, soft, flexible, full with mycelium, tender greenish, with an unique aroma of qu, and without an unpleasant smell. The physical and chemical requirements are as follows: moisture, 26% to 33%; qu protease activity, more than 1000 U/g (dry basis; Folin's method). Protease activity generally refers to the neutral protease activity. Currently, qu protease activity reaches up to 3000 U/g (dry basis).

An example of conventional qu preparation process: After the inoculated substrate is transferred to the incubation tank, substrate bed temperature is controlled in 28°C–30°C. Along with fermentation, mycelia grow over the surface of the substrate, and bed temperature rises gradually. As bed temperature reaches 33°C–34°C, forced aeration is applied; when bed temperature drops to 31°C–32°C, aeration is stopped. Repeat the aeration operation to maintain the temperature in the range of 31°C–34°C. First mixing (by agitator or other turning method) is needed to break the qu substrate to avoid aggregate formation in the bed. Choice of the right moment for the first mixing is very important and is a difficult task. After the first mixing operation, the mycelia grow quickly and vigorously, and increased rate of temperature speeds up. It is very important to aerate the qu bed to prevent the temperature rise. After 4–5 h, the qu bed is mixed again. The mycelia grow, become matured, and begin to produce spores. The temperature of the qu bed is no longer increased. Keep the temperature in the incubation room in the range of 30°C–32°C afterwards. Total incubation time is 48–50 h.

The mature qu is mixed with brine containing 18 Be or more salt at a weight ratio of 1:1.5 and is transferred to porcelain jars or kegs to create a mash for fermentation. The mash must be exposed to open air day and night for 6–12 months until full flavor is achieved, during which time the soy–wheat paste turns into a semiliquid, reddish-brown mature mash. This fermentation process creates over 200 different flavor compounds. The jars remain uncovered throughout the whole fermentation.

8.2.1.3.3 Bioreactors Used for Qu Cultivation Qu cultivation is a typical solid-state fermentation (Luh, 1995; Peng et al., 2007). The bioreactor commonly used in the qu industry is classified as an *intermittently-mixed forcefully-aerated bioreactor*. It has a *thick bed aeration tank* (Figures 8.24–8.25). The reactor is built from concrete with a length about 7–10 m, a breadth of 1.5–3 m, and an internal height of 0.7–1.2 m. A perforated stainless steel plate or bamboo mat (Figure 8.26), designed to support the bed, is fixed at a height of 0.4–0.6 m. The maximal depth of the substrate bed is 0.3 m. The capacity of the tank should be matched to the volume of the cooker. For a 5 m³ cooker, the thick bed aeration tank should contain about 1500 kg substrate (dry weight). The air system is not strictly aseptic, and air flow rate is 6000–7500 m³/h with the air pressure above 1 KPa. The air is humidified and warmed or cooled by an air conditioning apparatus, and the air is introduced through the air box below the substrate bed. The air box has a slope along the length of the tank at an incline of 8–10 degrees.

Another bioreactor used in the qu industry is called a *disc* or *cylinder qu-making reactor* (Figure 8.27), with a 6 m or 12 m diameter bed and a capacity for 3000–15000 kg of substrates (dry weight).

Figure 8.24 Aeration tank for solid-state fermentation of qu.

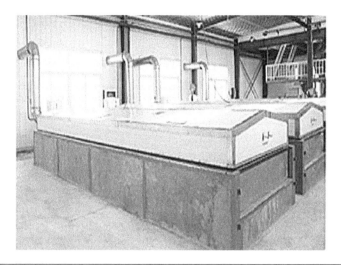

Figure 8.25 Aeration tank with a cover for solid-state fermentation of qu.

Figure 8.26 Perforated bottom mat made by a bamboo splint.

8.2.1.4 Soy Sauce Fermentation A representative soy sauce manufacturing process is as follows:

After mixing with roasted broken wheat and wheat bran, protein materials are steam cooked in a continuous pressure cooker. The seed qu, which is inoculated with *Aspergillus ozizae* and incubated in shallow vats with perforated bottoms, is added to the medium to make

Figure 8.27   Cylinder qu-making bioreactor.

qu. After several days of incubation under controlled conditions of temperature and moisture, mold growth covers the entire mass, which turns greenish from sporulation. The dry mash, called *koji*, is an essential ingredient in most Oriental fermented products because it is the source of amylolytic and proteolytic enzymes for the decomposition of starch and protein. Subsequently, the qu is mixed with brine and transferred to deep fermentation tanks. The high salt concentration effectively inhibits the growth of undesirable wild microorganisms. The starch is transformed to sugars, which are fermented to lactic acid and alcohol. The fermenting mash is held in fermentation tanks for 6–8 months. In the refining stage, the fermented mash is pressed with layers of filtration cloth to separate the sauce from the cake of wheat and soy residues. The sauce is filtered, clarified, and heated to 70°C–80°C. Heating is necessary to pasteurize the sauce and to develop the characteristic color and aroma. After the final clarification step, the sauce is bottled.

Soy sauce fermentation consists mainly of an enzymatic hydrolysis process as well as microbial fermentation and chemical reaction processes (Luh, 1995; Peng et al., 2007). The most important change is amino acid formation by fermentation. During fermentation, proteins and peptides in the qu are further hydrolyzed to small peptides and amino acids. Carbohydrates are hydrolyzed to glucose and other reducing sugar, then are further fermented to alcohol and lactic acid.

Various products from enzymatic reaction and fermentation may react with each other to form new products. They are fundamental substances of the color and aroma of soy sauce. Another important reaction is the formation of color substances from amino acid and glucose by the Maillard reactions.

Currently, the soy sauce fermentation processes employed are classified into three types:

Traditional fermentation process
Low-salt, solid-state fermentation process
High-salt, submerged fermentation process

8.2.1.4.1 Traditional Fermentation Process The raw material for the traditional fermentation process is mainly the soybean, and a small quantity of wheat meal. The soybean is soaked in water and then cooked with steam. After cooling, the steamed soybean is mixed with wheat meal and water. Microorganisms from nature or artificially prepared are inoculated to the mixed substrates. The qu-making procedures are the same as stated previously.

In the traditional fermentation process, the matured qu is mixed with brine containing 18 Be or more salt at a weight ratio of 1:1.5 and then transferred to earthen fermentation jars. The fermenting mash is usually exposed to open air day and night for 6–12 months until full flavor is formed, so the temperature of the mash changes with seasons.

The traditional fermentation of soy sauce in the open air is shown in Figure 8.28.

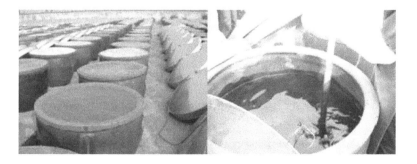

Figure 8.28 Traditional fermentation of soy sauce in open air.

8.2.1.4.2 Low-Salt, Solid-State Fermentation Process The low-salt, solid-state fermentation process is characterized by higher fermentation temperature, shorter fermentation time, and lower salt concentration. The raw materials for qu are defatted soybean meal, wheat meal, and wheat bran. Because fermentation is conducted at a higher temperature, the enzymes in the fermenting mash are easy to be denatured, resulting in a poor ability for further fermentation. Obviously, a higher fermentation temperature is not beneficial for production of amino acid and formation of flavor substances in the soy sauce. In this process, the mash is fermented for only 15–30 days, and the maturation of soy sauce is not enough due to the short fermentation time. The flavor and aroma of the soy sauce made by low-salt, solid-state fermentation is not as good as those made by the traditional fermentation process.

The standard procedures for low-salt, solid-state fermentation of soy sauce are as follows:

Brine preparation: Dissolve salt with water; concentration of the brine is adjusted in the range of 12–13 Be clarification.

Temperature of the brine: 45°C–50°C in summer and 50°C–55°C in winter.

Dosage of brine: Water moisture of the fermenting mash is controlled within 55%–58% for the process in which the fermentation of the mash, steeping, and filtration of soy sauce are carried out in the same fermentation vat, and 50%–53% for the process in which the fermentation of the mash, steeping, and filtration of soy sauce are carried out in different vessels. In order to prevent the superficial layer of fermentation mash from being oxidized by air, salt is spread onto the surface of the mash. It is called "cover salt."

Fermentation: The common fermentation vessels are cement pits with about 2 m depth and 1.5 m width. The cement pits have jackets surrounded and are warmed by hot water circulation in the jackets. Fermentation temperature is controlled in the range of 40°C–50°C.

Mash shift: The fermenting mash can be shifted to another pit. In this way, the fermenting mash is stirred and mixed. Shifting once or twice is enough with a time interval of 8–9 days. See Figure 8.29.

Figure 8.29   Fermenting sink for soy sauce.

8.2.1.4.3 Special Fermentation Process

8.2.1.4.3.1 Circumfluence Operation in Low-Salt, Solid-State Fermentation Process   In this operation, a pump is mounted in the sink at the bottom of the fermenting pit, and soy sauce drained from the fermenting pits is pumped back to the upper fermenting mash in the pits. The reflowed soy sauce is soaked into the mash gradually. The circulation can be repeated for several times during the fermentation of soy sauce. By circulating the operation, cultures of yeast or lactic acid bacteria can be added into the fermenting mash, and at the same time, oxygen can be supplemented into the mash while carbon dioxide can be removed from it. As soy sauce fermentation is static fermentation, the circulating operation is beneficial for substrate transfer in solid-state fermentation of soy sauce. Soy sauce flavor is improved by the circulating operation. Fermentation temperature can also be adjusted by the circumfluence process.

Oxidation layer in the upper surface of fermenting mash can be eliminated by circulation of soy sauce. As the fermenting pit is about 2 m in depth, and the liquid in the fermenting mash in the pit is always downwards, the resulting liquid lost is in the upper surface in the fermenting mash. As a result, the upper surface of the mash becomes porous. If the upper surface of the mash is not tightly sealed, the mash is oxidized easily and acquires an unpleasant flavor. Circulation of soy sauce to the surface can reduce the thickness of the oxidation layer.

8.2.2 Fermented Bean Paste

8.2.2.1 Brief Introduction   Fermented bean paste is a category of fermented foods typically made from ground soybeans, which are indigenous to the cuisines of East and Southeast Asia (Han et al., 2001). The pastes are usually salty and savory, but may also be spicy. Fermented bean pastes are usually in a semiliquid state.

Various types of fermented bean paste included *dajiang, doubanjiang,* sweet bean sauce, yellow soybean paste and miso, depend on the raw materials and microorganisms.

Dajiang is a kind of pasty product and made from soybeans. Usually, the steamed soybean is inoculated with *Aspergillus oryzae* to make the qu. The matured qu is soaked in brine for enzymatic hydrolysis for several months, and the product is ground to a pasty product.

8.2.2.2 Raw Materials for Fermented Bean Paste   The main raw materials for fermented bean pastes are bean, water, salt, and microorganisms. Flour is also used as a starting material for manufacturing sweet bean sauce and yellow soybean paste.

The soybean is the most popular starting material for fermented bean paste. The broad bean (or fava bean) and black bean are also used as raw materials for fermented bean paste in some cases. As beans are rich in proteins and flour is rich in starch, the ingredients of raw materials determine the growth of microorganisms and flavor of fermented bean paste.

8.2.2.3 Microorganisms   Microorganisms involved in fermented bean paste production are usually *Aspergillus niger, Aspergillus flavus,* and *Aspergillus oryzae.*

8.2.2.4 Production Procedures of Fermented Bean Paste   The production processes of fermented bean paste are divided into four stages:

Cooking or steaming of the raw materials;
Qu making: solid-state fermentation of the microorganism on the raw materials;
Enzymatic hydrolysis of the fermented substrate soaked in brine; and
Grinding and pasteurization of the paste.

The procedures of qu making are similar to those employed in qu making for soy sauce. The fermentation stage is actually the enzymatic hydrolysis of the proteins and starch in the qu soaked in brine.

8.2.2.5 Fermentation Processes for Various Types of Fermented Bean Paste

8.2.2.5.1 Dajiang

8.2.2.5.1.1 Brief Introduction   *Dajiang* is a typical fermented bean paste made from ground soybeans. Dajiang is used as a condiment to various flavor foods of East and Southeast Asia.

The production procedure for dajiang is similar to that for soy sauce. The soybean is the main raw material for making qu. The microorganism for qu making is typically *Aspergillus oryzae*. Matured qu is rich in hydrolysis enzymes. Then, the matured qu is steeped in brine and further fermented for several months, during which the proteins and starch in the qu are degraded to peptides, amino acids, and sugar. The fermented paste is further ground to a fine paste without filtration or squeezing.

8.2.2.5.1.2 Traditional Process for Dajiang   The traditional process for production of dajiang is still popular in small- and mid-sized factories. The fermented bean paste made by traditional processes is considered more savory.

The characteristics of the traditional techniques can be summarized as:

An open system for the traditional qu-making process
Natural fermentation process: The fermenting paste is exposed to open air for several months or even six months, day and night. Hence, in this process, the temperature of the fermenting paste fluctuates for a long time (see Figure 8.30).

Raw materials and pretreatment: Raw materials for dajiang are soybeans (100 kg) and salt (45 kg). The soybeans are rinsed with water and cooked in a pot. When the cooked soybeans can be smashed with a slight nip, stop the cooking and cool to an ambient temperature.

Qu making: Qu making is usually conducted in the summer in which the growth of molds can be promoted due to the relatively high temperature and humidity. The cooked soybeans are ground to wet

Figure 8.30   A typical production process for *dajiang* popular in countryside.

powder and shaped to bars (length × width × height = 10 × 5 ×5 cm). The bar is covered with paper and placed in the shallow basket or bamboo mats. The molds grow on the surface of the soybean bar gradually. The qu is cultivated in open air for several months. Before being used, the qu is rinsed with fresh water and dried under the sunlight.

Fermentation phase: The fermentation vessels are usually ceramic with straw mattress covers and mesh mantles. The straw mattress covers are used to maintain the temperatures of fermenting paste stable while the mesh mantles are used to discourage flies or insects.

Dissolve the salt (food grade) in water and place the brine into fermenting vessels. Put the qu into the brine. Cover the vessels with a straw mattress. The qu floats in the brine in the first 1 or 2 days, then will be gradually dissolved in the liquid. Three days later, stir the fermenting paste. After then, stir the paste in the morning and evening every day. The color of the fermenting paste will gradually become yellow, then brown-reddish. The fermentation time might be 1 month or 6 months.

The disadvantage of the traditional fermentation process is that the fermentation vessels are placed outside, resulting in the fermentation temperature fluctuating greatly. Another disadvantage is that the vessels occupy a large area.

In order to save energy, more and more factories use glass houses in which the fermentation vessels or the fermentation sink are placed or constructed. The quality and flavor of the dajiang paste manufactured in the glass house is superior to those in traditional methods.

8.2.2.5.1.3 Production of Dajiang in a Commercial Scale   In order to make delicious dajiang, the fermentation (actually hydrolysis of the qu in brine) time of fermented bean paste is 3 to 6 months for maturation at natural conditions. Commercial production of dajiang requires short fermentation time but basically guarantees the quality of the product.

The process flow of dajiang production on a large scale is shown in Figure 8.31.

> Raw materials: The raw materials for dajiang are 65% soybeans and 35% wheat flour. The soybeans are steeped in water for 3–4 hours. The soaked soybeans are cooked in a revolved cooker at a pressure of 0.13–0.14 MPa for about 20 min.
>
> Seed culture preparation: The seed is prepared in two stages. The raw materials for seed prepared in the first stage are 180 g soybeans, 20 g wheat bran, and 200 mL water. After sterilization, the substrate is cooled to 35°C, and inoculated with spores of *Aspergillus oryzae*. The culture is incubated at 30°C for 72 hours. The matured seed is used as an inoculum for the seed in the second stage.

The raw material for seed prepared in the second stage is soybeans or 80% soybean supplement with 20% wheat bran. The procedure of sterilization and cultivation is the same as that of the first stage.

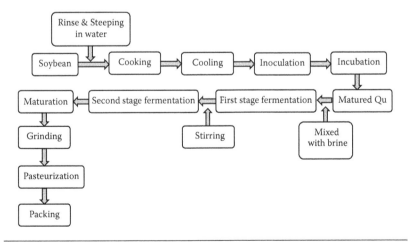

Figure 8.31   The flowchart of *dajiang* production on a large scale.

The seed prepared in second stage is mixed with wheat flour first. Then the mixture is mixed well with cooked and cooled soybeans. The temperature of the mixture is controlled in the range of 30°C–40°C.

Qu-making: Qu making is similar to that in soy sauce production. The depth of the substrate bed is about 20–25 cm.

Fermentation (enzymatic hydrolysis): The qu is mixed with brine of 17–19° Be. The temperature of brine is controlled at 40°C–45°C.

Fermentation is divided into three stages maintained at different temperatures: The first-stage fermentation lasts about 7 days, and the fermentation temperature is controlled at 41°C–43°C, while the second-stage fermentation lasts for about 8 days, and the fermentation temperature is controlled at 43°C–45°C. The third-stage fermentation lasts about 15 days at 35°C–38°C for maturation.

During fermentation, the fermenting dajiang is stirred intermittently.

Grinding and pasteurization of the fermented bean paste: After fermentation for about 30 days, the fermented bean paste is ground. Before packing, the paste is usually pasteurized at 60°C–65°C for 30–60 min. The heating temperature and time might affect the flavor of the dajiang products.

The fermented bean paste also might be seriously contaminated by thermophilic bacteria, which can survive in the product after pasteurization as the heating temperature is only 60°C–65°C. Recommended sterilization is by the method of ultra-high temperature (UHT) for a short time by using a continuous sterilizer. Various preservatives are used in the products to prolong their shelf life.

8.2.2.5.2 Sweet Sauce Paste

8.2.2.5.2.1 Brief Introduction   Sweet sauce or sweet bean sauce is manufactured with wheat flour as the main raw material. Roasted soybeans are sometimes used as a supplement to enhance the flavor of the sauce paste. Because the qu has a high concentration of starch, the final product contains a high sugar count, which accounts for the sweet flavor of the products. The typical production procedure for sweet sauce paste is the solid-state cultivation of qu, low-salt and solid-state fermentation process (Ma, 2001; Zhong and Feng, 2011).

For the production process flow of sweet sauce paste (Chinese style), see Figure 8.32.

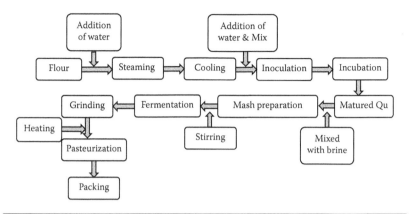

Figure 8.32 The flowchart of production of sweet sauce paste in Chinese style.

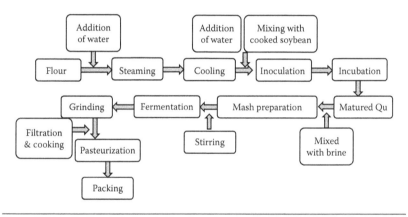

Figure 8.33 The flowchart of production of sweet sauce paste in Korean style.

For the production process flow of sweet sauce paste (Korean style), see Figure 8.33.

8.2.2.5.2.2 Raw Materials and Pretreatment   The principal raw material for production of sweet sauce paste is wheat flour, which requires a high starch content and moderate protein content. Whole-wheat flour or purified flour is not suitable for sweet sauce paste production.

The flour is mixed with water in a blender to allow the flour to absorb enough water. Usually the amount of water added to the flour is 28%–30% of the flour based on its weight. The dough is transferred to a steamer (constant pressure steamer or continuous steamer) for steaming. The starch is gelatinized and protein is denatured in the dough while the dough is sterilized during steaming. Steaming of the

Figure 8.34   Continuous steamer for flour.

dough actually allows the dough to absorb more water. The final moisture of the dough after steaming will increase to about 45% or more.
Steaming time is controlled in the range of 40–60 min. Continuous
steaming now is popular in large-scale factories and the steamer is
shown in Figure 8.34.

8.2.2.5.2.3 Qu Making   Before qu making, the seed culture is
incubated in flasks. The microorganism employed in qu making is
Aspergillus oryzae Huniang 3.042.

The formulation of the culture medium for the qu is wheat flour
(100–110 kg), water (30–36 kg), and seed culture (0.3 kg). The seed
culture is first mixed with some flour, then the mixture is blended
with steamed dough.

The qu cultivation follows the same procedures as that of soy sauce.
The detailed information was given previously. The qu cultivation

time is about 36–40 hours. The quality of the qu is determined by the amylase and proteinase activity.

8.2.2.5.2.4 Qu Mash Preparation   Matured qu (100 kg) is mixed with 13 °Be brine (80–85 kg). The brine is heated to 45°C. The mixture is transferred to fermenting tanks and steeped for 3 days. During this period, the temperature is controlled at about 50°C.

8.2.2.5.2.5 Fermentation of the Qu Mash   Fermentation of qu mash is actually enzymatic hydrolysis of proteins and starch in the qu mash by the enzymes in the qu under certain conditions. Due to unavoidable contamination during the qu-making period, the qu mash contains a large amount of microorganisms (bacteria and yeasts). Some microorganisms existing in the qu mash might be beneficial to the formation of the flavor substances in sweet sauce paste.

During the fermentation process, temperature control is very important. The temperature control is achieved by circulation of warm water in the jacket surround the fermentation tanks. The qu mash temperature is maintained at 40°C by a circulation of 50°C water in the first period, lasting for about 10 days. During this period, stir the mash twice every day. During the second period for about another 10 days, the circulation water is maintained at 45–50°C. During the third period for about 10–15 days, the temperature of the mash and circulation water is maintained at 40°C, and the mash is stirred. In large-scale fermentation tanks, the mash is stirred with compressed air.

8.2.2.5.2.6 Posttreatment of the Products   After fermentation for one month, the mash is ground, filtered, and pasteurized. Pasteurization temperature and time is recommended at 80°C for 15 min. Increasing the temperature and prolonging the pasteurization time might lead to a Maillard reaction in the sweet sauce paste because the paste is rich in amino acids and reduces sugar. But when the pasteurization temperature is too low and time is short, the sweet sauce paste might be easily contaminated.

Contamination of the sweet sauce paste might be due to following factors:

Qu cultivation and fermentation of the mash are carried out in open systems without a strict aseptic operation, which offers many chances for the contamination.

Sweet sauce paste is rich in reducing sugar and amino acids, which supply enough nutrition for the growth of bacteria.

The NaCl concentration in the mash is relatively low resulting in a reduced inhibition effect.

In order to prolong the shelf life of the products, preservatives should be added to the sweet sauce paste product before packing in factories.

8.2.3 Fermented Bean

Figure 8.35

8.2.3.1 Natto

8.2.3.1.1 Brief Introduction   Natto is very popular in Japan and the production method is similar with *touchi* fermented by *Bacillus subtilis* as the production techniques are believed to come from ancient China (Fujita et al., 1993; Okamoto et al., 1995). See Figure 8.36.

8.2.3.1.2 Raw Material and Microorganism   Natto is manufactured with soybeans as the raw material, which is fermented by *Bacillus subtilis natto*. *Bacillus subtilis natto* produces proteinases which convert proteins in soybeans into amino acids and peptides. A special enzyme called nattokinase which facilitates thrombolysis is also produced in natto (Choetal, 2010).

Figure 8.35   Fermented bean by *Aspergillus oryzae* (left) and *Aspergillus niger* (right).

Figure 8.36   *Natto.*

8.2.3.1.3 Fermentation Process   The process flow for natto production is shown in Figure 8.37.

Soybeans are steeped in water for about 16–24 hours, beyond which decrustation of the bean will occur. The bean is steamed at a constant pressure or pressure cooked at 121°C for 30 min.

The seed culture medium is composed of a soybean extract supplement with 5 g/L glucose. *Bacillus subtilis* natto seed culture is incubated for 24 h and is inoculated into the soybeans with the inoculum volume of 2%–4%. The fermentation method is a typical solid-state fermentation, which lasts about 24–40 h. Fermentation temperature is controlled in the range of between 34°C–37°C. Secondary fermentation is a very important step for formation of the special flavor of natto. The key variables are temperature and fermentation time, which are recommend to be 4°C and 24 h.

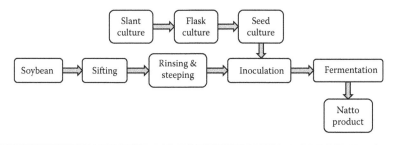

Figure 8.37   The process flow for *natto* production.

8.3 *Monascus* pigments

8.3.1 Introduction to Monascus *Pigments*

Monascus pigments are the most successful natural colorants manufactured by microbial fermentation on a large scale, which are widely employed in the food industry (Juzlová et al., 1996; De Carvalho et al., 2008; Lin et al., 2008). Commercial *Monascus* pigments include *Monascus* red rice (called "anka" or "red mold rice") and *Monascus* red pigments. *Monascus* red rice is manufactured from rice fermented by *Monascus* spp. by solid-state cultivation, while *Monascus* red pigment is a concentrated pigment that is extracted from fermentation broth by the liquid-state fermentation of *Monascus* (Johns and Stuart, 1991; Fabre et al., 1993; Nimnoi and Lumyong, 2011).

Monascus pigments have been widely used as food colorants, such as coloration of processed meat (sausage, hams), marine products like fish, fermented bean curd (*furu*), and rice wine and soy sauce since ancient times (Lee et al., 1995, 2002). The early descriptions of the red mold rice manufacture techniques in detail were shown in the literatures of the Ming Dynasty of China.

Monascus pigments are a group of fungal metabolites called *azaphilones*, which have similar molecular structures as well as similar chemical properties. At least six kinds of pigments are identified, among which ankaflavine and monascine are yellow pigments, rubropunctatine and monascorubrine are orange pigments, and rubropunctamine and monascorubramine are purple pigments. Two complexed pigments, glutamyl-monascorubrine and glutamyl-rubropunctatine, were isolated from the broth of submerged cultures (see Figures 8.38–8.39).

Traditionally, red mold rice production was conducted in family workshops in South China. Long-shaped-rice is used as the raw material for cultivation. The high-quality red mold rice is preserved as seed for next-batch fermentation. Fermentation facilities used for cultivation were very simple and included wooden boxes, bamboo mats, and even surface soil. The master's skill decides the success or failure of the products. Production of red mold rice by these old ways at family workshops still exists currently in the Fujian Province, South China.

Since the 1960s, a series of technological innovations have been put into practice in red mold rice enterprises. For example, steeping tanks

Figure 8.38   Three types of *Monascus* pigments.

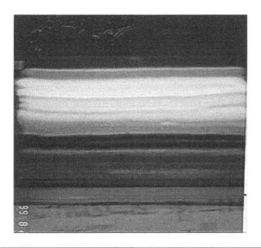

Figure 8.39   *Monascus* pigment bands by TLC (extracted from red mold rice).

for rice steeping, steaming cookers for rice cooking, inoculators for mixing seed with steamed rice, aeration tanks with air conditioners, hot air driers, and liquid seed cultivation tanks have been successfully used in red mold rice production on a large scale. The quality of the red mold rice, indicated by the color values (U/g, stand for pigment content in the red mold rice), was improved greatly. Currently, depth beds and aeration tanks with intermittently mixing are the major fermentation vessels for red mold rice cultivation in China.

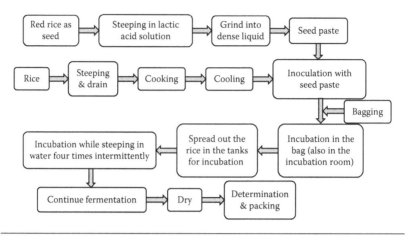

Figure 8.40   The flowchart of red mold rice production.

The flowchart of red mold rice production by depth bed aeration tanks with intermittent mixing is shown in Figure 8.40.

8.3.2 Monascus *Strains*

In general, *Monascus* can be identified as three species: *M. pilosus*, *M. purpureus*, and *M. ruber*. *Monascus* hypha is a multicellular coenocyte. In its asexual generation cycle, the hypha develops conidium, which is unicellular and multinuclear. In its sexual generation, cleistothecium is developed and ascospores are formed in the cleistothecium.

The suitable pH range for *Monascus* growth is 3.0–5.0, and it can even grow at pH 2.5. The optimal growth temperature ranged from 28°C–32°C. *Monascus* can tolerate a high ethanol concentration up to 10% and high concentration of glycerol up to 25% (see Figure 8.41).

8.3.3 Preparation of Monascus *Seed Cultures*

Currently, there are two processes for seed culture preparation: liquid-state cultivation of seed culture and solid-state cultivation of seed culture. Seed culture by liquid-state cultivation follows an ordinary operation in seed culture preparation. Whereas, seed culture by solid-state cultivation is actually high-quality red mold rice that is specially prepared as an inoculum for red mold rice cultivation.

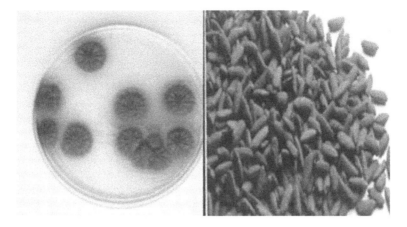

Figure 8.41 *Monascus ruber* and *Monascus* red rice.

8.3.3.1 Solid-State Cultivation of Monascus *Seed Culture* Solid-state cultivation of seed can be conducted in conical flasks or in aeration tanks. The seed incubated in conical flasks is free of contamination, so the pure cultures can be used as an inoculum to prepare more seed. The seed prepared in aeration tanks may be contaminated due to not strictly aseptic conditions during the cultivation period, so the seed prepared in this way can only be used as an inoculum for red mold rice fermentation, and it cannot be used as an inoculum further for seed preparation. Large-scale cultivation of solid-state seed culture can be carried out in "depth bed aeration tanks."

8.3.3.1.1 Seed Preparation in Conical Flask
1. Rice is steeped, and sterilized in conical flasks, then cooled as usual. For a 500 mL flask, add only 50 g rice.
2. Spore suspension is prepared from slant cultures, which have been incubated for 10 day in a PDA medium.
3. Inoculation 2 mL spore suspension to each flask, and the inoculated rice should be heaped up in a corner in the flask, and incubated at 35°C–36°C for 15–18 h. Until the mycelia grow and extend onto the steamed rice, shake the flask to let the rice spread across the bottom of the flask. The inoculated rice is incubated for 10 days at 30°C–32°C. During this period, add 2 mL sterilized water to each flask and shake if necessary.

Figure 8.42   Seed culture of *Monascus* red rice in conical flask.

4. Dry the red mold rice below 45°C to the moisture below 12%. This red mold rice is called *first-stage seed*, which should be used up within one year (see Figure 8.42).

8.3.3.1.2 Inoculation with Flask Seed

1. The first-stage seed (red mold rice) is steeped in an acetic acid solution (red mold rice: water: acetic acid = 1: 11: 0.2, by wet weight), and then is ground into seed paste.
2. The seed paste is mixed with steamed rice (seed paste: steamed rice = 6.2: 100, by wet weight), and this operation is actually an inoculation.
3. The inoculated rice is then bagged into a gunny sack.

8.3.3.1.3 Seed Cultivation in Aeration Tanks   The steamed rice is inoculated with the first seed and incubated in incubation room with well-controlled temperature and humidity. The inoculated rice may be incubated in a box or spread in the ground in an incubation room.

1. The inoculated rice in the gunny bag is cultivated in an incubation room for 20 h until the temperature of the fermenting rice reaches 45°C–50°C, and then the fermenting rice in the gunny bag is poured out into wooden box for further fermentation in an incubation room at a normal temperature.

2. The fermenting rice is steeped in an acetic acid solution again (acetic acid: rice = 0.2: 100, by wet). The fermenting rice is steeped in water a third and fourth time.

3. Drained, the fermenting rice is heaped up allowing the fermentation temperature to reach 36°C.

4. Then the fermenting rice is spread flat for further fermentation.

5. Repeat steps 2, 3, and 4 twice.

6. Dry the red mold rice at a lower temperature.

7. The dried red mold rice, which is served as first-stage seed, is preserved in dry and cold environments.

The first-stage seed is prepared in a laboratory in an aseptic operation, so the seed is free of contamination. The operation for second-stage seed is not at aseptic conditions. In order to avoid contamination during the operation and incubation, the acetic acid solution plays an important role when used in steeping the fermenting rice. Steeping the fermenting rice in an acetic acid solution not only favors *Monascus* growth in rice (optimal pH for *Monascus* growth is below 5), but also helps prevents the contaminant, especially, from killing the bacteria. Steeping in water intermittently also allows the fermenting rice to absorb water gradually and the *Monasucs* to grow inward gradually.

As stated previously, the inoculated rice in the gunny bag is precultivated for 20 h until the temperature reaches to 45°C–50°C. It seems the temperature is too high for the *Monascus* to grow. Actually at a high temperature, most of the bacterial contaminants died, but the spores of *Monascus* just begin to germinate during the period of precultivation (see Figure 8.43).

8.3.3.2 Seed Mash or Seed Paste Preparation In traditional processes, red mold rice is served as a starter culture. The red mold rice should be steeped in water acidified with vinegar before inoculating with steamed rice. This is called *seed mash* or *seed paste*.

1. Seed mash preparation (Method 1): Add 30 kg sterilized water (acidified with acetic acid to pH 3) to a pottery jar and cool to 30°C, then add 1 kg red mold rice and 1 g dry yeast into the water. Mix well and incubate the mixture at 28°C–30°C for about 24 h. The mixtures begin to ferment. Add about

Figure 8.43   Red mold rice incubated in a box.

3 kg steamed rice, seal the jar, and maintain the temperature at 28°C–30°C for 6–7 days. The mixtures prepared in this way are like red rice wine and can be served as seed mash for inoculating a large quantity of steamed rice.

2. Seed paste preparation (Method 2): The red mold rice is steeped in a water–acetic acid solution for about 2 h. (red mold rice: water: acetic acid = 1: 11: 0.2), then the mixture is ground to seed paste (see Figure 8.44).

Figure 8.44   Seed paste prepared by steeping red mold rice in an acetic acid-water solution.

8.3.3.3 Liquid-State Cultivation of Monascus *Seed Culture* Due to a nonaseptic operation in the period of solid-state cultivation of seed, it is obvious that the contamination of the red mold rice seed is unavoidable. So cultivation of the *Monascus* seed culture by liquid-state fermentation is necessary. A series of equipment for liquid cultivation are adopted in the red mold rice manufacture enterprises. The liquid-state cultivation of the seed culture shortens the cultivation cycle and reduces the risk of contamination. The procedures of liquid-state cultivation of the *Monascus* seed are the same as most submerged fermentation of microbial metabolites. The only difference is that the operation of inoculation of the *Monascus* seed in a liquid-state with steamed rice is still nonaseptic. The mixing operation of seed culture with steamed rice is conducted at operation desks in a clean room. After mixing uniformly, the inoculated rice is transferred to the aeration incubation tanks, which are not strictly aseptic.

Liquid seed cultivation conditions: Culture medium (in 1000 mL) for flask culture: rice meal 60–80 g; soybean powder 20–30 g; $MgSO_4 \cdot 7H_2O$ 2–3 g, KH_2PO_4 2–3 g, corn steep liquor 10–20 mL, $NaNo_3$ 2–3g; with initial pH 3.5–4.0. The medium is sterilized at 0.1 MPa for 30 min. Spore suspension is prepared from slant culture and inoculated into the shaker flasks. Cultivation of the seed is conducted at 32°C–35°C for 36–48 h with shaking at 180–200 r/m.

The flask seed is used for inoculating the seed tank, with an inoculum volume of 5%. The composition of the medium in seed tanks: rice powder 3%; soybean powder 2%–3%; $MgSO_4 \cdot 7H_2O$ 0.2%–0.3%, KH_2PO_4 0.2%–0.3%, corn steep liquor 1%–2%, $NaNo_3$ 0.2%–0.3%; with initial pH 3.5–4.0. Cultivation of the seed is conducted at 32°C–35°C for 36–48 h.

8.3.4 Raw Materials for Production of Red Mold Rice and Their Pretreatment

Various cereals, including rice, corn, millet, and wheat, are suitable for the production of red mold rice, among which rice is the most popular raw material. The rice should not be glutinous rice, which is too sticky to be used in the solid-state fermentation. Round-grained rice is also not suitable for red mold rice production. Only long-shaped rice is the best raw material for cultivation of *Monascus* red rice, as this kind of rice is fine and long with a greater ratio of surface area to the volume

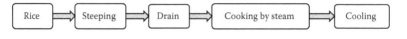

Figure 8.45 The flowchart of pretreatment of the raw materials.

than that of round-grained rice, which is favored for the growth of microorganisms on the surface of the rice. Most starch in long-shaped rice is amylose, so this kind of rice does not absorb too much water. The rice is still rigid after steaming, letting the steamed rice form the fermentation bed, and favoring air penetration through it.

The flowchart of pretreatment of the raw materials is seen in Figure 8.45.

8.3.4.1 Rice Steeping   The long-shaped rice is first steeped in water for soaking in a suitable amount of water. The moisture of the steeped rice is controlled within 28%. Steeping time is about 4–5 h in winter and spring, or 2–3 h in summer.

The proportion of main raw materials for red mold rice production should be in accordance with the seasons, which are listed below:

Springtime: Rice: red mold rice: acetic acid: water = 100: 0.5: 0.1: 56

Summertime: Rice: red mold rice: acetic acid: water = 100: 0.4: 0.08: 57

Wintertime: Rice: red mold rice: acetic acid: water = 100: 0.6: 0.15: 54

After steeping, the rice absorbs a suitable amount of water. The next step is rice cooking.

8.3.4.2 Rice Steaming   The steeped rice is sprinkled into the steamer many times during the steaming process. The traditional way for rice steaming is shown in Figure 8.46. When the steam goes off the top of steamed rice, keep steaming for another 20 min. Rice steaming on a large scale uses the cooker in a vertical type as shown in Figure 8.47. Rice steaming is processed continuously by this cooker.

Steeped rice is put into the cooker from upside of the cooker. From the top down, the rice goes through three sections till cooked: precooking happens in the first (top) section with the remaining steam coming out from the bottom; cooking in the middle section releases

Figure 8.46   Rice steaming in a traditional way.

steam from the jackets and sprayers; further cooking occurs in the bottom section.

The steam is sprayed into the rice from sprayers in the center and from the spouts in jackets around the cooker (upside, middle, and bottom). In this way, the steam is uniformly mixed with rice, ensuring the rice cooked well. If the rice is half-cooked, it can be returned into the cooker again for further steaming.

8.3.4.3 Cooling   The steamed rice is quickly cooled to about 36°C–38°C by ventilation.

When using a traditional steamer, the rice is sprayed onto the bamboo mat and cooled by a fan. If the steamed rice is caked, it should be stamped, breaking to pieces before cooling.

When using large-scale steaming, the rice is cooled by an air blower installed in the outlet of the steamer.

8.3.5 Inoculation

As stated previously, there are two types of seed cultures: liquid-state seed culture and solid-state seed culture that is actually well-chosen red mold rice. Here, the inoculation procedures with solid-state seed culture are given in detail below.

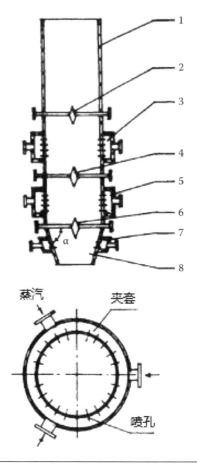

Figure 8.47 Continuous steamer in a vertical type. (1) cooker, (2) sprayer, (3) jacket, (4) sprayer, (5) jacket, (6) sprayer, (7) jacket, (8) cone-shaped bottom.

The traditional method of inoculation is shown in Figure 8.48. The steamed rice is spread out on a bamboo mat. The seed mash is quantitatively mixed with steamed rice.

In large-scale production, the inoculation operation system is composed of a screw conveyor, seed mash tank (seed cultivation tank if liquid-state seed is used), and a mixer with a stirrer. Steamed rice is transferred quantitatively by a screw conveyor. The proportions of rice to red mold rice are indicated in Section 8.3.4.

After mixing, the inoculated rice is poured to hollow-ware to release any unpleasant smell, and then the inoculated rice is loaded into gunny bags. One bag holds about 100 kg inoculated rice, which is equivalent to about 50–60 kg rice.

Figure 8.48   Inoculation with seed paste (in a traditional way).

8.3.6 Process Control in Solid-State Fermentation of Red Mold Rice

The gunny bags containing inoculated rice are carried to the incubation room and piled up in the incubation tanks. Fermentation of red mold rice in an incubation room lasts 5–8 days.

It can be divided into three stages:

Monascus spore germination stage
Monascus mycelium growth stage
Pigment production stage

8.3.6.1 Spore Germination Stage   The fermenting rice is incubated for 18–22 h for the spore germination. In the period of *Monascus* spore germination, the inoculated rice is put into gunny bags (one bag holds about 100 kg inoculated rice). The bags containing the inoculated rice are piled up on each other in one side in the aeration tanks. Furthermore, in order to keep warm, clean gunny bags are used as thermal insulation material to cover the bag piles. All the windows and door are closed in the incubation rooms. Within the bags, the *Monascus* spores begin to germinate in the inoculated rice. As one bag holds about 100 kg inoculated rice, large amounts of metabolic heat can not be released to the environment, resulting in the sharp increase of the temperature of the fermenting rice. The temperature increase in the fermenting rice favors the killing of contaminants.

As the inoculated rice is held in the gunny bags for about 20 hours, diffusion of oxygen into the bags is difficult, and large amounts of contaminated bacteria are killed due to depletion of oxygen and increase of the temperature.

Temperature control during the first period of fermentation: The temperature in fermenting rice held in the bags will rise gradually. Usually, the temperature of fermenting rice rises to 46°C–50°C. If the temperature exceeds 50°C, the fermenting rice in the gunny bag is poured out, occupying about a third of the area in the perforated bottom of the aeration tanks. The bed depth of fermenting rice in the fermentation tanks is reduced, so the temperature of the fermenting bed will drop naturally. But along with further fermentation, the temperature will rise gradually.

When the temperature rises to 45°C again, the fermenting rice should spread out and occupy all the areas in the perforated bottom. In this way, the depth of the fermenting rice is reduced to about 5 cm.

In conclusion, substrate temperature during red mold rice fermentation is determined by the bed depth of the fermenting rice, the growth state of microorganism, and climate change (see Figure 8.49).

8.3.6.2 Monascus *Growth Stage*   After spore germination, red spots appeared on the surface of the rice, and then the red spots extend to the whole rice granule. Gradually, the *Monascus* mycelia begins growing on the rice.

8.3.6.2.1 Moisture Control   Moisture control in red mold rice fermentation is very important. There are several steps that are involved with water addition in the whole process: steeping of rice and steaming

Figure 8.49   The bags containing the inoculated rice are piled up in one end of the aeration tank.

of rice, inoculation of seed culture (especially liquid-state seed culture), addition of water during fermentation period, and the elimination of water vapor by aeration during fermentation.

Monascus spp. are usually considered to be aerobic with pigments as the main metabolites. However, they also have the alcohol fermentation ability under anaerobic conditions. Moisture contents in fermenting rice are key parameters to determine the metabolic activity during fermentation. Higher moisture content of fermenting rice will result in an anoxic layer in the solid-state substrate, leading to ethanol fermentation, and formation of a hollow rice granule with little pigment. When the moisture content of fermenting rice is too low, *Monascus* growth is very poor, and its mycelium can not penetrate into the rice granule, resulting in the so-called "red surface and white core," which means only small pigments are produced on the surface of the rice and most of the starch in the rice is unchanged. Only when the moisture content is appropriate, will the *Monascus* grow normally; the main metabolites of fermentation are pigments, not ethanol. So the moisture control in the solid-state fermentation of red mold rice is very important and requires precise methods. The masters in ancient China have summarized such procedures. In short, the most precise method is the addition of water intermittently during fermentation after *Monascus* growth.

Figure 8.50   Fermenting rice is intermittently steeped in water.

There are two methods of adding water: steeping the fermenting rice into water (the traditional method as shown in Figure 8.50) and sprinkled water (in aeration tanks).

Traditionally, the red mold rice is incubated on the ground, a bamboo mat, or in a shallow tray (wooden). Adding of water intermittently is also necessary in the fermentation period. The fermenting rice is put into bamboo baskets, which is steeped in running water for about 30 s. Then the steeped rice is carried back to the incubation room for further incubation, with intermittently mixing, piling up, and spreading.

Addition of water at the right time and in suitable amounts can stimulate pigment production. In order to get higher pigment production yield, water addition four and five times is recommended.

8.3.6.2.2 First Addition of Water   Usually, addition of water the first time is carried out about 44 h later after inoculation. The amount of water is about 30% of the fermenting rice, but actually only a small amount of water is absorbed by the fermenting rice, and most of the sprinkled water is drained out of the hole in the perforated bottom.

After a sprinkling of water, the fermenting rice is heaped up. About 60 min later, the temperature of fermenting rice will rise gradually to 37°C, and the fermenting rice will be spread out uniformly in the perforated bottom. Aeration and mixing should be carried out intermittently.

8.3.6.2.3 Second Addition of Water   Addition of water a second time is carried out 8–10 h later after the first time, depending on the

Monascus growth. Usually, when a pink colony appeared on the fermenting rice uniformly, and the rice is loose and dry, addition of water is necessary. The amount of water is about 40% of the fermenting rice. The operation of heaping up and spreading is the same as with the first addition of water.

After the second addition of water, *Monascus* growth is sped up, and the temperature changes quickly. Aeration should be carried out almost continuously with intermittent mixing. The maximal temperature of the fermenting rice should be controlled below 36°C.

8.3.6.2.4 Third and Fourth Addition of Water   The amounts of water addition at the third and fourth time are 35% and 30% of the fermenting rice, respectively. The temperature of fermenting rice is controlled below 36°C by aeration and mixing intermittently.

8.3.6.2.5 Further Fermentation When Addition of Water Is Stopped   About 12 h later, after adding water a fourth time, the fermentation temperature is controlled below 34°C by opening the windows and doors of the incubation room. After fermentation for another 24 hours, the red mold rice can be dried and harvested (see Table 8.3).

Table 8.3   Process Control during Solid-State Fermentation of Red Mold Rice

TIME (h)	MAX. TEMP. (°C)	PROCESS CONTROL
0	38	Inoculated rice is loaded into bags
20	45–50	Fermenting rice spread onto perforated bottom
30	45	Stir
40	43	Stir
48	37	Stir, add water
56	37	Stir
58	36	Add water
64	36	Stir, aeration
68	36	Add water, stir, aeration
76	36	Stir
88	36	Add water
96	35	Stir
100	36	Stir
120	34	
144		Drying, stir every 2 hours
168		Red mold rice

8.3.6.3 Aeration and Agitation (Stirring) during Fermentation   The bio-reactor used for fermentation of red mold rice is an intermittently-mixed, forcefully-aerated bioreactor, each with overall dimensions of 6–9 m × 2–4 m × 0.6–1.0 m and a working capacity of about 10 t (20% dry matter) in an industrial scale-up.

Aeration time and frequency depend on the temperature of fermenting rice. Usually during the period of 58–68 h after inoculation, the temperature of the fermenting rice rises quickly. Accordingly, the aeration runs frequently with an interval time of 1–2 min.

Stirring of the fermenting rice is usually performed by a manual operation. Workers use both of their hands to turn out the fermenting rice gently (see Figure 8.51).

The depth of the fermenting rice is adjusted by piling up and spreading out the fermenting rice, so this operation is also a measure to control the temperature of fermenting rice.

Monascus has a relatively slow growth rate, and the fermentation time is usually 5–8 days. So it is better if the red mold rice is cultivated in clean conditions. Usually, it is considered that beyond 3 days' cultivation, it is very difficult to work in nonsterile conditions. The con-

Figure 8.51   Depth bed and aeration tanks for red mold rice fermentation.

taminated microbes come from air and the environment, especially if the aeration system is run in a non-aseptic condition.

8.3.7 Drying and Grinding

After fermentation, the red mold rice is dried in the aeration tank by hot air that is released from the perforated bottom. The depth of the red mold rice is controlled below 30 cm. Stir the red mold rice every 2 hours till the moisture of the red rice is below 12%.

8.3.8 Determination Methods for Pigment Content of Red Mold Rice

As *Monascus* pigments are mixtures of six or more pigments, it is difficult to gain a pure pigment substance. The method of determination of pigment contents of red mold rice is based on the optical densitometry.

The red mold rice is ground to powder in 80 meshes. Weighing 1.0000 g red rice powder, add 50 mL 65% ethanol to extract the pigments at 60°C for 1 h, shaking two times. The extract is filtered with filter paper, and 1 mL of the filtrate is diluted with 100 mL 65% ethanol solution. The absorbance of the diluted sample is measured at 505 nm (for red pigment) with 65% ethanol solution as reference.

The term "color value" is used to indicate the content of *Monasucs* pigments in red mold rice. The color value is defined as:

Color value (U/g) = $(OD_{505\ nm} \times D)/M$
D: dilution fold; D = $50 \times 100 = 5000$
M: weight of the red mold rice (g, dry)

8.3.9 Mixed Cultures of Monascus with Aspergillus

Two types of mixed cultures of *Monascus* with *Aspergillus* are popular in Southeast China: mixed culture of *Monascus* with *Aspergillus niger*, which is called *wu-yi-hong-qu*; and a mixed culture of *Monascus* with *Aspergilus flavus*, which is called *huang-yi-hong-qu* (Shin et al., 1998; Li et al., 2009). Wu-yi-hong-qu looks black outside and red inside while huang-yi-hong-qu looks yellow outside and red inside. They are all manufactured by solid-state fermentation and used for brewing special rice wine (see Figure 8.52).

Figure 8.52   Traditional production of wu-yi-hong-qu by *Monascus* and *Aspergillus niger* at home.

References

Bhumiratana, A., Flegel, T.W., Glinsukon, T., and Somporan, W. (1980). Isolation and analysis of molds from soy sauce koji in Thailand. *Applied and Environmental Microbiology.* **39**, 430–439.

Chen, F.S., Li, L., Qu, J., and Chen, C.X. (2009). Cereal vinegars made by solid-state fermentation in China. *Vinegars of the World.* **15**, 243–259.

Cho, Y.H., Song, J.Y., Kim, K.M., Kim, M.K., Lee, I.Y., Kim, S.B., Kim, H.S., Han, N.S., Lee, B.H., and Kim, B.S. (2010). Production of nattokinase by batch and fed-batch culture of *Bacillus subtilis*. *New Biotechnology.* **27**, 341–346.

Couto, S.R. and Sanromán, M. (2006). Application of solid-state fermentation to food industry: A review. *Journal of Food Engineering.* **76**, 291–302.

De Carvalho, J.C., Soccol, C.R., Babitha, S., Pandey, A., and Wojciechowski, A.L. (2008). Production of pigments. In *Current Developments in Solid-State Fermentation*, eds. A. Pandey, C.R. Soccol and C. Larroche, pp. 337–355. India: Springer.

Fabre, C.E., Santerre, A.L., Loret, M.O., Baberian, R., Pareilleux, A., Goma, G., and Blanc, P.J. (1993). Production and food applications of the red pigments of *Monascus rubber*. *Journal of Food Science.* **58**, 1099–1102.

Fujita, M., Nomura, K., Hong, K., Ito, Y., Asada, A., and Nishimuro, S. (1993). Purification and characterization of a strong fibrinolytic enzyme (nattokinase) in the vegetable cheese natto, a popular soybean fermented food in Japan. *Biochemical and Biophysical Research Communications.* **197**, 1340–1347.

Han, B.Z., Rombouts, F.M., and Nout, M.J.R. (2001). A Chinese fermented soybean food. *International Journal of Food Microbiology.* **65**, 1–10.

Huang, D.M., Yang, Y., and Zhang, Z.C. (2008). Changes of aroma components of Zhenjiang savory vinegar during different fermentation periods and aging periods. *China Brewing.* **9**, 56–61.

Johns, M.R. and Stuart, D.M. (1991). Production of pigments by *Monascus purpureus* in solid culture. *Journal of Industrial Microbiology and Biotechnology.* **8**, 23–28.

Juzlová, P., Martínková, L., and Křen, V. (1996). Secondary metabolites of the fungus *Monascus*: A review. *Journal of Industrial Microbiology and Biotechnology.* **16**, 163–170.

Lee, B.K., Piao, H.Y., and Chung, W.J. (2002). Production of red pigments by *Monascus purpureus* in solid-state culture. *Biotechnology and Bioprocess Engineering.* **7**, 21–25.

Lee, Y.K., Chen, D.C., Chauvatcharin, S., Seki, T., and Yoshida, T. (1995). Production of *Monascus* pigments by a solid-liquid state culture method. *Journal of Fermentation and Bioengineering.* **79**, 516–518.

Li, K., Chen, M.B., and Zhu, Z.J. (2009). Production of koji by black-skin-red-koji. *China Brewing.* **4**, 124–126.

Lin, Y.L., Wang, T.H., Lee, M.H., and Su, N.W. (2008). Biologically active components and nutraceuticals in the *Monascus*-fermented rice: A review. *Applied Microbiology and Biotechnology.* **77**, 965–973.

Liu, D.H., He, W.H., Xiang, L.Y., and Wang, Y.J. (2008). Application of functional *Monascus* in vinegar brewing. *China Condiment.* **8**, 70–73.

Liu, D.R., Zhu, Y., Beeftink, R., Ooijkaas, L., Rinzema, A., Chen, J., and Tramper, J. (2004). Chinese vinegar and its solid-state fermentation process. *Food Reviews International.* **20**, 407–424.

Luh, B.S. (1995). Industrial production of soy sauce. *Journal of Industrial Microbiology and Biotechnology.* **14**, 467–471.

Ma, X.C. (2001). Making process of sweet soybean jam. *Chinese Condiment.* **8**, 28–30.

Nimnoi, P. and Lumyong, S. (2011). Improving solid-state fermentation of *monascus* purpureus on agricultural products for pigment production. *Food and Bioprocess Technology.* **4**, 1384–1390.

Okamoto, A., Hanagata, H., Kawamura, Y., and Yanagida, F. (1995). Antihypertensive substances in fermented soybean, natto. *Plant Foods for Human Nutrition.* **47**, 39–47.

Peng, T., Yang, X.X., and Chen, S.H. (2007). Current situations and future applications of China soy sauce. *China Condiment.* **9**, 26–29.

Röling, W.F.M., Apriyantono, A., and Van Verseveld, H.W. (1996). Comparison between traditional and industrial soysauce (kecap) fermentation in Indonesia. *Journal of Fermentation and Bioengineering.* **81**, 275–278.

Shin, C.S., Kim, H.J., Kim, M.J., and Ju, J.Y. (1998). Morphological change and enhanced pigment production of *Monascus* when cocultured with *Saccharomyces cerevisiae* or *Aspergillus oryzae*. *Biotechnology and Bioengineering.* **59**, 576–581.

Sievers, M., Sellmer, S. and Teuber, M. (1992). *Acetobacter Europaeus* sp. Nov., a main component of industrial vinegar fermenters in central Europe. *Systematic and Applied Microbiology.* **15**, 386–392.

Tesfaye, W., Morales, M.L., García-Parrilla, M.C., and Troncoso, A.M. (2002). Wine vinegar: Technology, authenticity and quality evaluation. *Trends in Food Science and Technology.* **13**, 12–21.

Wicklow, D.T., Mcalpin, C.E., and Yeoh, Q.L. (2007). Diversity of *Aspergillus oryzae* genotypes (RFLP) isolated from traditional soy sauce production within Malaysia and Southeast Asia. *Mycoscience.* **48**, 373–380.

Wu, J.J., Ma, Y.K., Zhang, F.F., and Chen, F.S. (2012). Biodiversity of yeasts, lactic acid bacteria and acetic acid bacteria in the fermentation of "Shanxi aged vinegar," a traditional Chinese vinegar. *Food Microbiology.* **30**, 289–297.

Yong, F.M. and Wood, B.J.B. (1976). Microbial succession in experimental soy sauce fermentations. *International Journal of Food Science and Technology.* **11**, 525–536.

Yue, Y.Y., Zhang, W.X., Yang, R., Zhang, Q.S. and Liu, Z.H. (2007). Design and operation of an artificial pit for the fermentation of Chinese liquor. *Journal of the Institute of Brewing.* **113**, 347–380.

Zhang, L. and Li, Z.X. (2009). Effects of wheat brans on functional characteristics of vinegar. *Journal of Triticeae Crops.* **29**, 915–918.

Zhong, S.R. and Feng, Z.P. (2011). Study on production technology of sweet soybean paste. *The Food Industry.* **1**, 43–45.

9

SOLID-STATE FERMENTED ALCOHOLIC BEVERAGES

XIAOQING MU, YAN XU, WENLAI FAN,
HAIYAN WANG, QUN WU, AND DONG WANG

*State Key Laboratory of Food Science and Technology, School
of Biotechnology, Jiangnan University, Wuxi, China*

Contents

9.1 Introduction

Chinese traditional solid-state fermented beverages are produced using natural microflora in empirical processes based on the spontaneous fermentation of different raw materials. All these beverages usually contain ethanol and are divided into two general classes: distilled spirits and rice wines. Chinese distilled spirit, also named *baiju* (means "transparent alcoholic drink") or Chinese liquor, is distilled mainly from fermented cereals. Chinese rice wine, also named *huangjiu* (means "yellow alcoholic drink"), yellow wine or yellow liquor, is a type of Chinese alcoholic beverage brewed directly from fermented grains and contains less than 20% alcohol. These wines are traditionally sterilized, aged, and filtered before their final bottling for sale to consumers.

9.2 Chinese Distilled Spirit

9.2.1 Introduction

Chinese distilled spirit is one of the oldest distillates in the world. Compared with other spirits such as vodka, whisky, and brandy, Chinese liquor has a higher ethanol content (normally 40%–55% by volume). Although the trend today is toward lower strength, it will still mostly be found in the 46%–52% range, with exceptions up to 60%. This high alcohol-containing product is normally consumed straight or "neat," not mixed with water, soda, fruit juice, or some other liquid. Up to now, Chinese liquor has an annual consumption of approximately 4 million tons.

Chinese liquor is mainly produced in the provinces of Sichuan, Shandong, Jiangsu, and Guizhou. Especially, about one-fourth of the total liquors in China are produced in Sichuan, which is mainly the strong aroma-style liquor, and the representative liquors are Wuliangye liquor, Jiannanchun liquor, and Luzhoulaojiao liquor. Liquors produced in Shandong are mainly sesame aroma-style liquors, and the representative liquors are Jingzhi liquor, Bandaojing liquor, and Taishan liquor (Figure 9.1). Most of the liquors produced in Guizhou are soy sauce aroma-style liquors, including the famous Maotai liquor and Xi liquor. Other liquors, including Fen liquor and

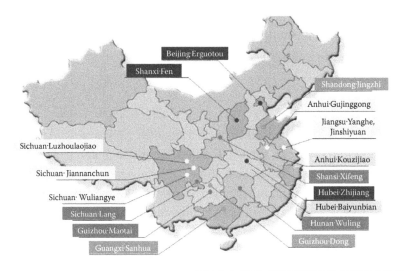

Figure 9.1 The distribution of Chinese distilled spirits in China.

Erguotou liquor, are light aroma-style liquors, which are produced in Shanxi and Beijing, respectively.

Because of differences in manufacturing practices, the aroma profiles of various Chinese liquors are quite different. On the basis of aroma characteristics, Chinese liquor can be classified into five categories: strong aroma style, light aroma style, soy sauce aroma style, sweet honey style, and miscellaneous style. Of these, the strong aroma-style accounts for about 70% of total liquor production. Strong aroma-style liquors typically have strong fruity, pineapple- and banana-like aromas (Fan and Chen, 2001). Within this category, Wuliangye and Jiannanchun are two of the most famous brands, followed by Yanghe Daqu and a few others. Chinese Moutai, a most famous spirit named the "national liquor," belongs to the soy sauce aroma-type liquor.

According to the alcoholic content, Chinese liquors can be classified into three types: high alcoholic level (50%–60%), middle alcoholic level (40%–50%), and low alcoholic level (18%–40%) liquors. Some 20 years ago, most of Chinese liquors had a high alcoholic content. But now, about 40% of Chinese liquor production is comprised of the low alcoholic level liquors.

Chinese liquors are typically fermented using the solid-state technique, which means the main processes including saccharifying, fermentation, and distillation were carried out in the fermenting grains (mash) with 60% moisture content. Most premium Chinese liquors, including Moutai, Wuliangye, and Jiannanchun, must be manufactured using the solid-state technique.

9.2.2 Production Processes

Most Chinese liquor is traditionally fermented from grains. After the fermentation, the fresh spirit is distilled out and then aged under controlled conditions. The aged distillate is adjusted to the designated ethanol concentration and blended to ensure the quality of the finished product and maintain brand consistency (Shen, 1996). The raw materials for making Chinese liquor can be quite different, depending upon availability and the cost of the raw materials. In general, Chinese liquor is made from sorghum or a mixture of sorghum, wheat, corn, rice, and sticky rice. Rice hull is typically used as the fermentation aid (Shen, 1996; Fan and Teng, 2001).

Daqu or *xiaoqu* is starter culture used for Chinese liquor (Fan and Qian, 2006b). Daqu is the most widely used culture, made from wheat or a mixture of wheat, barley, and pea. The raw materials of daqu are typically milled and pressed into a mold of different sizes of bricks, according to the manufacturer. The daqu bricks are put into a room for fermentation under controlled conditions. On the basis of the maximum temperature at which the daqu is incubated, the types of daqu can be classified into low temperature (<45°C), moderate temperature (45°C–60°C), and high temperature (>60°C) daqu. Daqu is rich in various microorganisms including bacteria, yeast, and fungi (Shen, 1996). In addition, complex enzyme systems are accumulated in the finished daqu (Fan and Xu, 2000).

The grains used for liquor fermentation are first milled, cooked, and then mixed with husk. Hot water (about 95°C) is added to the cooked grains to adjust the moisture content to 55% (w/w). The cooked grains are cooled to 13°C–16°C and mixed with the daqu powder. The mixture is fermented in a special fermentor (3.4 m length, 1.8 m width, and 2.0 m height); the inside of the fermentor is coated with a layer of fermentation mud made of clay, spent grain, bean cake powder, and fermentation bacteria. The fermentation is typically carried out at 28°C–32°C for 60 days under anaerobic conditions in a solid state. After fermentation, the liquor is distilled out with steam and aged in sealed pottery jars to develop the balanced aroma. While most of the liquors are aged for about 1 year, some of them are aged for more than 3 years. The aged liquor is diluted with water and blended to give an ethanol content of 40%–55% (v/v) for constant quality in the finished product.

9.2.3 Functional Microorganisms and Enzymes

Many enzymes are accumulated in Chinese daqu during fermentation, including liquefying amylase, saccharifying amylase, acid protease, cellulase, lipanase, and esterase. Fan and coworkers (Fan and Xu, 2000; Fan et al. 2002, 2003) studied these enzymes in Chinese daqu. The results are listed in Table 9.1. Liquefying amylase in the daqu is mainly produced by *Aspergillus*, *Rhizopus*, and *Endomycopsis*, whereas saccharifying amylase is produced from *Rhizopus*, *Aspergillus*, and *Monascus*. Acidic protease is made by *Aspergillus*, *Rhizopus*, *Rhizomucor*, and other molds. The microorganism-producing cellulase includes *Trichoderma (Tri.)*, *Aspergillus*, *Mucor*, such as *Tri. reesei*, *Furium oxysporum*, *Tri. viride*, and *Tri. koningii oud*, which are all well known for yielding cellulase. It is reported that some bacteria can also produce cellulose (Shen and J., 1991). Lipases widely present in daqu, are produced by bacteria and epiphytes. Esterases present in daqu are produced by *Rhizopus*, *Saccharomyces*, and *Monascus*. In general, molds

Table 9.1   The Activities of Enzymes and Microorganisms in Chinese Daqu

	SURFACE OF DAQU	CENTER OF DAQU	CORNERS OF DAQU	WHOLE OF DAQU
Activities of enzymes				
Liquefying amylase (g oluble starch/g dry daqu·h)	3.949	0.905	2.056	1.342
Saccharifying amylase (mg glucose/g dry daqu·h)	201.54	66.51	179.82	109.89
Acid protease (µg tyrosine/g dry daqu·min)	100.21	50.33	80.56	61.06
Cellulase (mg glucose/g dry daqu·h)	68.25	35.67	45.48	39.90
Lipanase (mg xylose/g dry daqu·h)	180.24	120.55	145.51	128.43
Esterase (decrease hexanoic acid,%)	66.64	42.39	—	—
Microorganisms				
Bacteria	6.25×10^5	1.12×10^5	—	—
Yeasts	2.88×10^5	1.35×10^4		
Molds	3.21×10^6	2.55×10^5	—	—

account for a large proportion of microorganisms yielding liquefy-ing amylase, saccharifying amylase, acid protease, cellulase, lipase, and esterase.

Molds are influenced by oxygen in the environment. Under an aer-obic condition, molds boost and produce a large amount of enzymes. During the fermentation of the daqu, more oxygen is dissolved in the surface of the daqu than in the center. Compared with the center of the daqu, the number of microorganisms on the surface of the daqu is almost 10 times higher. The activities of the hydrolyzing enzymes from different parts of the daqu, including a liquefying enzyme, sacchari-fying amylase, acidic protease, cellulase, and lipanase, are also quite different. The activities of hydrolyzing enzymes produced from the surface of the daqu are the highest, followed by the enzymes from the corners of the daqu. The enzyme activities in the center of the daqu are the lowest. It is said the activities of these enzymes are tightly cor-related with the amount of microorganism. Also, activities of differ-ent enzymes produced by microorganisms growing in the same part of the daqu are different. The activity of liquefying and saccharifying amylases from the surface of the daqu is three times of that from the center of the daqu, whereas the activities of acidic protease and cellu-lase are only double. Furthermore, the difference between the activity of lipase from the surface of the daqu and that from the center of the daqu is not evident as other kinds of hydrolyzing enzymes.

A fungal strain capable of synthesizing ethyl esters of short-chain fatty acids, which is the key flavor compound, is isolated from daqu ("moldy grains") samples through combined screening strategy of lipid hydrolysis and with esterification ability, and is identified to be *Rhizopus chinesis* CCTCCM201021 (Xu et al., 2002). When compared with 10 other commercial lipases for ester synthesis, the whole-cell lipase of *R. chinesis* CCTCCM201021 (RCL) shows a much higher ability in the synthesis of ethyl hexanoate with a maxi-mum yield of 96.5% after 72 h conversion using 0.5 M at optimized reaction conditions (Table 9.2).

9.2.4 Micraobial Community of Some Typical Chineses Liquors

Solid-state fermentation initiated by starters (daqu or xiaoqu in Chinese) is a major characteristic of Chinese liquor. *Qu* is prepared

Table 9.2 Effect of Various Sources of Lipases on the Synthesis of Ethyl Hexanoate in the Heptane Phase

LIPASE SOURCE[a]	MOLAR CONVERSION (%)	RELATIVE ACTIVITY (%)
Rhizopus chinensis lipase (whole-cell)	96.5	100.0
Rhizopus chinensis lipase (cell-free)	84.9	88.0
Rhizomucor miehei lipase	91.5	94.8
Candida rugosa lipase	75.4	78.1
Triacylglycerolacylhydrolase	87.7	90.9
Candida lypolytica lipase	32.6	33.8
Candida Antrarctica lipase	61.2	63.4
Mucor javanicus lipase	16.0	16.6
Rhizopus arrhizus lipase	80.0	82.9
Aspergillus niger lipase	18.5	19.2
Pseudomonas Cepacia Lipase	87.1	90.3

[a] Reaction condition: equal molar ethanol and hexanoic acid concentration of 0.5 M and 6 g/L of enzyme at 150 rpm and 30°C for 72 h.

by using the natural inoculation of mold, bacteria, and yeast or the artificial inoculation of a special species isolated from liquor production. Microorganisms play important roles in the production of liquor and are provide by qu. Different types of qu have specific microbial compositions. Many intrinsic and extrinsic factors of qu production influence the richness and structure of the microbial community, including raw materials variety, moisture content, temperature control, geographical location, climate, and water.

A number of researchers have been working on liquor microorganisms ever since the beginning of the 1960s, and their work has been focused mainly on microorganisms in daqu (Shi et al., 2001; Wang et al., 2008), pit mud, and fermented grains (Xiong, 1994; Wu et al., 2009) through traditional culture-dependent methods. Culture-dependent methods are time consuming due to long culture periods and elaborate culture techniques. Moreover, species occurring in low numbers are often out-competed *in vitro* by numerically more abundant microbial species, and some species may be unable to grow *in vitro*. In contrast, molecular biology approaches have been proven to be powerful tools in providing a more complete inventory of the microbial diversity in environmental samples (Gich et al., 2001). Following the development of molecular ecological technology, it has generated a keen interest from Chinese researchers to investigate the microflora of Chinese liquor using culture-independent methods. Polymerase chain reaction–denaturing gradient gel electrophoresis (PCR-DGGE) has recently been shown to be a useful tool for studying microbial community structure. The great potential shown in analyzing samples from natural environments has stimulated food microbiologists to investigate the suitability of PCR-DGGE to study microbial fermentations in food and food-related ecosystems. Now PCR-DGGE has successfully been applied to analyze the microflora in various traditional fermented foods, such as Chinese soybean paste (Zhao et al., 2009), Japanese *shochu* (Endo and Okada, 2005), and Korean *kimchi* (Chang et al., 2008).

Daqu is a mixture of grains, microorganisms, enzymes, and aroma precursors. To date, very few studies have employed culture-independent approaches to investigating the microflora in daqu. The whole composition and functions of the microbial group are largely unknown. The purpose of this study was to analyze the microbial diversity in daqu using PCR-DGGE and obtain a better knowledge of microbial composition.

There are rich LABs in daqu and the identified species were *Lactobacillus brevis*, *L. helveticus*, *L. panis*, *L. fermentum*, *L. pontis*, and *Weissella cibaria*. LABs are the main producers of lactic acid and provide substrates for esterification by yeasts. Though the functions of LAB in the fermented foods such as sourdough and wine were revealed gradually, further investigation is required to obtain detailed function of LAB in Chinese liquor. Another dominant species, *Staphylococcus xylosus*, was first reported in daqu. *S. xylosus* was supposed to be used as a single-strain culture or associated with lactic acid bacteria for the production of fermented sausages (Fiorentini et al., 2009). The coexistence and high richness of LAB and *S. xylosus* in daqu may help them to act synergistically during the fermentation process. The detected *Pseudomonas sp.* in DGGE profile was also isolated from the fermented grains of liquor and proved to reduce the concentration of higher alcohols in liquor (Wu, 2006). Liquor production factories try to reduce these concentrations, and utilization of the suitable microorganism may be an effective measure.

It is a well-established fact that the saccharification and alcoholic fermentation of Chinese liquor result from the combined actions of several yeast and mold species, and they grow more or less in succession throughout the brewing process. Many studies have described the isolation and identification of yeasts and molds in daqu, and concluded that the culture-dependent isolation process could not fully assess the total fungal population of daqu (Wang et al., 2008; Shi et al., 2009).

To investigate the community of yeast and molds in daqu, the 26S rRNA gene of yeast and 18S rRNA gene of fungi were amplified, and the resulting PCR products were used for DGGE. *Saccharomycopsis fibuligera* and *Pichia anomala* were observed in 10 daqu samples analyzed. *P. anomala* showed a lower intensity in the Langjiu sample and the Kouzijiaogaowen sample than other samples. *Saccharomyces cerevisiae* could be detected in all samples, but its band density was weak. *Saccharomyces exiguus* was also mostly detected in samples with low operation temperature. *Saccharomyces bulderi* was only predominant in samples B, C, and D, which were produced in the same liquor plant with a different technological process. There were several non-*Saccharomyces* yeasts in daqu including *Hanseniaspora guilliermondii*, *Debaryomyces hansenii*, *Trichosporon asahii*, and *Pichia kudriavzevii (Issatchenkia orientalis)*. *Rhizomucor miehei*, *Absidia blakesleeana*, and *Aspergillus terreus* were the dominant molds. *R. miehei* was detected in medium-temperature daqu samples such as E, F, G, and I (Table 9.3).

In the DGGE profile *Saccharomycopsis fibuligera* and *Pichia anomala* existed in all of the daqu and their predominance were in agreement with previous findings by culture-dependent methods (Cui et al., 2002). *S. fibuligera* possessed high α-amylase and glucoamylase activities and could utilize raw starch as a carbon source (Van Rensburg et al., 1998). *P. anomala* is known as an ester-producing yeast. The pilot tests proved that adding more *P. anomala* into the production could increase the ester content of liquor. Most detected yeasts in this research were reported in many kinds of fermented food especially in fermented alcoholic beverage (Aidoo et al., 2006; Beh et al., 2006). In recent years the significance of non-*Saccharomyces* species in wine making attracted the interest of wine-making researchers (Mendoza et al., 2009; Takush and Osborne, 2009), and they contribute to the final taste and flavor of wines. However, there are few reports published about the function of non-*Saccharomyces* yeast in Chinese liquor. The detected main molds in daqu were *Absidia sp.* and *Aspergillus sp.*, and this result accorded with the previous reports (Zhang et al., 2007).

Table 9.3 Samples of Different Chinese Liquor Daqu

SAMPLE NO.	NAME OF DAQU	WHEAT (%)	BARLEY (%)	PEA (%)	HIGHTEST MANUFACTURE TEMPERATURE (°C)	REGION (CITY, PROVINCE)	DATE OF RIPENING	DATE OF SAMPLING
A	Laobaigan (LBG)	100	0	0	50	Hengshui, Hebei	2008.05	2008.11
B	Fenjuqingcha (FQC)	0	60	40	44–46	Xinghuacun, Shanxi	2008.04	2008.11
C	Fenjiuhouhuo (FHH)	0	60	40	47–48	Xinghuacun, Shanxi	2008.03	2008.11
D	Fenjiuhongxin (FHX)	0	60	40	45–47	Xinghuacun, Shanxi	2008.04	2008.11
E	Jinshiyuan (JSY)	100	0	0	50–55	Huaian, Jiangsu	2008.04	2009.05
F	Tanggou (TG)	60	30	10	55–60	Lianyungang, Jiangsu	2008.04	2008.07
G	Kouzizhongwen (KZJH)	60	30	10	50–55浓	Huaibei, Anhui	2008.09	2009.03
H	Kouzigaowen (KZJL)	100	0	0	65	Huaibei, Anhui	2008.04	2009.03
I	Jiannanchun (JNC)	90	10	0	55–60	Mianzhu, Sichun	2008.04	2008.01

They can secrets hydrolyzing enzymes to decompose macromolecular materials and some metabolites to affect liquor flavor.

Although many studies have clearly demonstrated the broad applicability of the PCR-DGGE method, the ability of DGGE to discriminate among target rRNA genes is dependent on the PCR primers used. Thus, the use of appropriate consensus primers has a critical influence on the resolution of DGGE analysis in mixed microbial systems (Fontana et al., 2005). We used two primers to fully assess the fungal community, including yeasts and molds, and DGGE profiles of the 26S rRNA and 18S rRNA gene revealed different compositions. As assessment of the total microbial diversity by molecular means has been hampered by the lack of sufficiently specific primers, it is predicted that the use of several primers will enable more accurate microbial community analysis (Schabereiter-Gurtner et al., 2001).

Predominant species of bacteria, yeast, and mold in different typical daqu were revealed for the first time. The microbial community of daqu was correlated not only with production techniques, but also with grains and the local geographic environment around liquor production factories. For the microbes reported in this study, further investigation needs to be carried out for obtaining more detailed information on the specific function of each individual microorganism and its contribution to the final formation of the unique aroma of Chinese liquors.

Although xiaoqu is named because of its small volume, it has several advantages such as low usage amount and high utilization ratio of the starch substrate. There was lower microorganism diversity in xiaoqu than in daqu. *Staphylococcus, Micrococcus, Leuconostoc, Pediococcus,* and *Lactobacillus* were the main bacteria in xiaoqu. Because pH in the fermented grains remained about pH 4, lactic acid bacteria was dominant in fermented grains, and other bacteria quickly died. *S. cerevisiae, Pichia anomala, Saccharomycopsis fibuligera, Issatchenkia orientalis,* and *Pichia fabianii* were isolated from xiaoqu, and the former two were dominant species. *Rhizopus oryzae* was absolutely dominant in both xiaoqu and fermented grains. Other molds only included small amounts of *Aspergillus* and *Mucor,* which were just detected in the beginning of the fermentation process.

9.2.5 Aroma Compounds in Chinese Liquors

With the introduction of gas chromatography (GC) in the 1960s, it became possible to study the trace volatile compounds in Chinese liquors. GC equipment was widely applied to analyze the volatiles of Chinese liquors in the 1980s. Recently, some new techniques, such as aroma extraction dilution analysis (AEDA) and time-intensity method (Ruth, 2001; Fan and Qian, 2005; Chen and Ji, 2006; Fan and Qian, 2006b,c), have been developed to evaluate the relevance of detected volatiles to odor perception in Chinese liquors. Acids seem to play important roles in Chinese liquor, including hexanoic, butanoic, 3-methylbutanoic, pentanoic, acetic, 2-methylpropanoic, octanoic, heptanoic, propanoic, 4-methylpentanoic, and nonanoic acids. Volatile esters introduce fruity flavor notes and are considered highly positive flavor attributes of Chinese liquors. Much more important esters in Chinese liquors include ethyl acetate, ethyl propanoate, ethyl butanoate, ethyl pentanoate, ethyl hexanoate, ethyl octanoate, ethyl nonanoate, ethyl decanoate, ethyl 2-methylpropanoate, ethyl 2-methylbutanoate, ethyl 3-methylbutanoate, methyl hexanoate, propyl hexanoate, butyl hexanoate, pentyl hexanoate, hexyl acetate, hexyl butanoate, hexyl hexanoate, hexyl octanoate, 2-methylpropyl acetate, 2-methylpropyl hexanoate, 3-methylbutyl butanoate, 3-methylbutyl hexanoate, 3-methylbutyl octanoate, ethyl 2-hydroxyhexanoate, ethyl 2-hydroxypropanoate, ethyl 2-hydroxy-3-methylbutanoate, diethyl butanedioate, and ethyl cyclohexanecarboxylate. Ethyl hexanoate has a fruity, floral, and sweet aroma. The OSME value of this compound is very high among the aroma compounds identified in Chinese liquor, especially in strong aroma-type liquors (Fan and Qian, 2006b,c). In Chinese strong aroma-style liquor, important alcohols consist of 2-methylpropanol, 3-methylbutanol, 2-ethyl-1-hexanol, 1-butanol, 1-pentanol, 1-hexanol, 1-heptanol, 1-octanol, 2-butanol, 2-pentanol, 2-heptanol, and 2-octanol. In Chinese liquors, guaiacol (2-methoxyphenol), 4-methylguaiacol (4-methyl-2-methoxyphenol), and 4-ethylguaiacol (4-ethyl-2-methoxyphenol) could be important to the aroma basis on their high OSME values or FD values detected on the DB-Wax column (Fan and Qian 2006). These compounds contribute to strong clove, spicy, and smoky odors. Phenol, o-cresol (2-methylphenol), p-cresol (4-methylphenol), 4-ethylphenol, and

4-vinylphenol are also identified in Chinese liquors. These compounds contribute herbal and animal odors. Some acetals are identified in the neutral/basic fraction. 1,1-Diethoxy ethane has a strong fruity aroma. 1,1-Diethoxy-3-methylbutane gives a fruity aroma and has an extremely high FD value. 1,1-Diethoxynonane, 1,1,3-triethoxy-propane, 1,1-diethoxy-2-methylpropane, 1,1-diethoxy-2-methylbu-tane, and 1,1-diethoxy-2-phenylethane are also identified in Chinese liquors. They contributed a fruity aroma. Two sulfur compounds, dimethyl disulfide and dimethyl trisulfide, are detected in Chinese liquors. Aldehydes are detected in liquors, including acetaldehyde, 2-methylbutanal, 3-methylbutanal, 1-pentanal, and so on. Several aromatic esters are identified, including ethyl benzoate, ethyl 2-pen-ylacetate, ethyl 3-phenylpropanoate, 2-phenylethyl acetate, 2-phenyl-ethyl butanoate, and 2-phenylethyl hexanoate.

In 2007, Fan and coworkers identified the pyrazines of Chinese liquors using liquid–liquid extraction, and quantified using headspace SPME coupled with a GC–flame thermionic detector (Fan et al., 2007). A total of 27 pyrazines are identified in Chinese liquors, mainly alkyl- and acetylpyrazines. 2,6-Dimethylpyrazine was detected in all samples and ranged from 20.34 to 1057.46 µg/L. Several pyrazines, including 2-methylpyrazine (39.19–1011.55 µg/L), 2,5-dimethyl-pyrazine (0.63–182.15µg/L), 2-ethylpyrazine (19.52–101.7 µg/L), 2-ethyl-3-methylpyrazine (47.21–897.49 µg/L), 2,3,5-trimethylpyra-zine (0.41–2327.95 µg/L), 2,6-diethylpyrazine (84.05–1621.30 µg/L), 3,5-diethyl-2-methylpyrazine (41.12–545.58 µg/L), 2-methyl-6-propenylpyrazine (41.86–450.76 µg/L), and 2-acetyl-3,5-dimethyl-pyrazine (38.56–337.90 µg/L), were detected and determined in most of the Chinese liquor.

9.2.6 Kweichou *Moutai*

Kweichow Moutai (or *Guizhou Maotai* in mandarin Chinese), the first and a typical representative of the soy sauce aroma-style liquor made from grains and *daqu* (also shortened as *qu*, a kind of starter for mak-ing liquor, like Japanese *koji*), boasts both its long history and mysteri-ous origin. Legend has it that the aboriginals of Maotai Town on both banks of the *Chishui* River or the Red River (marking the boundary between Sichuan and Guizhou Province of Southwest China) had been adept at making liquor during the reign of Emperor Dayu living in remote antiquity.

According to the historical literature, Kweichow Moutai originated in 135 BC of the Western Han Dynasty (206 BC–25 AD). At that time, Maotai Town and its neighboring areas abounded in wolfberry sauce aroma liquor, the predecessor of present-day Kweichow Moutai, which had been the tribute liquor to the royal palace across dynasties as one of the best alcoholic beverages ever since Emperor Wu (156–87 BC) of the Han Dynasty (206 BC–220 AD) sampled it and acclaimed it as "sweet and delicious," as recorded in *Shi Ji* or *Records of the Grand Historian*. In the 1915 Panama-Pacific International Exposition held in San Francisco, America, Kweichow Moutai won the gold award, sharing the top rating with Scottish whiskey and French cognac as the three most world-renowned distillates, which represented the pioneering steps taken by Chinese national industry and commerce to go international.

From a little primitive still in ancient times to a contemporary supersized enterprise group of China, Kweichow Moutai's success can be attributed to its unique production processes, consistent high quality, peculiar attributive properties, distinctive brand culture, and unique liquor-making eco-environment, as well.

Kweichow Moutai is produced at Maotai Town, Renhuai County, in northwest Guizhou Plateau. Kweichow Moutai Distillery Co. Limited is situated on the eastern bank of the Red River with an altitude of 450 m, where a winding range of mountains are bordered by crystal waters. With long summers and short winters, the region enjoys an agreeable climate, high humidity, and rare frost weather, its temperature ranging from –2°C to 40°C with an annual average of 18°C. Its relative humidity is about 78% while its wind speed is about 1.2 m/s on average per year. All these peculiar natural elements constitute the unique ecological environment for making Kweichow Moutai (Xu et al., 2010b).

The unique flavor of Kweichow Moutai relies largely on the unique climate, peculiar water, exclusive eco-environment, and superior natural conditions of Maotai Town, which renders the impossibility of distilling Kweichow Moutai elsewhere in the world. Speaking of the desirable natural conditions of Maotai Town, a diplomat once remarked after his visit to Kweichow Moutai Distillery, "It turns out to be even more difficult to make Kweichow Moutai than to make atomic bombs as the later can be made all over the world while the former is confined to here in this region" (Chen and Ji, 2006). The production of Kweichow Moutai involves a perfect combination of three mysterious factors, namely, the mysterious nature, the mysterious land, and the mysterious people here. By the "mysterious nature," we refer to the very fact that there dwells a vast community of active microorganisms in the environment of the Maotai area, without which no Kweichow Moutai could come into being. In terms of the mysterious land, Maotai Town is blessed with unique red soil, a plentiful supply of peculiar water in the Red River, and the advantaged climate, all fitting for distilling Kweichow Moutai. As for the mysterious people, they are the ancestors of Maotai Town who derived the scientific liquor-making processes with unique taste from their long-term practices of liquor distillation. It is the three mysterious factors that contribute to the birth of the unique Kweichow Moutai, whose esoteric characteristics and distinctive flavor cannot be imitated elsewhere, thus "working a miracle" in the world's liquor industry.

Basic raw materials for making Kweichow Moutai: Raw materials are *hongyingzi* (meaning "with red-tasselled ears") glutinous sorghums, and wheat, which are local varieties and nourished with natural organic fertilizers, without chemical fertilizers and pesticides used in their growth at all. The sorghums locally grown in Maotai Town or its neighboring areas well meet the peculiar demands for raw materials to make Kweichow Moutai through unique processing. For one thing, they are thick in husk and small in kernels with a high proportion of germ and amylopectin starch. For another, they are resistant to steaming and boiling and slow to gelatinize due to a low rate of water absorption as they have a starch content of more than 60% and an amylopectin starch content of above 88%.

Qu-making raw materials are all selected from the locally grown wheat in Maotai Town or its surrounding areas. The wheat is compactly-built, evenly-ripened and plump-kerneled without worm infestation or moldy spoil, and it is thin in hull, high in purity, and light yellow in color, without any brown hue at the end of each kernel. Besides, its starch content is over 60% with mealy sections.

The water used for distilling Kweichow Moutai is from the sources of the Red River and it is clean, colorless, and transparent with high

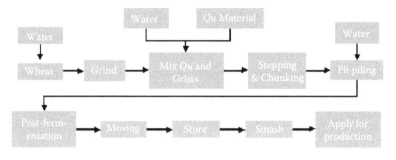

Figure 9.2   Qu-making process of Kweichow Moutai.

quality and purity, free from deposited solids and flavor taints. Its temperature is on the low side and its pH range fluctuates between 6.5 and 7.5.

Qu-making process of Kweichow Moutai (Figure 9.2): Qu-making raw materials are to be ground roughly into scoured grist with pieces of husks yet without too much fine powder and without an obviously coarse or smooth feeling. The amount of qu inoculum and water accounts for about 6%–8% and 40%, respectively, of all the qu materials in the mixing process to ensure the perfect combination of water, qu materials, and wheat powder so that no dry powder can be spotted, and the mixture can easily be molded into a ball or scattered if flung. The qu is trodden by feet and shaped into brick chunks without fracture and dust, tight along the side edges while loose in the central part. Qu chunks are then stacked on edge in the fermentation barn with two horizontal and three vertical chunks interlacing with each other. During the fermentation period of 40 days or more, chunks are to be turned upside down twice and the temperature in the barn must be monitored so as to clear up the moisture. The fermented qu chunks are profuse in aroma, thin in skin, evenly distributed in color, and chrysanthemum-shaped in their centers without any greasy, sour smell, or moldy spoil or rot from *Penicillium* and *Mucor.*

Distillate-making process of Kweichow Moutai (Figure 9.3): Grains are added twice (namely, *xiasha* and *zhaosha*, meaning the first and second deposits) during the whole process. Raw materials are to be ground to such a degree that its content of fine powder reaches 27%–30%. The moistening temperature must be above 90°C and the water used here is about 50%–54% of the total amount of raw materials. The pressure for steaming grains ranges from 0.08–0.12 MPa and 80 to 100 minutes

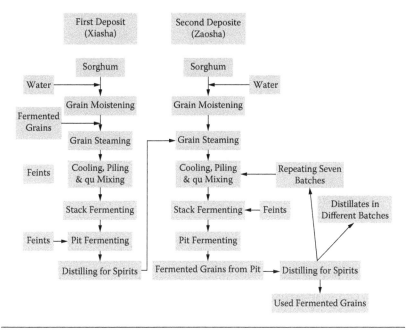

Figure 9.3   Distillate-making process of Kweichow Moutai.

are needed for steaming the second deposits. Fermented grains should be 8%–12% of the total amount of raw materials. The mixing process needs a temperature between 28°C and 32°C with a proper dose of qu according to different batches and 3%–5% feints. The mixture was stacked for about 2 days and moved into a stone pit for 30 d fermentation. The suitable temperature of stack fermentation for yeast and other strains propagated on the ground is between 50°C–53°C, while fermentation temperature in pits should be around 30°C to 43°C. As for the distilling process, the pressure should remain at 0.08 MPa while its duration is more than 40 minutes. A spirit distillation is divided into three fractions: *foreshots, middle cut,* and *feints.* The main runnings of spirit distillation are middle cut, named as the basic distillate and stored for blending. Basic distillates in each batch can then be collected on the basis of incoming strength drops of distillate (from 70% to 48%). Such main distillates are high in characteristic flavor compounds of Moutai-flavor liquor, making future blending easier in universal quality.

Blending process of Kweichow Moutai (Figure 9.4): Prior to elaborate blending and packaging processes, the freshly made Kweichow Moutai basic distillates will be stored in jars for at least 3 years according to

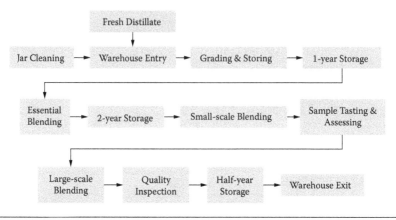

Figure 9.4　Blending process of Kweichow Moutai.

their round of distillation, type, and alcoholic strength. It will altogether take at least five years from the very beginning of production to the very end of making Kweichow Moutai ready for sale. During the storage of the basic distillates, processes such as essential blending, small-scale blending, and large-scale blending are to be implemented. Every batch of Kweichow Moutai blends is finished by using 80 to 100 or even more basic distillates.

Jars are to be cleaned before the fresh distillate is stored. As a rule, jars will first be filled with clean water and then left alone for 2 to 3 days to check whether there are any leakages, cracks, impurities, or dusts. Only when they are confirmed clean and sanitary can fresh distillates be filled in. After the new distillates are put into the barns for preservation, such labels and signs are to be made on the jars such as the year of production, date, barn number, jar number, workshop, team staff, round, and quantity. After they are stored for a whole year, distillates of the same round, aroma, and class can be used for essential blending. Two years after that, small-scale blending can be carried out in accordance with the production standards of Kweichow Moutai. And then large-scale blending is proceeded with the same proportional parameters as small-scale blending. After that, the blended liquor will be stored for another half a year before filling and packaging for sale.

Generally speaking, the blending of Kweichow Moutai is done with basic distillates of various batches of production, types, alcoholic strength, and duration of storage in a certain proportion on the

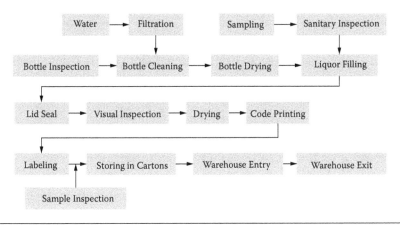

Figure 9.5   Filling and packaging process of Kweichow Mouai liquor.

condition that the standards of the blended samples meet the requirements of Kweichow Moutai.

Filling and packaging process of Kweichow Moutai liquor blends from large-scale blending are stored for half a year before filling and packaging for ex-work sale (see Figure 9.5). The filling and packaging process includes procedures like blend intake, filling, marking, etc. Bottles are expected to meet the requirements for uniform thickness and size without any skews, spots, bubbles, or flaws. Prior to the filling process, bottles are to be cleansed with clean, clear, and residue-free water and inspected to guarantee both the interior and exterior of bottles are sanitary and taintless without impurities and excess water left. Liquor blends to be filled in bottles should be first sampled for quality inspection to ensure they are pure, and free from turbidity and suspended solids, and then checked against the specifications required for each specific variety and batch. The net weight of liquor needs to meet the requirement as well. The lid for the seal should be in accordance with standards as follows: smooth, clean and tight without dusts, skews, or leaks. They should fit the bottle so well that no water oozes out even though the bottle stands upside down. After the lid is sealed, a ribbon is to be tied with its scripts facing outside in alignment. Then such marks as the brand name "Kweichow Moutai," date of production, batch number, and serial number of the bottle and production venue codes are to be coated evenly on the lid, followed by the labeling process, during which a specified tag coupled with an anticounterfeit label is to be attached fast and symmetrically without any inversions, folds, bubbles, or flushes. Bottles of liquor are then put into cartons or boxes complete with accessories according to the stipulated variety and quantity without inversion or mistake before the carton openings are sealed with tapes neatly and smoothly. In the end, the cartons are hooped with straps tightly and stably.

Kweichow Moutai is produced with daqu as the saccharifying and fermenting agents, simultaneously fermented together with saccharifying as the technique and sorghum starch as the raw material, which differs from both Scotland whiskey and French cognac in that the former is made with malts as the saccharifying agent while the latter is made with grapes, containing mainly sucrose and other sacchrides as the raw materials by adding yeasts for fermentation.

Apart from the discrepancies in raw materials and production processes, Kweichow Moutai distinguishes itself from other spirits with its unique production settings and ecological environment. The qu-making and liquor-making processes of Kweichow Moutai give full play to the ecological advantages and the unique natural conditions and resources. Such processes as qu making, stack fermentation, and anaerobic fermentation all take place under the ambience of high temperatures, which constantly domesticate the liquor-making microorganisms. Varieties of microorganisms, through their heredity, mutation, and evolution, have contributed to the abundance and gathering of the extreme microcoenosis around Kweichow Moutai workshops that are resistant to high temperature, acidity, and alcoholicity. The metabolism of the functional microorganisms gives birth to heat- and acid-resistant enzymes of high stability in the course of making Kweichow Moutai, such as amylases, proteases, glucoamylases, cellulases, glucosidases, and xylanases as well as various dehydrogenases and phosphoenolpyruvate carboxykinases involved in the oxidation-reduction reactions, etc. (Yuan, 2007). All kinds of extreme microorganisms and enzymes thus occurred to help bring about rich flavor compounds for the liquor with the prolonged and repeated unique production processes desirable for the growth of microorganisms and enzymes over a long period of time. These flavor compounds enter into the basic distillates in different processes and further form the unique taste and perfect quality of Kweichow Moutai. Up till now, such functional microorganisms as *Bacillus licheniformis* (Zhang and Xu, 2010) and *Bacillus subtilis* (Zhu et al., 2009a,b; Zhu and Xu,

2010c) for the purpose of producing soy sauce aroma have already been isolated from high-temperature-resistant Moutai daqu and their fermentation and metabolism can further produce flavor substances such as acetoin and 2, 3, 5, 6-tetramethylpyrazine, furaneol (Huang et al., 2006; Fan et al., 2007; Zhu and Xu, 2010a,b). This special kind of fermentation eco-environment of Kweichow Moutai provides it with the unique, distinctive, and exclusive attributes distinguishing it from all the other spirits and liquors, and giving it its core competitiveness.

According to the analysis and comparison of microcomponents in China's soy sauce aroma spirits (or Kweichow Moutai-flavor liquor), strong aromatic spirits (or Luzhou-flavor liquor), faint-scented spirits (or Fen-flavor liquor), brandy and whiskey via the highly-sensitive full two-dimensional GC×GC/TOFMS, there are 963, 674, 484, 440, and 264 peaks detected, respectively, confirming that Kweichow Moutai is the distillate with the richest flavor microcomponents worldwide. Qualitative analysis also shows that there are 873 chemical components in Moutai including 380 esters, 85 acids, 155 alcohols, 96 ketones, 73 aldehydes, 36 nitrogenous compounds, and 48 other chemicals. In particular, there are larger varieties of nitrogenous compounds in Kweichow Moutai as compared to the other aroma-type liquors, with up to 3 and 7 times as many as those in Luzhou-flavor liquor and Fen-flavor liquor, respectively (Ji and Guo, 2006; Xu et al., 2010a; Ji et al., 2007).

The findings from researchers on Kweichow Moutai's flavor impact compounds by GC-O, GC-MS together with OSME and AEDA show that there are over 300 flavor-impact compounds that contribute to the unique taste of Kweichow Moutai, 126 of which have a major influence on the formation of Kweichow Moutai's aroma while 65 of which also play important roles in the process.

Research done on Kweichow Moutai and Luzhou flavored liquors, regarding their "aroma diffusion" compounds and aroma compounds lingering in empty glasses, show that there are up to 163 kinds of compounds in the latter, while there are 28 kinds in Kweichow Moutai; their key components have been defined already.

Studies have also shown that Kweichow Moutai liquor abounds in nutritious and restorative microcomponents. For example, over 17 of Kweichow Moutai's flavor microcomponents have been found beneficial to human health. At the Fifth International Conference on Metallothionein, M. George Cherian, chief scientist on metallo-thionein research at Andrew University in Canada, and Dr. Liu Jie, an American scientist of Chinese origin, confirm that a sufficient intake of Kweichow Moutai can contribute to health as the liver can be stimulated to produce quite a large quantity of metallothionein if the liquor is consumed in moderation.

9.3 Chinese Rice Wine

9.3.1 Introduction

Chinese rice wine is one of the oldest drinks in the world, along with beer and wine. It is one of the most popular alcoholic beverages with an annual consumption of more than 2 million kiloliters in China, especially in the south (Zhu et al., 2004). The techniques for making rice wines were known to the Chinese before 5000 BC. The Chinese have always taken great pride in their rice wines. Nowadays, Zhejiang Province, Jiangsu Province, and Shanghai are the central regions where Chinese rice wines are produced and sold popularly. Chinese rice wines are also popular in Fujian Province, Shandong Province, and Taiwan. Their market even includes Anhui, Hubei, Hunan, Jiangxi, Shaanxi, Gansu, Ningxia, Hebei, and some other regions (Figure 9.6).

Shaoxing, located in the Zhejiang Province, has been the most famous rice wine-brewing center from ancient times. Jimo Laojiu produced in Shangdong and red rice wine in Fujian are also famous because of their unique flavor and brewing technique. The best rice wines are thought to be produced by traditional natural fermenta-tion methods and sold without additives. However, rice wines are now manufactured commercially in modernized rice breweries by pro-fessional craftsmen on a large scale. Many varieties of Chinese rice wines are now available throughout the world. Rice wines are being marketed in Hong Kong, Japan, and other countries and regions in Southeast Asia.

Comparing sake, which is typically described as having "caramel," "burnt," "heavy," and "complicated" characteristics (Isogai et al., 2005), Chinese rice wine has the characteristics of "yellow," "sweet aroma," and "abundant nutrition." According to its concentration of sugar, Chinese rice wine could be classified into four categories:

Figure 9..6 The distribution of Chinese rice wines in China.

dry type (sugar content <15 g/L), semidry type (sugar content of 15–40 g/L), semisweet type (sugar content 40–100 g/L), and sweet type rice (sugar content >100 g/L) wines. Chinese rice wine is saccharified in a solid-state, and fermented in a semisolid state or liquid state.

Chinese rice wine is typically fermented from sticky rice with wheat qu used as saccharifying and with yeast as a fermenting agent. Wheat qu, similar to the *koji* of sake, is made of wheat. First, the wheat, the raw material of making wheat qu, is typically milled and pressed into a mold of different sizes with appropriate natural water. Then, it is incubated at 28°C–30°C for 48 h in a special room, and dried at 45°C as far as its humidity is fewer than 12% (wt/wt). The fermented wheat qu is rich of various microorganisms, including bacteria, yeast, and mold. In addition, complex enzyme systems inside wheat qu are also accumulated in it during this process.

Like making sake, in the production processes of Chinese rice wine, the sticky rice is first polished before it is used. Then, the polished sticky rice is washed by natural water at room temperature, and immersed in the same water at 12°C–15°C for 40 h. After that, the steeped sticky rice used for Chinese rice wine fermentation is cooked with steam in a steaming machine to gelatinize the starch contained in the rice kernel. Finally, the steamed rice is cooled by cold water till its temperature falls to 30°C.

Preparation of seed mash is a special process of Chinese rice wine fermentation. The steamed sticky rice, as just described, is mixed with 10% (w/w, weight ratio of wheat qu and sticky rice) wheat qu and 200% (w/w) spring water at 25°C. Then, 3% (w/w) selected pure yeast is added to the mixture after adjusting the pH of the fermenting mash to 4 with lactic acid. Subsequently, the mixture is cultured at about 25°C for 48 h.

After those processes, steamed sticky rice is mixed with 10% (w/w) wheat qu, 6% (w/w) seed mash, and 150% (w/w) spring water in the pottery vats. The most typical characteristic of Chinese rice wine fermentation is that this process is simultaneously saccharifying and fermenting. The main fermentation is typically carried out at 28°C–30°C for 7 days in a semisolid state, and the postfermentation is carried out at room temperature for 15 days. After postfermentation, the rice wine mash is first filtered by a filter press, and then the slightly turbid rice wine is pumped to the fining tanks for clarification. The caramel, which determines the particular color of Chinese rice wine, is added to the clarified rice wine, and then the fresh rice wine is cooked by steam at 88°C–90°C for 3 minutes. Finally, the cooked rice wine is aged in sealed pottery jars to develop the balanced aroma. Most of Chinese rice wines are aged for 1–3 years, but some of them are aged for 3–5 or more years. The aged rice wine is generally blended to yield an ethanol content of 14%–17% (v/v) for constant quality in the finished product.

9.3.2 Wheat Qu and Yeast-Making Rice Wine

The main characteristic of Chinese rice wine brewing is the use of several types of wheat qu. Wheat qu is mainly a culture of molds and yeasts grown on and within grounded rice meal. Wheat qu, which is comparable to koji used in sake brewing, is used for liquefaction and saccharification of the starch contained in the rice. Wheat qu also contributes to aroma and flavor substance formation. Without wheat qu, the rice wine could lose its typically characteristic aroma and flavor.

Some researchers have studied the molecular identification and classification of the main molds from wheat qu (Fang et al., 2006). In this study, 18 molds are separated from wheat starter by a potato agar medium, wort agar medium, and CDA medium. The five main strains (the order of magnitude above 10^5) were identified as *Absidia corymbifera*, *Rhizomucor pusillus*, *Aspergillus oryzae*, *Aspergillus fumigatus*, and *Rhizopus oryzae* through sequencing and sequence contrast (DNA extracted by benzyl chloride method, then amplification of the whole ITS sequence including ITS1 and ITS2 by primer ITS1 and ITS4 by use of ITS rDNA).

The α-amylase from wheat qu of Chinese rice wine has been studied. A strain to produce α-amylase was isolated from the wheat qu of Chinese rice wine. It is identified as *Asp. oryzae* by molecular methods and sequence alignment. Ion-exchange chromatography and gel filtration chromatography were performed for purification of α-amylase from the wheat qu of Shaoxing rice wine and α-amylase produced by *Asp. oryzae*. In addition, the properties of α-amylase were studied.

Yeast converts sugar to ethanol and carbon dioxide. But different yeast strains will produce different products, like esters, alcohols, acids, and other chemical compounds that affect the nuances of aroma and flavor of rice wine. These compounds will be present in various levels, depending on the choice of yeast and the processing parameters of fermentation. The fermentation temperature highly affected the product of yeast, for example, fatty acids esters, high alcohols, and other compounds.

Several kinds of wine yeast, including rice-sprinkling yeast, fast-fermenting yeast, saccharifying yeast, and active dry yeast and so on, have been applied to the producing of different rice wine.

The tolerance of major Chinese rice wine yeasts (HJ1, HJ2, HJ3, and HJ4) and sake yeast (K-7) to high sugar concentration, high temperatures, low temperatures, and low pH, are compared. The results indicated that the tolerance of rice wine yeast showed statistically significant differences among yeasts. HJ4 rice wine yeast had good tolerance to high sugar concentration, whereas HJ3 rice wine yeast had the highest fermentation efficiency when sugar concentration was up to 40%. K-7 sake yeast had good tolerance to temperature, whereas HJ4 rice wine yeast showed better tolerance to low pH than other yeasts.

9.3.3 Characteristic Aroma Compounds of Chinese Rice Wine

The investigations on the volatile components of Chinese rice wine have been performed for a few years. More than 50 volatile and semi-volatile compounds were detected by liquid–liquid extraction (LLE), static headspace, and direct-injection gas chromatography (Luo et al., 2007). In 2008, Luo and coworkers (Luo et al., 2008) first employed headspace solid phase microextraction (HS–SPME) followed by gas chromatography–mass spectrometry (GC–MS) for analyzing the volatile and semivolatile trace compounds of Chinese rice wines. A total of 97 volatile and semivolatile compounds were detected and identified in several typical Chinese rice wines, including alcohols, acids, esters, aldehydes and ketones, aromatic compounds, lactones, phenols, sulfides, furans, and nitrogen-containing compounds. Up to now, no studies were published on aroma compounds in Chinese rice wine. The present work was intended for identifying the aroma compounds in typical Chinese rice wine by gas chromatography–olfactometry (GC–O), and for determining the concentrations of volatile compounds so as to find important aroma compounds.

To facilitate the identification of aromas, the extract of a rice wine sample was separated into three fractions: acidic/water-soluble, basic, and neutral. GC–O and GC–MS were performed on each fraction. A total of 57 aroma compounds were identified on DB-Wax and DB-5 columns in these two samples (Table 9.4), including 9 alcohols, 9 esters, 8 fatty acids, 10 aromatic compounds, 7 phenolic derivatives, 5 furans, 1 lactone, 2 sulfur-containing compounds, 4 nitrogen-containing compounds, 1 aldehyde, and 1 ketone. Among these, five aroma compounds, unknowns, were detected by GC–O but could not be identified by GC–MS.

Like other alcoholic beverages, alcohols were the main volatile aroma compounds in Chinese rice wines. On the basis of the OSME values, the potentially important alcohols were 2-methylbutanol and 3-methylbutanol, and imparted alcoholic and nail polish notes, respectively. 1-Propanol and 2-methylpropanol could also be important aroma compounds because they had medium OSME values. 1-Propanol had fruity and alcoholic aromas, whereas 2-methylpropanol contributed wine and solvent notes. Higher alcohols could be formed during the fermentation, under aerobic conditions from

Table 9.4 Aroma Compounds in JF12 and GY30 Detected by GC–O

						OSME VALUE	
RI_{Wax}	RI_{DB-5}	AROMA COMPOUND	DESCRIPTOR	FRACTION[a]	BASIC OF IDENTIFICATION[b]	JF12	GY30
ALCOHOLS							
1035	530	1-Propanol	Fruity, alcoholic	A/W	MS, RI, aroma	2.67	2.17
1087	618	2-Methylpropanol	Wine, solvent	A/W	MS, RI, aroma	2.67	3.50
1137	643	1-Butanol	Rancid	A/W	MS, RI, aroma	2.17	2.33
1195	753	2-Methylbutanol	Alcoholic	A/W	MS, RI, aroma	3.50	4.00
1201	783	3-Methylbutanol	Nail polish, rancid	A/W	MS, RI, aroma	3.50	4.00
1268	798	1-Pentanol	Fruity, balsamic	A/W	MS, RI, aroma	2.00	2.50
1341	888	1-Hexanol	Floral, green	A/W	MS, RI, aroma	ND[c]	1.67
1448	986	1-Octen-3-ol	Mushroom	A/W	RI, aroma	2.50	ND
1443	984	1-Heptanol	Alcoholic, fruity	A/W	RI, aroma	2.00	2.00
ESTERS							
892	584	Ethyl acetate	Pineapple	N	MS, RI, aroma	3.00	2.67
953	705	Ethyl propanoate	Fruity, banana	N	MS, RI, aroma	3.00	2.00
961	754	Ethyl 2-methylpropanoate	Fruity, sweet	N	MS, RI, aroma	2.17	1.17
1031	800	Ethyl butanoate	Pineapple	N	MS, RI, aroma	3.00	3.00
1102	875	3-Methylbutyl acetate	Fruity	N	MS, RI, aroma	2.00	2.83
1128	900	Ethyl pentanoate	Apple	N	MS, RI, aroma	2.67	1.17
1235	1010	Ethyl hexanoate	Fruity, floral, sweet	N	MS, RI, aroma	3.50	3.17
1409	1196	Ethyl octanoate	Fruity	N	RI, aroma	1.50	1.67
1655	1176	Diethyl butanedioate	Fruity, sweet	N	MS, RI, aroma	3.17	1.50
FATTY ACIDS							
1424	582	Acetic acid	Acidic, vinegar	A/W	MS, RI, aroma	3.50	4.33
1555	789	2-Methylpropanoic acid	Rancid, acidic	A/W	MS, RI, aroma	3.67	3.33
1602	802	Butanoic acid	Rancid, cheesy	A/W	MS, RI, aroma	4.33	4.50
1655	877	3-Methylbutanoic acid	Rancid, acidic	A/W	MS, RI, aroma	4.33	4.67
1727	911	Pentanoic acid	Sweat, rancid	A/W	MS, RI, aroma	2.17	2.67
1846	1019	Hexanoic acid	Sweat, cheesy	A/W	MS, RI, aroma	2.17	2.33

continued

Table 9.4 (continued)   Aroma Compounds in JF12 and GY30 Detected by GC–O

RI_{Wax}	RI_{DB-5}	AROMA COMPOUND	DESCRIPTOR	FRACTION[a]	BASIC OF IDENTIFICATION[b]	OSME VALUE JF12	OSME VALUE GY30
1955	1103	Heptanoic acid	Sweat	A/W	MS, RI, aroma	0.83	2.33
2060	1171	Octanoic acid	Sweat, cheesy	A/W	MS, RI, aroma	2.17	0.67
AROMATIC COMPOUNDS							
1501	963	Benzaldehyde	Fruity, berry	N	MS, RI, aroma	3.00	3.67
1620	1047	Phenylacetaldehyde	Floral, rose	N	MS, RI, aroma	3.00	3.33
1625	1035	Acetophenone	Sweet, fruity, floral	N	MS, RI, aroma	2.00	2.00
1640	1175	Ethyl benzoate	Fruity	N	MS, RI, aroma	1.50	2.00
1694		1-Phenyl-1-propanone[d]	Pungent, floral	N	MS, aroma	3.33	3.17
1768	1247	Ethyl 2-phenylacetate	Rosy, honey	N	MS, RI, aroma	1.67	2.17
1801	1260	2-Phenylethyl acetate	Rosy, floral	N	MS, RI, aroma	ND	2.50
1873	1353	Ethyl 3-phenylpropanoate	Rose, floral	A/W	MS, RI, aroma	3.00	2.00
1906	1116	2-Phenylethanol	Honey, rose	A/W	MS, RI, aroma	4.00	3.83
1916	1276	Z-2-phenyl-2-butenal	Cocoa, sweet, rum	N	MS, RI, aroma	1.17	2.67
PHENOLIC DERIVATES							
1858	1090	Guaiacol	Spicy, clove, animal	A/W	MS, RI, aroma	1.50	2.67
1952	1195	4-Methylguaiacol	Smoke	A/W	MS, RI, aroma	2.33	ND
2007	987	Phenol	Phenol, medicinal	A/W	MS, RI, aroma	3.17	3.17
2080	1082	4-Methylphenol	Animal, phenol	A/W	MS, RI, aroma	2.67	ND
2185	1181	4-Ethylphenol	Smoky	A/W	MS, RI, aroma	2.50	ND
2200	1323	4-Vinylguaiacol	Spicy, clove	A/W	MS, RI, aroma	3.67	3.67
2208	1345	2,6-Dimethoxyphenol[d]	Smoke	A/W	RI, aroma	1.67	ND
FURANS							
1456	831	Furfural	Almond, sweet	N	MS, RI, aroma	4.50	4.50
1489	917	2-Acetylfuran	Sweet, caramel	N	MS, RI, aroma	2.50	2.33
1555	967	5-Methyl-2-furfural	Green, roasted	N	MS, RI, aroma	ND	1.50
1603	1058	Ethyl 2-furoate	Balsamic	N	MS, RI, aroma	1.83	2.00
1647	854	2-Furanmethanol	Burnt sugar	A/W	MS, RI, aroma	1.50	1.33

Table 9.4 (continued) Aroma Compounds in JF12 and GY30 Detected by GC–O

RI$_{Wax}$	RI$_{DB-5}$	AROMA COMPOUND	DESCRIPTOR	FRACTION[a]	BASIC OF IDENTIFICATION[b]	OSME VALUE JF12	GY30
LACTONES							
2018	1363	γ-Nonalactone	Coconut, peach	N	MS, RI, aroma	4.83	4.83
ALDEHYDES AND KETONES							
1073	797	Hexanal	Green, grass, apple	N	MS, RI, aroma	1.67	0.83
1300	980	1-Octen-3-one[d]	Mushroom, earthy	N	RI, aroma	2.17	2.17
SULFUR-CONTAINING COMPOUNDS							
1360	976	Dimethyl trisulfide	Sulfur, rotten cabbage	N	Aroma, RI	2.67	2.50
1702	978	3-(Methylthio)propanol	Cooked vegetable	A/W	MS, RI, aroma	2.67	3.00
NITROGEN-CONTAINING COMPOUNDS							
1315	915	2,5-Dimethylpyrazine	Baked, nutty	B	MS, RI, aroma	2.67	2.67
1330	910	2,6-Dimethylpyrazine	Nutty	B	MS, RI, aroma	2.67	2.67
1430	1089	2,5-Dimethyl-3-ethylpyrazine[d]	Roasted, baked	B	RI, aroma	2.67	1.67
1972	1024	2-Acetylpyrrole	Herbal, herbal	B	MS, RI, aroma	1.17	2.33
UNKNOWNS							
1314		Unknown	Cooked rice	B		2.33	2.67
1417		Unknown	Fatty, oily	N		2.50	ND
1762		Unknown	Acid, sour	A/W		2.00	ND
1783		Unknown	Sweet, flower	N		ND	2.17
1949		Unknown	Caramel	N		1.50	2.50

[a] A/W, acidic/water-soluble fraction; N, neutral fraction; B, basic fraction.
[b] MS, compounds were identified by MS spectra; aroma, compounds were identified by the aroma descriptors; RI, compounds were identified by a comparison to the pure standard.
[c] ND: not detected by GC–O.
[d] Tentatively identified.

sugar and anaerobic conditions from amino acids (Fan and Qian, 2006a; Luo et al., 2008). Since the raw materials (rice and wheat) are rich sources of amino acids, amino acids could be converted to higher alcohols by yeast via the Ehrlic metabolic pathway. A small amount of higher alcohols could also be made by yeast through reduction of corresponding aldehydes (Li, 2001).

Esters of fatty acids especially ethyl esters in both rice wines had medium OSME values. The aroma intensities of ethyl acetate, ethyl propanoate, ethyl butanoate, and ethyl hexanoate were strong and contributed pineapple, floral, and fruity notes to Chinese rice wines. The aroma intensity of diethyl butanedioate in JF12 rice wine, which had fruity and sweet aromas, was much higher than in GY30. 3-Methylbutyl acetate, which gave a fruity aroma, was identified with weak aroma intensities in both JF12 and GY30 rice wine. The aroma intensities of other esters were weak that gave fruity, sweet, and banana aromas. Esters were mostly formed through esterification of alcohols with fatty acids during fermentation and aging process. Ester formation can be influenced by many factors such as fermentation temperature, oxygen availability, and fermentation strains (Fan et al., 2003; Belitz et al., 2004).

A total of eight fatty acids were identified in these two Chinese rice wines by GC–O and GC–MS. On the basis of OSME values, the potentially important fatty acids in both samples were butanoic acid and 3-methylbutanoic acid (OSME values > 4.33) and contributed rancid, acidic, and cheesy notes. Acetic, propanoic acid, and 2-methylpropanoic acid were found in both JF12 and GY30 rice wine, but they had lower OSME values. The aroma intensities of fatty acids in GY30 rice wine were stronger than in JF12 rice wine. Most of the fatty acids in Chinese rice wines were produced by yeast metabolism (Luo et al., 2008). And some would be from raw material and wheat qu.

There were 10 aroma-active aromatic compounds identified in both rice wines. The aroma intensity of 2-phenylethanol produced by *Saccharomyces cerevisiae* (Ledauphin et al., 2003) was strongest among these all aromatic compounds, and it contributed to honey and rose aromas. Some other aromatic compounds, including benzaldehyde, phenylacetaldehyde, and 1-phenyl-1-propanone (propiophenone, tentatively identified), had medium aromas (OSME values = 3–3.67). They contributed to fruity, berry, floral, rose, and pungent aromas, and could be important to the aroma of Chinese rice wine. Acetophenone, ethyl benzoate, ethyl 2-phenylacetate, Z-2-phenyl-2-butenal, and ethyl 3-phenylpropanoate had low aroma intensities (OSME values = 1–2.67). Acetophenone gave musty, almond, and glue aromas, and ethyl benzoate gave a fruity aroma, whereas ethyl 2-phenylacetate had fruity and sweet aromas. 2-Phenylethyl acetate was only identified in GY30 rice wine and gave rosy and floral aromas. Aromatic compounds were mainly formed through aromatic amino acids metabolism (Diaz et al., 2001).

A total of seven phenolic derivatives were detected in these two Chinese rice wines. Only three phenolic derivates were identified in GY30 rice wine. 4-Vinylguaiacol and phenol had OSME values >3 in both rice wines and contributed to strong clove, spicy, and smoky aromas. 4-Methylguaiacol, 4-methylphenol, 4-ethylphenol, and 2,6-dimethoxyphenol (tentatively identified) were only identified by GC–O in JF12 rice wine. These compounds contributed smoky, animal, and phenolic aromas. The OSME value of phenolic derivatives in these two samples was obviously different. Phenolic derivatives in Chinese rice wines would belong to the secondary plant constituents, mainly derived from lignin degradation (Mart Nez et al., 2001).

Furans were also found to be important to both Chinese rice wine aroma profiles. There were five furans identified in both Chinese rice wines. Furfural had the strongest aroma intensity (OSME value = 4.50) among all furans and contributed to almond and sweet aromas. 2-Furanmethanol had a medium aroma intensity (OSME value = 3.33–3.50). 2-Acetylfuran and ethyl 2-furoate were detected with weak aroma intensities in both samples. 5-Methylfufural was only detected in GY30 rice wine with very low aroma intensities. As a result of cooking process, furans were formed by thermal degradation and rearrangement of carbohydrate and protein in nonenzymatic browning reactions (Maillard reaction; Limacher et al., 2008).

Two aroma-active sulfur-containing compounds were identified in both Chinese rice wines. Dimethyl trisulfide and 3-(methylthio) propanol (methionol) were identified in both rice wines. The aroma intensities of these two compounds were more than 2.5. However, due to their very low aroma threshold, sulfur-containing compounds may be important for the Chinese rice wine aroma. Sulfur-containing compounds probably came from the degradation of sulfur-containing amino acids (Mestres et al., 2000).

Only 1-lactone (γ-nonalactone) was identified in both samples with strong aroma intensities (OSME value = 4.83). It mainly contributed to coconut and peach aromas, and could be mainly produced by bacteria (Muller et al., 1973). Four nitrogen-containing compounds were detected by GC–O in this study with very low aroma intensities. 2-Acetylpyrrole had herbal and medicine aromas, while others gave nutty and roasted aromas. Pyrazines could be less important, but they were important in Chinese liquors (Fan et al., 2007). Only one aldehyde and one unsaturated ketone were detected by GC–O and GC–MS in both JF12 and GY30 rice wine. Hexanal gave green grass and apple aromas. 1-Octen-3-one, tentatively identified, had mushroom and earthy aromas, identified in both rice wines.

HS–SPME and GC–MS were applied to analyze the volatile and semivolatile compounds in Chinese rice wines (Luo et al., 2008). The results of quantitative analysis showed that the three largest group were fatty acids, alcohols, and aromatic compounds, whereas the three highest concentrations of the aroma compounds quantified in two samples were 3-methylbutanol (122043.16–129286.17 µg/L), 2-phenylethanol (76160.94–133472.45 µg/L), and acetic acid (44722.55–265223.95 µg/L). The short-chain fatty alcohols were the main alcohols in these Chinese rice wines; especially, the concentrations of 3-methylbutanol were higher than 120000 µg/L in both samples. The concentrations of fatty acids in GY30 were much higher than in JF12 rice wine. Due to the different brewing technique, the concentration of acetic acid in GY30 was much higher than that in JF12 rice wine. According to the concentrations of these compounds in Chinese rice wines, the most abundant aromatic compound were 2-phenylethanol. Ethyl esters of fatty acids were the mainly esters in Chinese rice wines. Furfural and 3-(methylthio)propanol were the other two high concentration compounds in these two samples. Due to some compounds' low concentrations in Chinese rice wines, these compounds identified by GC–O were not detected.

Data in Table 9.5 indicate that the concentrations of 23 aroma compounds quantified in Chinese rice wines could be higher than their corresponding odor thresholds, among which, the OAVs of 15 and 19 aroma compounds quantified in JF12 and GY30, respectively, were more than 1.

Among all aroma compounds quantified in both rice wines, dimethyl trisulfide had the highest OAV, followed by ethyl octanoate, ethyl butanoate, phenylacetaldehyde, and ethyl hexanoate with OAVs from 10 to 50. Some aroma compounds with OAVs >10 were only detected in GY30 or JF12 rice wines. For example, 3-(methylthio)propanol (methionol), 2-phenylethanol, and γ-nonalactone were only detected in GY30, whereas ethyl

Table 9.5 Quantitative Data and OAVs of Aroma Compounds in Both JF12 and GY30 (n = 3)

AROMA COMPOUNDS	ODOR THRESHOLD (µg/L)	JF12			GY30			REFERENCES
		CONCN (µg/L)	RSD%	OAV	CONCN (µg/L)	RSD%	OAV	
ALCOHOLS								
1-propanol	306000	11312.40	1.97	<1	30015.02	4.65	<1	Peinado et al. 2004
2-methylpropanol	40000	42709.37	0.11	1.1	33123.41	1.87	<1	Guth 1997
1-butanol	150000	1237.36	2.42	<1	2758.82	0.71	<1	Tominaga et al. 1998
2-methylbutanol		19931.98	8.99	—	23242.20	3.18	—	
3-methylbutanol	30000	129286.17	0.42	4.3	122043.16	0.76	4.1	Guth 1997
1-pentanol		<q.l.[a]	—	—	<q.l.	—	—	
1-hexanol	8000	197.53	0.28	<1	524.69	0.10	<1	Guth 1997
1-octen-3-ol	40	19.28	0.39	<1	20.43	0.87	<1	Guerche et al. 2006
1-heptanol	3000	<q.l.	—	—	48.86	9.31	<1	Guadagni et al. 1963
ESTERS								
ethyl acetate	7500	31512.13	0.79	4.2	29302.84	0.83	3.9	Guth 1997
ethyl propanoate	1800	403.89	0.83	<1	304.94	0.42	<1	Peinado et al. 2004
ethyl 2-methylpropanoate	15	279.42	1.21	18.6	<q.l.	—	—	Ferreira et al. 2000
ethyl butanoate	20	636.49	0.59	31.8	444.57	0.56	22.2	Guth 1997
3-methylbutyl acetate	30	65.69	2.55	2.2	24.88	9.33	<1	Guth 1997
ethyl pentanoate	10	34.72	1.74	3.5	<q.l.	—	—	Maarse 1991
ethyl hexanoate	5	97.60	0.68	19.5	52.17	7.79	10.4	Guth 1997
ethyl octanoate	2	91.98	0.21	46.0	105.08	1.02	52.5	Guth 1997
diethyl butanedioate	200000	7113.25	1.46	<1	23715.23	3.13	<1	Tominaga, Murat et al. 1998

continued

Table 9.5 (continued)   Quantitative Data and OAVs of Aroma Compounds in Both JF12 and GY30 (n = 3)

AROMA COMPOUNDS	ODOR THRESHOLD (μg/L)	JF12 CONCN (μg/L)	RSD%	OAV	GY30 CONCN (μg/L)	RSD%	OAV	
FATTY ACIDS								
acetic acid	200000	44722.55	3.28	<1	265223.95	6.01	1.3	Guth 1997
2-methylpropanoic acid	200000	1082.06	0.75	<1	1033.96	0.99	<1	Guth 1997
butanoic acid	10000	<q.l.	—	—	2818.68	1.86	<1	Guth 1997
3-methylbutanoic acid	3000	2546.61	0.42	<1	5820.41	2.83	1.9	Guth 1997
pentanoic acid	3000	432.07	1.09	<1	<q.l.	—	—	Fazzalari 1978
hexanoic acid	3000	2257.73	1.01	<1	2040.75	1.88	<1	Guth 1997
heptanoic acid	3000	<q.l.	—	—	66.22	10.26	<1	Fazzalari 1978
octanoic acid	500	111.05	2.14	<1	144.38	5.28	<1	Ferreira, Lopez et al. 2000
AROMATIC COMPOUNDS								
benzaldehyde	990	648.15	0.47	<1	1799.75	1.92	1.8	Isogai, Utsunomiya et al. 2005
phenylacetaldehyde	1	29.01	2.45	29.0	39.17	4.56	39.2	Fazzalari 1978
acetophenone	65	59.77	0.52	<1	274.22	2.03	4.2	Buttery et al. 1988
ethyl benzoate	575	29.47	0.13	<1	96.17	2.99	<1	Guth 1997
ethyl 2-phenylacetate	100	57.72	3.29	<1	238.88	1.24	2.4	Isogai, Utsunomiya et al. 2005

2-phenylethyl acetate	250	30.38	0.32	<1	77.07	2.99	<1	Guth 1997
ethyl 3-phenylpropionate		<q.l.	—	—	45.69	2.65	—	
2-phenylethanol	10000	76160.94	4.08	7.6	133472.45	21.71	13.4	Guth 1997
PHENOLS								
guaiacol	10	<q.l.	—	—	62.80	3.71	6.3	Guth 1997
4-methylguaiacol		<q.l.	—	—	<q.l.	—	—	
phenol	30	<q.l.	—	—	49.99	5.56	1.7	Maarse 1991
4-methylphenol	68	<q.l.	—	—	<q.l.	—	—	Ferreira, Lopez et al. 2000
4-ethylphenol	440	51.68	5.31	<1	43.51	1.47	<1	Ferreira et al. 2002
4-vinylguaiacol	40	<q.l.	—	—	<q.l.	—	—	Guth 1997
FURANS								
furfural	14100	4156.65	2.31	<1	20532.92	1.70	1.5	Ferreira, Lopez et al. 2000
2-acetylfuran		<q.l.	—	—	<q.l.	—	—	
5-methyl-2-furfural	20000	67.29	0.53	<1	97.50	5.14	<1	Tominaga, Murat et al. 1998
ethyl 2-furoate	16000	<q.l.	—	—	21.81	1.70	<1	Ferreira, Ortin et al. 2002
2-furanmethanol	2000	<q.l.	—	—	<q.l.	—	—	Culleré et al. 2004
LACTONES								
γ-nonalactone	30	130.36	3.76	4.4	303.64	2.95	10.1	Ferreira, Lopez et al. 2000

continued

Table 9.5 (continued)   Quantitative Data and OAVs of Aroma Compounds in Both JF12 and GY30 (n = 3)

AROMA COMPOUNDS	ODOR THRESHOLD (µg/L)	JF12			GY30			
		CONCN (µg/L)	RSD%	OAV	CONCN (µg/L)	RSD%	OAV	
ALDEHYDES								
hexanal	5	21.16	2.38	4.2	19.14	4.43	3.8	Buttery, Turnbaugh et al. 1988
SULFUR-CONTAINING COMPOUNDS								
dimethyl trisulfide	0.2	86.13	0.35	430.7	83.92	0.06	419.6	Guth 1997
3-(methylthio)propanol	500	4233.06	2.09	8.5	31069.19	2.24	62.1	Guth 1997
NITROGEN-CONTAINING COMPOUNDS								
2,5-dimethylpyrazine		<q.l.	—	—	<q.l.	—	—	
2,6-dimethylpyrazine		<q.l.	—	—	<q.l.	—	—	
2-acetylpyrrole	170000	<q.l.	—	—	48.19	12.83	<1	Buttery, Turnbaugh et al. 1988

[a] Quantatitive limit.

Table 9.6 General Composition of Chinese Rice Wines Obtained after Fermentation by Yeast Strain under Different Conditions

PARAMETERS	30°C		18°C	
	WHEAT QU	ENZYMES	WHEAT QU	ENZYMES
pH	3.89 ± 0.02	4.06 ± 0.04	4.31 ± 0.03	4.27 ± 0.01
Ethanol (% vol)	18.91 ± 0.33	18.83 ± 0.25	19.53 ± 0.21	19.45 ± 0.17
Residual sugar (g/L)[a]	4.69 ± 0.09	5.13 ± 0.13	4.70 ± 0.21	4.43 ± 0.15
Total acid (g/L)[b]	4.81 ± 0.07	4.58 ± 0.10	4.21 ± 0.15	4.3 ± 0.09
Volatile acidity (g/L)	0.42 ± 0.03	0.39 ± 0.05	0.35 ± 0.02	0.30 ± 0.03
Glycerol (g/L)	5.32 ± 0.12	5.47 ± 0.08	4.85 ± 0.25	4.63 ± 0.14
α-amino nitrogen (g/L)	0.65 ± 0.05	0.42 ± 0.03	0.58 ± 0.03	0.35 ± 0.02

[a] As glucose.
[b] As lactic acid.

2-methylpropanoate was only in JF12. These results were similar to the research of aroma compounds in aged sake, a kind of brewed alcoholic beverage in Japan. Isogai and coworkers (Isogai et al., 2005) studied the changes of aroma compounds during sake aging. They showed that 3-(methylthio) propoanal (methional) and dimethyl trisulfide were present in aged sake at concentrations exceeding their odor thresholds, and the highest OAV was observed for dimethyl trisulfide. Although the formation of dimethyl trisulfide in Chinese rice wine (Isogai et al., 2005) had not been studied, it had been well studied in beer (Gijs et al., 2000). There were considered to be two formation pathways of dimethyl trisulfide: The reaction between methanesulfenic acid and hydrogen sulfide and the oxidation of methionol derived from the degradation of methional (Gijs et al., 2000, 2002).

There were 9 and 11 other aroma compounds with OAVs from 1 to 10 identified in the samples of JF12, and GY30 rice wines respectively. These compounds mainly were short-chain alcohols, ethyl esters of fatty acids, and aromatic compounds.

In summary, GC–O is a suitable method to fast screen potent aroma compounds in Chinese rice wines. GC–O and OAVs showed that dimethyl trisulfide, phenylacetaldehyde, ethyl octanoate, and ethyl butanoate were the most important aroma compounds in Chinese rice wine. Other components, such as γ-nonalactone, 2-phenylethanol, 3-methylbutanol, and some other aromatic compounds were also determined to be powerful odorants. These odorants are associated with fruity, flora, and sweet odor descriptions, which are closely related to the aromas of Chinese rice wine.

9.3.4 Changes in Volatile Compounds of Chinese Rice Wine Wheat Qu during Fermentation and Storage

Chinese rice wine is fermented from glutinous rice or rice with wheat qu made from wheat in the solid state. Wheat qu is incubated and fermented under spontaneous conditions, and it can be rich in various microorganisms and complex enzyme systems during their fermentation (Wang et al., 2004). Therefore, wheat qu plays a key role in Chinese rice wine production. It is not only a source of enzymes necessary for the breakdown of carbohydrates and proteins in grains, a supplier of varieties of microorganisms, but also a source of flavor and aroma substances in the Chinese rice wine.

In general, the quality classification of Chinese rice wines is based on Chinese rice wine wheat qu, which determines the organoleptic characteristics of the finished Chinese rice wine (Wang, 2004). During Chinese rice wine making using wheat qu, various enzymes, varieties of microorganisms, and different substances of Chinese rice wine wheat qu are present in the mash of Chinese rice wine. So ingredient characteristics of wheat qu directly influence the quality classification of Chinese rice wines. Research has generally focused on enzymes and microorganisms of wheat qu (Xie et al., 2007; Shou et al., 2008), and the flavor of wheat qu is evaluated only by its simple sensory analysis. However, the volatile compounds of wheat qu are the most important factors influencing the flavor of Chinese rice wine (Shou et al., 2007). Therefore, it is necessary for analysis of the flavor of wheat qu to study volatile components instead of sensory evaluation.

Chinese rice wine wheat qu is made of wheat through spontaneous fermentation under solid-state fermentation. This fermentation is a complex process, involving a concerted series of microbiological, biochemical, and chemical reactions. Primary degradation of wheat constituents by glycolysis, lipolysis, and proteolysis leads to the formation of a wide range of precursors of flavor compounds (Shou et al., 2007). It is well known that these changes are widely found in cheese and curd-fermented meat products. During cheese ripening and curd-fermented meat products there are many reactions in which carbohydrate fermentation, lipid breakdown, and proteolysis are the main pathways involved. These reactions are then followed and/or overlapped by a series of secondary catabolic reactions, for example, transformation of the free amino acids and fatty acids into important volatile compounds such as methyl ketones, alcohols, esters, aldehydes, lactones, and sulfur compounds, which are responsible for the flavor characteristics of a variety of cheese and fermented-meat products (Ordonez et al., 1999; Marilley and Casey, 2004). Therefore, it can also be suggested that Chinese rice wine wheat qu could accumulate a diversity of volatile compounds during fermentation and storage. Chinese rice wine wheat qu is usually fermented for 15 days in one room and then removed to another room for storage for at least 2 months. During the fermentation and storage period, changes of enzymes, microorganisms, and substances occur, and these changes depend on many factors such as temperature, humidity, and time of fermentation and storage (Wang, 2004). Because of differences in time of fermentation and storage, enzymes and microorganisms of wheat qu are quite different. Thus, substances, especially, volatile compounds of wheat qu are changed. In order to understand the optimum time of using wheat qu for Chinese rice wine brewing on the basis of the contents of volatile compounds, the evolution of the volatile compounds of wheat qu during their fermentation and storage needs analyzing. However, no study on the volatile profile of wheat qu during fermentation and storage has been reported up to now.

Wheat qu is a solid matrix, and the concentrations of its volatile compounds are present in trace amounts. Therefore, it is necessary to have a sample preparation technique that can be used in the analysis of volatiles of wheat qu. SPME is a solvent-free extraction technique and a rapid sample preparation. This technique can extract a wide range of aroma compounds and has been widely used for the analysis of volatile and semivolatile compounds in solid food samples such as mushrooms (Guedes De Pinho et al., 2008), beef (Moon and Li-Chan, 2004), and breakfast cereal (Klensporf and Jeleń, 2008), as well as in other food matrices such as olive oil (Koprivnjak et al., 2009) and alcoholic beverages (Wang et al., 2004; Fan and Qian, 2005; Xu et al., 2007). SPME has also been shown to be a very suitable technique for the analysis of volatile and semivolatile compounds in Chinese rice wines (Luo et al., 2008). So in this study SPME is

Table 9.7 Volatile Compounds Concentration of Chinese Rice Wines Obtained from Different Conditions

COMPOUNDS	30°C		18°C	
	WHEAT QU ± SD	ENZYMES ± SD	WHEAT QU ± SD	ENZYMES ± SD
2-Methylpropanol (mg/L)	126.37 ± 5.95	85.54 ± 4.93	114.70 ± 7.29	81.19 ± 4.55
1-Butanol (mg/L)	8.00 ± 0.61	7.84 ± 0.72	6.12 ± 0.58	6.57 ± 0.63
2-Methylbutanol (mg/L)	37.85 ± 1.78	26.31 ± 2.39	38.62 ± 3.31	23.90 ± 2.47
3-Methylbutanol (mg/L)	240.90 ± 15.54	177.42 ± 11.36	238.40 ± 17.49	184.82 ± 8.87
1-Pentanol (µg/L)	184.73 ± 19.12	133.48 ± 15.48	158.57 ± 14.78	135.49 ± 15.02
1-Hexanol (µg/L)	631.92 ± 60.01	509.18 ± 50.59	763.86 ± 65.71	538.68 ± 36.72
1-Hentanol (µg/L)	72.76 ± 6.85	53.13 ± 4.42	69.15 ± 5.79	60.35 ± 5.61
2-Phenylethyl alcohol (mg/L)	106.15 ± 5.88	85.35 ± 5.17	103.90 ± 7.54	79.61 ± 6.10
Ethyl acetate (mg/L)	23.15 ± 1.30	17.51 ± 1.01	35.71 ± 1.27	28.17 ± 1.15
2-Methylpropyl acetate (µg/L)	79.85 ± 7.11	62.35 ± 6.21	98.17 ± 9.70	86.39 ± 7.59
3-Methylbutyl acetate (µg/L)	84.17 ± 5.39	73.29 ± 5.05	95.95 ± 7.52	99.24 ± 4.57
2-Phenylethyl acetate (µg/L)	16.31 ± 1.03	16.49 ± 1.54	10.53 ± 0.96	11.39 ± 0.78
Ethyl propanoate (µg/L)	85.67 ± 7.39	72.60 ± 6.31	97.28 ± 7.57	87.85 ± 7.14
Ethyl 2-methylpropanoate (µg/L)	63.24 ± 6.62	40.11 ± 3.11	91.07 ± 5.83	77.90 ± 9.34

Ethyl butanoate (µg/L)	171.09 ± 11.57	109.23 ± 9.98	327.10 ± 20.18	296.73 ± 21.36
Ethyl pentanoate (µg/L)	55.34 ± 5.34	48.28 ± 3.35	70.45 ± 7.18	57.47 ± 5.84
Ethyl hexanoate (µg/L)	125.45 ± 11.20	102.89 ± 7.26	260.55 ± 16.47	214.03 ± 19.01
Ethyl octanoate (µg/L)	104.45 ± 10.06	72.91 ± 5.74	254.66 ± 19.25	198.57 ± 16.77
Ethyl decanoate (µg/L)	786.81 ± 51.34	613.17 ± 43.31	1375.88 ± 108.61	1080.77 ± 75.05
Eiethyl succinate (µg/L)	711.24 ± 58.72	692.20 ± 39.69	785.12 ± 58.41	760.75 ± 68.26
Ethyl benzoate (µg/L)	43.93 ± 4.31	27.22 ± 2.02	74.76 ± 7.17	63.96 ± 3.75
Ethyl 2-phenylacetate (µg/L)	48.05 ± 4.55	24.92 ± 1.86	49.74 ± 3.53	33.27 ± 2.89
Acetic acid (mg/L)	507.19 ± 43.59	425.76 ± 31.64	366.20 ± 24.41	317.44 ± 17.84
2-Methylpropanoic acid (mg/L)	1853.35 ± 149.17	1531.69 ± 123.20	2435.39 ± 185.89	2619.20 ± 289.15
Butanoic acid (µg/L)	949.56 ± 101.29	891.29 ± 71.05	732.29 ± 70.99	703.78 ± 72.60
3-Methylbutanoic acid (µg/L)	962.97 ± 56.44	811.27 ± 55.98	1486.57 ± 101.49	1324.77 ± 104.48
Hexanoic acid (µg/L)	1067.03 ± 101.54	827.86 ± 86.39	1141.26 ± 101.12	1280.44 ± 115.27
Octanoic acid (µg/L)	390.06 ± 17.54	417.31 ± 36.28	770.80 ± 57.12	915.65 ± 75.84
3-Methylthiopropanol (mg/L)	5.90 ± 0.38	4.59 ± 0.30	4.90 ± 0.23	3.79 ± 0.24
Guaiacol (µg/L)	226.33 ± 24.99	59.22 ± 16.72	193.28 ± 23.74	40.15 ± 14.67
4-Vinylguaiacol (mg/L)	2.79 ± 0.26	0.55 ± 0.09	2.09 ± 0.24	0.49 ± 0.09

applied to analyze volatile compounds of Chinese rice wine wheat qu during fermentation and storage.

The primary aim of this research effort was to determine the volatiles in Chinese rice wine wheat qu during their fermentation and storage by headspace solid phase microextraction, and then analyze the concentration changes in volatile compositions of wheat qu during their fermentation and storage within 30 days. It is expected that the findings of this research will help to better understand the optimum time of using wheat qu for Chinese rice wine brewing on the basis of the contents of volatile compounds.

A total of 58 volatile compounds were identified and measured in wheat qu samples during fermentation and storage within 30 days in this study, including 11 alcohols, 6 acids, 3 esters, 7 ketones, 5 aldehydes, 9 aromatics, 1 lactone, 4 phenols, 2 sulfides, 3 furans, 4 nitrogen-containing compounds, and 3 terpenes. Most of these compounds had been previously detected in Chinese rice wine (Luo et al., 2008). The evolution of each group of volatile compounds during fermentation and storage of wheat qu is represented in Figure 9.7. The different groups of compounds generally behaved predictably, and the volatile compounds except menthol have been increased from the first day of fermentation.

Alcohols identified in Chinese rice wine wheat qu during fermentation and storage were 2-methylpropanol, 1-penten-3-ol, 3-methylbutanol, 1-pentanol, 2-heptanol, 1-hexanol, 1-octen-3-ol, 1-heptanol, 2-ethyl-1-hexanol, 1-octanol, and 1-nonanol. There was a clear trend for these compounds in Chinese rice wine wheat qu during fermentation and storage. The levels of alcohols rose from the first day to the fourth day of fermentation, decreased gradually, and then tended to stabilize. The concentrations of 3-methylbutanol, 1-pentanol, and 1-hexanol, increased sharply from the first day to the fourth day of fermentation, decreased from the sixth day to the fifteenth days of fermentation, and then tended to stabilize gradually. While 2-methylpropanol, 1-penten-3-ol, 2-heptanol, 1-octen-3-ol, 1-heptanol, 2-ethyl-1-hexanol, 1-octanol, and 1-nonanol, were present in very small amounts in Chinese rice wine wheat qu during fermentation and storage, and their concentrations gradually rose, decreased, and then tended to stabilize. Alcohols have been described with fruity, floral,

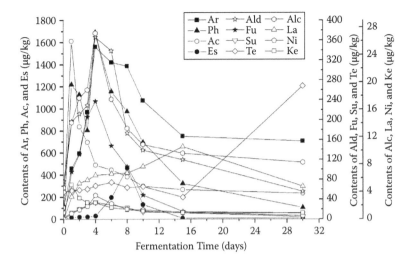

Figure 9.7 Changes of the volatile compounds in the Chinese rice wheat qu during fermentation and storage within 30 days. Key: Ar, aromatic compounds; Ph, phenols; Ac, acids; Es, esters; Ald, aldehydes; Fu, furans; Su, sulfides; Te, terpenes; Alc, alcohols; La, lactones; Ni, nitrogen-containing compounds; Ke, ketones.

and alcohol-like aromas. It is well known that many metabolic pathways are involved in the biosynthesis of the alcohols in cheese. These metabolic pathways include amino acid metabolism, lactose metabolism, and methyl ketone reduction, as well as degradation of linoleic and linolenic acids (Curioni and Bosset, 2002). Chinese rice wine wheat qu was made of wheat through spontaneous fermentation under solid-state fermentation, and it was rich in various microorganisms and complex enzyme systems. Cramer et al. (2005) reported that the lipid content was about 1.70%, and the protein content ranged from 15.5% to 16.8% in wheat varieties. So alcohols in wheat qu could be produced in lipid oxidation catalyzed by lipoxygenase originating from microbiota present in the medium or coming from a microbial catabolism amino acid, as well as methyl ketone reduction.

The evolution of aromatic compounds was similar to those of alcohols. The sum of aromatic compounds showed a sharp increase between the first day and the fourth day of fermentation, and then decreased gradually. As an exception, the concentrations of 1,2,3-trimethoxybenzene and 2-phenylethanol of this group increased from the first day to the eighth day of fermentation, and then decreased gradually from the ninth day of fermentation. Among all the aromatic compounds, 1,2,3-trimethoxybenzene and 4-ethenyl-1,2-methoxybenzene were identified only in wheat qu but not in wheat. Benzaldehyde, phenylacetaldehyde, benzyl alcohol, ethyl benzoate, acetophenone, 1,2-dimethoxybenzene, and 2-phenylethanol, which were detected in wheat qu throughout all fermentation and storage stages studied, were also detected in Chinese rice wine (Luo et al., 2008). 1,2,3-Trimethoxybenzene, which is associated with smoke and musty odors, presented a higher concentration on the fourth day of fermentation. 2-Phenylethanol contributed to rosy and honey aromas, which is important in wheat qu; it also had the highest concentration in Chinese rice wine (Luo et al., 2008).

The volatile phenolic compounds identified in the Chinese rice wine wheat qu during fermentation and storage, including phenol, guaiacol, 4-ethylguaiacol, and 4-vinylguaiacol, have been formed from lignin degradation of raw materials wheat (Mart Nez et al., 2001). Different studies have been performed to determine the capacity of certain microorganisms to produce volatile phenols (Couto et al., 2006; Vanbeneden et al., 2008). These volatile phenolic compounds, except guaiacol, have been reported in Chinese rice wine (Luo et al., 2008). The evolution of these compounds was somewhat more erratic than others, but the tendency was for an increase until the fourth day of fermentation, from which time the concentration also started to fall. The levels of these volatile phenolic compounds increased sharply on the first day of fermentation, and then deceased slowly from the second day to the third day of fermentation, but tended to increase from the third day to the fourth day of fermentation, subsequently tending to fall. The concentration of 4-vinylguaiacol increased markedly from the first day to the fourth day of fermentation, therefore, 4-vinylguaiacol presented a maximum concentration on the fourth day of fermentation in this study sample, but its concentration gradually decreased, and then remained stable. While the concentration of 4-ethylguaicol achieved a peak after 2 days of fermentation, it decreased continuously until zero. It is for this reason that the change of phenols was erratic.

Aldehydes and nitrogen-containing compounds were present in small amounts, and sulfides and furans were found in trace amounts in the studied samples. So small fluctuations of these groups of volatile compounds were observed in Chinese rice wine wheat qu during fermentation and storage. The tendency was for a slight increase from the first day to the fourth day of fermentation, and then for a slight decrease. As an exception, the concentrations of (E,E)-2,4-decadienal and (E)-2-heptenal of aldehydes achieved a peak on the second day and the third day of fermentation, respectively, and then decreased continuously. Most aldehydes such as hexanal and (E,E)-2,4-decadienal are the expected oxidation products of linoleic acid (Valim et al., 2003). A total of four nitrogen-containing compounds were detected only in Chinese rice wine wheat qu but not in wheat. Wheat qu was generally exposed to 45°C–55°C for at least 2 days, so there was a high fermentation temperature in wheat qu. Nitrogen-containing compounds could be produced through both nonenzymatic pathways such as the Maillard reaction and enzymatic pathways produced by *Bacillus subtilis* (Fan et al., 2007) in wheat qu during fermentation, and a high temperature would benefit the Maillard reaction. Similar to nitrogen-containing compounds, a high temperature also would facilitate furan formation through thermal degradation of carbohydrates followed by cyclation in Maillard-type systems (Ben Tez et al., 2003). Two sulfides appear in wheat qu at low concentrations. Sulfur-containing compounds often have very low sensory thresholds, and these came from the degradation of sulfur-containing amino acids (Fan and Qian, 2005).

Acids and ketones peaked after one day of fermentation, decreased gradually from the second day of fermentation, and then stabilized. The concentrations of acids increased sharply on the first day of fermentation, and then decreased gradually. Acids were found at a high concentration in wheat qu at all stages. Acids can be produced either by yeasts during alcoholic fermentation (Ugliano and Moio, 2005) or by lactic acid bacteria (LAB) genera during fermentation of cheese (Gobetti et al., 1997). Many LAB isolated from cheese are known to contain lipases that can hydrolyze lipids, resulting in the formation of volatile fatty acids. Fatty acids are not only aroma compounds by themselves, but also serve as precursors of alcohols, esters, lactones, and methyl ketones. Most ketones are produced by lipid oxidation and the β-oxidation of free fatty acids by microbial fermentation in fermented-meat products, and ketones may be reduced to alcohols

(Ordonez et al., 1999). Thus, most acids and ketones in wheat qu were speculated to be derived from lipid oxidation by microbial fermentation. These two groups of volatile compounds are intermediate compounds which may be transformed into the other volatile compounds. Therefore, the levels of these volatile compounds in wheat qu increased sharply on day 1 of fermentation and then decreased.

Only 3 esters were detected in wheat qu. Esters tended to increase slowly from the first day to a peak on the eighth day of fermentation, decreased gradually, and then stabilized. Esters, which impart fruity and floral odors to Chinese rice wine wheat qu, could be formed either via the esterification of alcohols with fatty acids or through the synthesis in the microorganisms cells by alcohol acetyltransferase using acetyl-CoA and higher alcohols as substrates during the fermentation. Generally speaking, the latter played a more important role in formation of esters during the fermentation (Fan and Qian, 2006c). Therefore, ester formation can be influenced by the concentration of the two substrates (acetyl-CoA and higher alcohols) and the activity of alcohol acetyl transferase. All factors that influence substrate concentrations or enzyme activity will affect ester production. Many factors such as fermentation temperature, fatty acid, and nitrogen and oxygen levels can affect ester production. In the fermentation processes of the Chinese rice wine wheat qu, suitable conditions such as temperature and carbon dioxide and oxygen levels were maintained, thus favoring the formation of esters.

Lactones tended to increase slowly from the first day to a peak on the fifteenth day of fermentation, and then decreased. In cheese, lactones are generated by hydrolysis of hydroxy-fatty acid triglycerides, followed by lactonization (Curioni and Bosset, 2002). Only one lactone, namely γ-nonalactone, was detected in wheat qu in this study. This compound, which probably came from bacteria fermentation (Pinto et al., 2006), may be formed by other volatiles such as acids, that is, it is a secondary metabolite. So the concentration of γ-nonalactone increased slowly up to 15 days of fermentation. γ-Nonalactone has also been presented in Chinese rice wine (Luo et al., 2008).

The evolution of terpenes was somewhat erratic because the concentrations of trans-1,10-dimethyl-trans-9-decalinol tended to increase slowly within the 15 days of fermentation, and then increased fast during the storage period studied. Terpenes are usually exited in free and glycosylated form in grapes. During alcoholic fermentation, the content of free terpenes often increases due to the β-glucosidase activity of yeasts (Izquierdo Canas et al., 2008). For this reason, the increase of some terpenes observed in wheat qu may be attributed to the β-glucosidase activity originating from microbiota present in the medium. Trans-1,10-dimethyl-trans-9-decalinol is produced by a lot of microorganisms, including most *Streptomyces* and several species of myxobacteria, cyanobacteria, and fungi (Jiang et al., 2007). Terpenes, which were present in small amounts in wheat qu, were not detected in Chinese rice wine (Luo et al., 2008).

As analyzed above, most groups of volatile compounds including alcohols, aldehydes, aromatic compounds, phenols, sulfides, furans, and nitrogen-containing compounds are primary metabolites. So the levels of these groups of volatile compounds peaked on the fourth day of fermentation, and then tended to decrease. Acids and ketones are not only primary metabolites, but also the precursors of other aromatic compounds. So the levels of these volatile compounds in wheat qu increased sharply on the first day of fermentation and then decreased. A minority group of volatile compounds such as esters and lactones, which were the secondary metabolites, could be produced via other volatiles such as alcohols, fatty acids, or other substances. Thus, these volatile compounds attained a peak on 8 days and 15 days, respectively, afterward tending to decrease. Nonetheless, the levels of these volatile concentrations were low, and they may not have an important impact in Chinese rice wine. Terpenes increased slowly within the 15 days of fermentation, and then increased fast during storage. These volatiles were also present in small amounts in wheat qu but not in Chinese rice wine. According to the analysis above, during fermentation there was a marked change in all these groups of volatiles except terpenes, however, no significant change was shown during storage. Hence, it can be speculated that wheat qu incubated for four days should be used for brewing of Chinese rice wine in terms of the general evolution of volatile compounds during fermentation and storage. The results can provide a theoretic reference for shorting the wheat qu-making periods of Chinese rice wine production on the basis of the contents of volatile compounds.

9.3.5 *Effect of Temperature on the Fermentation Processes and Volatile Flavor Characteristics of Chinese Rice Wine*

Fermentation temperature is one of the most important parameters that can drastically affect the alcohol fermentation process in Chinese rice wine brewing. Two-step temperature control strategy is carried out in Chinese rice wine fermentation; at the beginning the fermentation temperature always holds at 30°C–35°C for about 3–4 days, as the main fermentation, over 80% of sugar has been consumed at this stage. Then the temperature of fermentation mash was decreased below 18°C for about 15 days or longer, as postfermentation, in which reducing sugar consumption and flavor maturation are occurring (Zhou, 1996).

Because some of the raw materials used in Chinese rice wine fermentation are unsterile and have a high ethanol level in fermentation mash, higher fermentation temperature always results in fermentation spoilage. And there is an increasing interest in conducting beverage fermentation at a lower temperature, for developing better taste and aroma. In fact most white wine and Japanese *sake* fermentations are carried out at or below 18°C. Fermentation temperature can dramatically affect the growth, fermentation, and metabolism of *saccharomyces cerevisiae*. Research carried out about wine fermentation show that fermentation temperature can regulate the expression of yeast genes involved in aroma compounds metabolism, and then regulate the concentration of volatile flavor compounds in wines (Beltran et al., 2006; Molina et al., 2007).

The overall fermentation kinetic profiles, that is, yeast cell growth, the variation in CO_2 production, and the concentration of residual sugar versus fermentation process, were shown in Figures 9.8 and 9.9. In Chinese rice wine brewing, high yeast pitching rate (about 2×10^7 CFU/mL) was always used to ensure the inoculated yeast strain was the dominant strain in the fermentation. So the yeast growth curve showed that there was almost no lag phase at 30°C, and the yeast cell reached maximal population (about 3.2×10^8 CFU/ml) within 20 hours. In contrast, yeast cells grew slower at 18°C. There was a clear growth delayed (about 20 hours) and yeast cell growth to maximal after 72 hours. Except for fermentation temperature, saccharifying agents also affected the yeast cell growth rate. The use of wheat qu as saccharifying agent

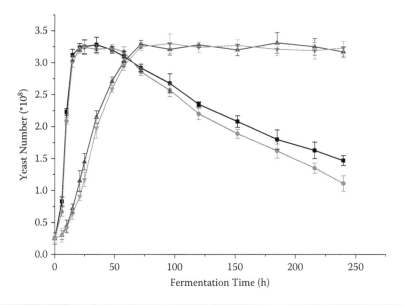

Figure 9.8 Growth kinetics of the yeast strain used: viable plate counting of cells inoculated in the YPD medium and incubated statically at 30°C, under different conditions. ■: 16% wheat qu at 30°C, ●: enzyme at 30°C, ▲: 16% wheat qu at 18°C, ▼: Enzyme at 18°C.

appears in a promotion role in yeast growth, and the promotion effect appears more obviously at lower temperature fermentation.

The raw material used for ethanol fermentation was rice starch, so there was little sugar at the beginning of fermentation. As the rice starch was degraded by enzymes, the residual sugar gradually accumulated in fermentation mash. The results showed that at the beginning of fermentation, the sugar liberation rate was faster than the yeast assimilation rate (Figures 9.8–9.9). The concentration of residual sugar reached a maximum, 110, 120g/L at 30°C and 18°C, and then gradually decreased when yeast cell population reached to maximal. Temperature is one of the most important parameters that can significantly affect the alcohol fermentation process. The fermentation kinetics showed that fermentation rates significantly decreased from 30°C to 18°C. After 4 days of fermentation at 30°C, the concentration of residual sugar in fermentation mash decreased lower than 10 g/L, and the ethanol content was about 14% v/v (date not shown here); then the temperature of fermentation mash was reduced to 15°C to going on postfermentation to complete fermentation of the residual sugar in mash. The fermentation rate was slower at low temperature

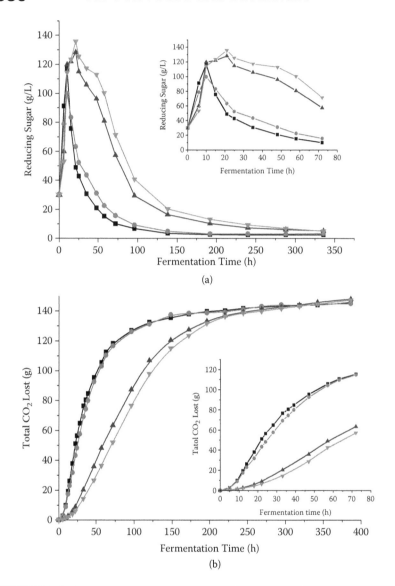

Figure 9.9 Fermentation kinetics of yeast strain under different conditions. A: Reducing sugar consumption throughout the alcohol fermentation; B: Total CO_2 production during alcohol fermentation. ■: 16% wheat qu at 30°C, ●: Enzymes at 30°C, ▲: 16% wheat qu at 18°C, ▼: Enzymes at 18°C.

fermentations. After 20 days of fermentation, all fermentations were completed with a concentration of residual sugar below 5 g/L.

As the fermentation continued, a high amount of yeast cell death was observed at 30°C fermentation. We also observed that when commercial enzymes were used as a saccharifying agent, the rate of dead yeast cells was higher than wheat qu used as a saccharifying agent. However, at low temperature fermentation there was no clear yeast cell decline phase at the first 10 days of fermentation.

At the end of the alcohol fermentation, Chinese rice wine samples were taken to evaluate the final physicochemical characteristics (Table 9.6). The results were within the normal range of values expected. The amount of final ethanol in all the fermentations ranged from 18.8% to 19.5% (v/v), and the ethanol content was a little higher in Chinese rice wine samples obtained at a lower fermentation temperature. pH was lower in Chinese rice wine samples fermented at 30°C; this agreed with the higher total acid detected at 30°C. A higher level of glycerol was produced at a higher fermentation temperature.

Aroma is a most important distinguishing characteristic of Chinese rice wine. To assess how Chinese rice wine aroma composition could be effected by the fermentation condition evaluated in this study, all Chinese rice wine samples were analyzed by HS-SPME-GC-MS after fermentation.

Thirty-three volatile flavor compounds including 8 alcohols, 14 esters, 8 fatty acids, 1 aldehyde, 1 sulfur-containing compounds, and 2 volatile phenols were identified and quantified. The results obtained herein were in line with those obtained in previous studies conducted with Chinese rice wine (Luo et al., 2009; Chen and Xu, 2010). The aromatic profiles of the Chinese rice wines evidenced significant differences for those obtained by different fermentation temperature conditions, as well as those obtained from different saccharifying agents.

The total concentration of higher alcohols in the Chinese rice wine samples represented important variables varied from 389.22 mg/L to 520.15 mg/L in different treatments. 2-Methylpropanol (fusel—spiritous), 3-methylbutanol (harsh, nail polish), and 2-phenylethyl alcohol (floral—rose) are the major alcohols in Chinese rice wine, which all showed a marked increase in those fermentations when wheat qu is used, compared to enzymes used. The concentration of 2-phenylethyl alcohol was about 1.3-fold higher in Chinese rice wine samples obtained with wheat qu. 1-Pentanol, 1-hexanol, and 1-hentanol showed low concentrations in Chinese rice wines, and showed little change in all fermentations. Compared with saccharifying agents, fermentation temperature showed lesser influence on the concentrations of higher alcohols in Chinese rice wines.

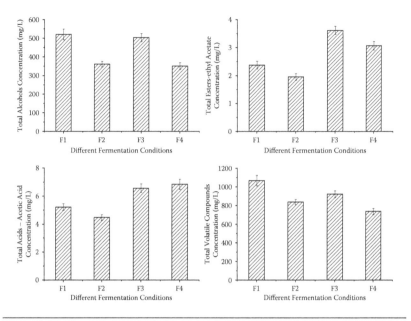

Figure 9.10 Concentration of volatile compounds in the Chinese rice wine samples obtained from different fermentation conditions. F1: 16% wheat qu at 30°C; F2: enzymes at 30°C; F3: 16% wheat qu at 18°C; F4: enzymes at 18°C.

Esters were one of the largest and most important groups of volatile flavor compounds in Chinese rice wines due to their low threshold concentrations and desirable fruity aromas. Fourteen esters were quantified in our Chinese rice wine samples. The results showed that the formations of esters were greatly influenced by the fermentation temperature and the use of wheat qu (Table 9.7, Figure 9.10). The highest concentration of total esters was formation in Chinese rice wine samples fermented with wheat qu as a saccharifying agent at 18°C. The concentration of total esters (except ethyl acetate) was 1.5-fold higher in samples obtained at 18°C than 30°C.

Ethyl acetate (fruity–nail polish) was found to be the major ester in Chinese rice wine, ranging from 20.51–33.71 mg/L, contributing over 90% of total esters. Lower fermentation temperature strongly promoted the formation of ethyl acetate, more than 1.5–fold higher at 18°C than at 30°C. Wheat qu also played a prominent role in the formation of ethyl acetate, so the highest concentration of ethyl acetate was found in Chinese rice wine samples fermented at 18°C with wheat Qu as a saccharifying agent (35.71 mg/L). While 2-methylpropyl acetate (banana–fruity) and 3-methylbutyl acetate (banana–pear) only correlated closed with the fermentation temperature other than saccharifying agents, and they all showed higher levels at a lower fermentation temperature.

Ten ethyl esters were quantified in this study. In general, all the ethyl esters showed a marked increase at lower fermentation temperature. Among them, ethyl butanoate (floral, fruity), hexanoate (apple), and octanoate (sweet) were the most important ethyl esters in Chinese rice wine for their fruity odor and low threshold (Luo et al., 2009; Chen and Xu, 2010), which increased 1.5-2 fold at a lower temperature. The use of wheat qu in fermentation also showed a contribution to the formation of ethyl esters.

Although the total acid contents excluding acetic acid, was about 1.3–1.5 fold higher at 18°C than 30°C, the production of acetic acid was much higher at 30°C than 18°C. The final concentration of acetic acid was 507.19 mg/L in the Chinese rice wine sample fermented at 30°C with wheat qu used, and was nearly twofold higher than the sample fermented at 18°C with enzymes as a saccharifying agent. The concentrations of acetic acid in our study were all within the range of common value in Chinese rice wines. The concentrations of other organic acids were showed higher values in a lower fermentation temperature except for butanoic acid, which showed a higher concentration at a higher fermentation temperature. There was no statistically significant difference observed in the concentration of organic acids between different saccharifying agents.

Two volatile phenols were quantified in our study, which is 4-vinylguaiacol (clove-like) and guaiacol (smoky, spicy). 4-Vinylguaiacol was the major volatile phenol found in Chinese rice wines, with concentrations ranging from 0.49 mg/L to 2.79 mg/L. It is interesting to find that the concentrations of volatile phenols were markedly higher in Chinese rice wines when wheat qu was used. The highest concentrations of volatile phenols were found in a Chinese rice wine sample obtained with wheat qu used at 30°C, which was 5.5-fold higher than a Chinese rice wine sample obtained with enzymes used at 18°C.

3-Methylthiopropanol (methionol) was found in a high concentration in Chinese rice wines, varying from 3.79 to 5.90 mg/L. The formation of 3-methylthiopropanol was also regulated by the saccharifying agents used, the use of wheat qu contributing to the formation of 3-methylthiopropanol.

References

Aidoo, K.E., Rob Nout, M., and Sarkar, P.K. (2006). Occurrence and function of yeasts in Asian indigenous fermented foods. *FEMS Yeast Research.* **6**, 30–39.

Beh, A.L., Fleet, G.H., Prakitchaiwattana, C., and Heard, G.M. (2006). Evaluation of molecular methods for the analysis of yeasts in foods and beverages. *Advances in Food Mycology.* **571**, 69–106.

Belitz, H.D., Grosch, W., and Schieberle, P. (2004). *Food Chemistry.* Heidelberg, Germany: Springer.

Beltran, G., Novo, M., Leberre, V., Sokol, S., Labourdette, D., Guillamon, J.M., Mas, A., Francois, J., and Rozes, N. (2006). Integration of transcriptomic and metabolic analyses for understanding the global responses of low-temperature winemaking fermentations. *Fems Yeast Research.* **6**, 1167–1183.

Ben Tez, P., Castro, R., and Barroso, C.G. (2003). Changes in the polyphenolic and volatile contents of "fino" sherry wine exposed to ultraviolet and visible radiation during storage. *Journal of Agricultural and Food Chemistry.* **51**, 6482–6487.

Chang, H.W., Kim, K.H., Nam, Y.D., Roh, S.W., Kim, M.S., Jeon, C.O., Oh, H.M., and Bae, J.W. (2008). Analysis of yeast and archaeal population dynamics in kimchi using denaturing gradient gel electrophoresis. *International Journal of Food Microbiology.* **126**, 159–166.

Chen, S. and Xu, Y. (2010). He influence of yeast strains on the volatile flavour compounds of Chinese rice wine. *Journal of the Institute of Brewing.* **116**, 190–196.

Chen, X. and Ji, K. (2006). General introduction to the individualities of moutai. *Liquor-Making Science and Technology.* **2**, 79–84.

Couto, J.A., Campos, F.M., Figueiredo, A.R., and Hogg, T.A. (2006). Ability of lactic acid bacteria to produce volatile phenols. *American Journal Enology and Viticulture.* **57**, 166–171.

Cramer, A.C.J., Mattinson, D.S., Fellman, J.K., and Baik, B.K. (2005). Analysis of volatile compounds from various types of barley cultivars. *Journal of Agricultural and Food Chemistry.* (53), 7526–7531.

Cui, Z.Y., Ren, Y.L., and Yan, F.X. (2002). Method of adding mold in fen-flewour daqu. *Liquor Making.* (29), 79–79.

Curioni, P.M.G. and Bosset, J.O. (2002). Key odorants in various cheese types as determined by gas chromatography-olfactometry. *International Dairy Journal.* **12**, 959–984.

Diaz, E., Ferrandez, A., Prieto, M.A., and Garcia, J.L. (2001). Biodegradation of aromatic compounds by *Escherichia coli. Microbiology and Molecular Biology Reviews.* **65**, 523–569.

Endo, A. and Okada, S. (2005). Monitoring the lactic acid bacterial diversity during shochu fermentation by pcr-denaturing gradient gel electrophoresis. *Journal of Bioscience and Bioengineering.* **99**, 216–221.

Fan, W. and Chen, X. (2001). Study on increase of ratio of famous product in luzhou-flavor baijiu by sandwich fermentation with mud. *Niangjiu.* **28**, 71–73.

Fan, W. and Qian, M.C. (2005). Headspace solid phase microextraction (hs-spme) and gas chromatography-olfactometry dilution analysis of young and aged Chinese "yanghe daqu" liquors. *Journal of Agricultural and Food Chemistry.* **53**, 7931–7938.

Fan, W. and Qian, M.C. (2006a). Characterization of aroma compounds of Chinese "wuliangye" and "jiannanchun" liquors by aroma extraction dilution analysis. *Journal of Agricultural and Food Chemistry.* **54**, 2695–2704.

Fan, W. and Qian, M.C. (2006b). Identification of aroma compounds in chinese "yanghe daqu" liquor by normal phase chromatography fractionation followed by gas chromatography/olfactometry. *Flavour and Fragrance Journal.* **21**, 333–342.

Fan, W. and Teng, K. (2001). Brewing technology of yanghe daqu liquor. *Niangjiu.* **28**, 36–37.

Fan, W. and Xu, Y. (2000). Research progress of enzyme in daqu. *Niangjiu.* **27**, 35–40.

Fan, W., Xu, Y. and Dao, Y. (2002). Study on hydrolyzing enzymes and its determination method from daqu for Chinese strong flavor liquor making. *Niangjiu.* **29**(25–31).

Fan, W., Xu, Y., Lu, H. and Dao, Y. (2003). Study on esterifying power and the rate of breaking-up ester from daqu of Chinese strong flavored liquor making. *Niangjiu.* **30**, 10–12.

Fan, W.L. and Qian, M.C. (2006c). Characterization of aroma compounds of Chinese "wuliangye" and "jiannanchun" liquors by aroma extract dilution analysis. *Journal of Agricultural and Food Chemistry.* **54**, 2695–2704.

Fan, W.L., Xu, Y. and Zhang, Y.H. (2007). Characterization of pyrazines in some Chinese liquors and their approximate concentrations. *Journal of Agricultural and Food Chemistry.* **55**, 9956–9962.

Fang, H., Cao, Y., Lu, J. and Xie, G. (2006). The molecular identification and classification of the main mold from wheat starter. *Liquor-Making Science and Technology.* **3**, 45–47.

Fiorentini, M., Sawitzki, M.C., Bertol, T.M. and Sant'anna, E.S. (2009). Viability of *Staphylococcus xylosus* isolated from artisanal sausages for application as starter cultures in meat products. *Brazilian Journal of Microbiology.* **40**, 129–133.

Fontana, C., Vignolo, G. and Cocconcelli, P.S. (2005). Pcr-dgge analysis for the identification of microbial populations from Argentinean dry fermented sausages. *Journal of Microbiological Methods.* **63**, 254–263.

Gich, F., Garcia-Gil, J. and Overmann, J. (2001). Previously unknown and phylogenetically diverse members of the green nonsulfur bacteria are indigenous to freshwater lakes. *Archives of Microbiology.* **177**, 1–10.

Gijs, L., Chevance, F., Jerkovic, V. and Collin, S. (2002). How low pH can intensify β-damascenone and dimethyl trisulfide production through beer aging. *Journal of Agricultural and Food Chemistry.* **50**, 5612–5616.

Gijs, L., Perpete, P., Timmermans, A. and Collin, S. (2000). 3-methylthiopropionaldehyde as precursor of dimethyl trisulfide in aged beers. *Journal of Agricultural and Food Chemistry.* **48**, 6196–6199.

Gobetti, M., Fox, P.F. and Stepaniak, L. (1997). Isolation and characterization of a tributyrin esterase from *Lactobacillus plantarum. Journal of Dairy Science.* **80**, 3009–3019.

Guedes De Pinho, P., Ribeiro, B., Goncalves, R.F., Baptista, P., Valent, O.P., Seabra, R.M. and Andrade, P.B. (2008). Correlation between the pattern volatiles and the overall aroma of wild edible mushrooms. *Journal of Agricultural and Food Chemistry.* **56**, 1704–1712.

Huang, Y., Xu, H. and Huang, P. (2006). Extreme liquor-making environment and extreme liquor-making microbes of moutai liquor. *Liquor-Making Science & Technology.* **12**, 47–50.

Isogai, A., Utsunomiy, H., Kanda, R. and Iwata, H. (2005). Changes in the aroma compounds of sake during aging. *Journal of Agricultural and Food Chemistry.* **53**, 4118–4123.

Izquierdo Canas, P.M., Garcia Romero, E., Gomez Alonso, S. and Palop Herreros, M.L.L. (2008). Changes in the aromatic composition of tempranillo wines during spontaneous malolactic fermentation. *Journal of Food Composition and Analysis.* **21**, 724–730.

Ji, K. and Guo, K. (2006). Investigation on microconstituents in moutai liquor. *Liquor-Making Science and Technology.* **10**, 98–100.

Ji, K., Guo, K., Zhu, S., Lu, X. and Xu, G. (2007). Analysis of microconstituents in liquor by full two-dimensional gas chromatography/time of flight mass spectrum. *Liquor-Making Science and Technology.* **3**, 100–102.

Jiang, J.Y., He, X.F. and Cane, D.E. (2007). Biosynthesis of the earthy odorant geosmin by a bifunctional streptomyces coelicolor enzyme. *Nature Chemical Biology.* **3**, 711–715.

Klensporf, D. and Jeleń, H.H. (2008). Influence of the addition of raspberry seed extract on changes in the volatile pattern of stored model breakfast cereal. *Journal of Agricultural and Food Chemistry.* **56**, 3268–3272.

Koprivnjak, O., Bubola, K.B., Majetić, V. and Škevin, D. (2009). Influence of free fatty acids, sterols and phospholipids on volatile compounds in olive oil headspace determined by solid phase microextraction-gas chromatography. *European Food Research and Technology.* **229**, 539–547.

Ledauphin, J., Guichard, H., Saint-Clair, J.F., Picoche, B. and Barillier, D. (2003). Chemical and sensorial aroma characterization of freshly distilled calvados. 2. Identification of volatile compounds and key odorants. *Journal of Agricultural and Food Chemistry.* **51**, 433–442.

Li, J. (2001). Sources of color components, aroma components and taste components in yellow rice wine. *Liquor-Making Science and Technology.* **105**, 48–50.

Limacher, A., Kerler, J., Davidek, T., Schmalzried, F. and Blank, I. (2008). Formation of furan and methylfuran by Maillard-type reactions in model systems and food. *Journal of Agricultural and Food Chemistry.* **56**, 3639–3647.

Luo, T., Fan, W.L. and Xu, Y. (2007). The review of volatile and non-volatile compounds in Chinese rice wine. *Liquor Making.* **34**, 44–48.

Luo, T., Fan, W.L. and Xu, Y. (2008). Characterization of volatile and semi-volatile compounds in Chinese rice wines by headspace solid phase microextraction followed by gas chromatography-mass spectrometry. *Journal of the Institute of Brewing.* **114**, 172–179.

Luo, T., Fan, W.L., Xu, Y. and Zhao, G.A. (2009). Aroma components in Chinese rice wines from different regions. *China Brewing.* **203**, 14–19.

Marilley, L. and Casey, M.G. (2004). Flavours of cheese products: Metabolic pathways, analytical tools and identification of producing strains. *International Journal of Food Microbiology.* **90**, 139–159.

Mart Nez, A.T., Camareroa, S., Guti Rrezb, A., Bocchinic, P. and Gallettic, G.C. (2001). Studies on wheat lignin degradation by pleurotus species using analytical pyrolysis. *Journal of Analytical and Applied Pyrolysis.* **58/59**, 401–411.

Mendoza, L.M., De Nadra, M.C.M., Bru, E. and Farias, M.E. (2009). Influence of wine-related physicochemical factors on the growth and metabolism of non-saccharomyces and saccharomyces yeasts in mixed culture. *Journal of Industrial Microbiology and Biotechnology.* **36**, 229–237.

Mestres, M., Busto, O. and Guasch, J. (2000). Analysis of organic sulfur compounds in wine aroma. *Journal of Chromatography A.* **881**, 569–581.

Molina, A.M., Swiegers, J.H., Varela, C., Pretorius, I.S. and Agosin, E. (2007). Influence of wine fermentation temperature on the synthesis of yeast-derived volatile aroma compounds. *Applied Microbiology and Biotechnology.* **77**, 675–687.

Moon, S.Y. and Li-Chan, E.C.Y. (2004). Development of solid-phase microextraction methodology for analysis of headspace volatile compounds in simulated beef flavour. *Food Chemistry.* **88**, 141–149.

Muller, C.J., Kepner, R.E. and Webb, A.D. (1973). Lactones in wines: A review. *American Journal of Enology and Viticulture.* **24**, 5–9.

Ordonez, J.A., Hierro, E.M., Bruna, J.M. and De La Hoz, L. (1999). Changes in the components of dry-fermented sausages during ripening. *Critical Reviews in Food Science and Nutrition.* **39**, 329–367.

Pinto, U.M., Viana, E.D.S., Martins, M.L. and Vanetti, M.C.D. (2006). Detection of acylated homoserine lactones in gram-negative proteolytic psychrotrophic bacteria isolated from cooled raw milk. *Food Control.* **18**, 1322–1327.

Ruth, S.M.V. (2001). Methods for gas chromatography of actometry. *Biomolecular Engineering.* **17**(121–128).

Schabereiter-Gurtner, C., Pinar, G., Lubitz, W. and Rolleke, S. (2001). Analysis of fungal communities on historical church window glass by denaturing gradient gel electrophoresis and phylogenetic 18s rDNA sequence analysis. *Journal of Microbiological Methods.* **47**, 345–354.

Shen, T. and J., W. (1991). *Biochemistry*. Beijing, China: Higher Education Press.

Shen, Y. (1996). *Manual of Chinese Liquor Manufactures Technology*. Beijing, China: Light Industry Publishing House of China.

Shi, A., Guan, J., Zhang, W., Xu, E. and Xu, C. (2001). Analysis of microbial species in xufang daqu & determination of the dominant microbes. *Liquor Making Science and Technology*. **6**, 26–28.

Shi, J.H., Xiao, Y.P., Li, X.R., Ma, E.B., Du, X.W. and Quan, Z.X. (2009). Analyses of microbial consortia in the starter of fen liquor. *Letters in Applied Microbiology*. **48**, 478–485.

Shou, H.Z., Ling, Z.Y., Yang, X. and Xie, G.F. (2007). Relationship between metabolites of wheat qu microorganisms and flavor of Chinese rice wine. *China Brewing*. **8**, 55–67.

Shou, Q.H., Zao, G.A. and Wei, T.Y. (2008). Study on saccharifying properties of wheat starter of yellow rice wine. *Liquor-Making Science and Technology*. **8**, 95–98.

Takush, D.G. and Osborne, J.P. (2009). Impact of saccharomyces and non-saccharomyces yeast on the flavor and aroma of pinot noir. *American Journal of Enology and Viticulture*. **60**, 391–391.

Ugliano, M. and Moio, L. (2005). Changes in the concentration of yeast-derived volatile compounds of red wine during malolactic fermentation with four commercial starter cultures of *Oenococcus oeni*. *Journal of Agricultural and Food Chemistry*. **53**, 10134–10139.

Valim, M.F., Rouseff, R.L. and Lin, J.M. (2003). Gas chromatographic-olfactometric characterization of aroma compounds in two types of cashew apple nectar. *Journal of Agricultural and Food Chemistry*. **51**, 1010–1015.

Van Rensburg, P., Van Zyl, W.H. and Pretorius, I.S. (1998). Engineering yeast for efficient cellulose degradation. *Yeast*. **14**, 67–76.

Vanbeneden, N., Gils, F., Delvaux, F. and Delvaux, F.R. (2008). Formation of 4-vinyl and 4-ethyl derivatives from hydroxycinnamic acids: Occurrence of volatile phenolic flavour compounds in beer and distribution of pad1-activity among brewing yeasts. *Food Chemistry*. **107**(221–230).

Wang, C., Shi, D. and Gong (2008). Microorganisms in daqu: A starter culture of Chinese maotai-flavor liquor. *World Journal of Microbiology and Biotechnology*. **24**, 2183–2190.

Wang, J.G. (2004). Function of traditional wheat qu in Chinese rice wine and its characteristics. *China Brewing*. **10**, 29–31.

Wang, L., Xu, Y., Zhao, G., and Li, J. (2004). Rapid analysis of flavor volatiles in apple wine using headspace solid-phase microextraction. *Journal of the Institute of Brewing*. **110**, 57–65.

Wu, Y.Y. (2006). Exploration and application of microbial resources in liquor-making industry. *Liquor-Making Science and Technology*. **11**, 111–115.

Wu, Z.Y., Zhang, W.X., Zhang, Q.S., Hu, C., Wang, R., and Liu, Z. (2009). Developing new sacchariferous starters for liquor production based on functional strains isolated from the pits of several famous luzhou-flavor liquor brewers. *Journal of the Institute of Brewing*. **115**, 111–115.

Xie, G.F., Li, W.J., Lu, J., Cao, Y., Fang, H., Zou, H.J., and Hu, Z.M. (2007). Isolation and identification of representative fungi from shaoxing rice wine wheat qu using a polyphasic approach of culture-based and molecular-based methods. *Journal of the Institute of Brewing.* **113**, 272–279.

Xiong, C. (1994). Research on changes of microbes and materials in Zaopei during fermentation of strong aroma style liquor. *Liquor-Making Science and Technology.* **62**, 25–27.

Xu, Y., Fan, W., Wang, H., and Wu, Q. (2010a). Advance in flavor-directed analytic technology for Chinese liquor (baijiu). *Liquor-Making Science and Technology.* **11**, 73–76.

Xu, Y., Fan, W.L., and Qian, M.C. (2007). Characterization of aroma compounds in apple cider using solvent-assisted flavor evaporation and headspace solid-phase microextraction. *Journal of Agricultural and Food Chemistry.* **55**, 3051–3057.

Xu, Y., Wang, D., Fan, W., Mu, X.Q., and Chen, J. (2010b). *Traditional Chinese Biotechnology Advances in Biochemical Engineering/Biotechnology.* Springer.

Xu, Y., Wang, D., Mu, X.Q., Zhao, G.A., and Zhang, K.C. (2002). Biosynthesis of ethyl esters of short-chain fatty acids using whole-cell lipase from *Rhizopus chinesis* CCTCC M201021 in non-aqueous phase. *Journal of Molecular Catalysis B: Enzymatic.* **18**, 29–37.

Yuan, R. (2007). More difficult to produce moutai than atomic bombs. http://finance.qq.com/a/20070421/000086.htm.

Zhang, R. and Xu, Y. (2010). Research on *Bacillus licheniformis* from daqu of soy sauce type and its metabolites with flavor characteristics. *Industrial Micorbiology.* **40**, 1–7.

Zhang, W.X., Qiao, Z.W., Tang, Y.Q., Hu, C., Sun, Q., Morimura, S., and Kida, K. (2007). Analysis of the fungal community in Zaopei during the production of Chinese Luzhou-flavour liquor. *Journal of the Institute of Brewing.* **113**, 21–27.

Zhao, J., Dai, X., Liu, X., Chen, H., Tang, J., Zhang, H., and Chen, W. (2009). Changes in microbial community during Chinese traditional soybean paste fermentation. *International Journal of Food Science and Technology.* **44**, 2526–2530.

Zhou, J., Ed. (1996). *Chinese Rice Wine Brewing Process.* Beijing: China Light Industry Press.

Zhu, B.F. and Xu, Y. (2010a). A feeding strategy for tetramethylpyrazine production by *Bacillus subtilis* based on the stimulating effect of ammonium phosphate. *Bioprocess and Biosystems Engineering.* **33**, 953–959

Zhu, B.F. and Xu, Y. (2010b). High-yield fermentative preparation of tetramethylpyrazine by *Bacillus* sp. Using an endogenous precursor approach. *Journal of Industrial Microbiology and Biotechnology.* **37**, 179–186.

Zhu, B.F. and Xu, Y. (2010c). Production of tetramethylpyrazine by batch culture of *Bacillus subtilis* with optimal pH control strategy. *Journal of Industrial Microbiology and Biotechnology.* **37**, 815–821.

Zhu, B.F., Xu, Y., and Fan, W.L. (2009a). Study of tetramethylpyrazine formation in fermentation system from glucose by *Bacillus subtilis* XZ1124. *New Biotechnology*. **25**, 237–239.

Zhu, B.F., Xu, Y., and Fan, W.L. (2009b). Tetramethylpyrazine production by fermentative conversion of endogenous precursor from glucose by *Bacillus* sp. *Journal of Bioscience and Bioengineering*. **108**, 122–122.

Zhu, Y., Zhang, J., Shi, Z., and Mao, Z. (2004). Optimization of operating conditions in rice heat blast process for Chinese rice wine production by combinational utilization of neural network and genetic algorithms. *Journal of the Institute of Brewing*. **110**, 117–123.

10

SOLID-STATE FOOD FERMENTATION AND SUSTAINABLE DEVELOPMENT

XIAOMING LIU AND PENG ZHOU

State Key Laboratory of Food Science and Technology, School of Food Science and Technology, Jiangnan University, Wuxi, China

Contents

10.1 Introduction

Our nutritional, mental, and physical status depends significantly on the food we eat and how we eat it. The food industry in developing countries is constantly facing the problems of growing population, food security, food safety, and hygiene. To address current and future threats to food security and food safety, as well as seek opportunities for increasing food availability, access, stability, and utilization, sustainable development of the food industry is urgent to provide a solution to save people in developing countries from potential food quality and safety risks. Moreover, demand for more animal protein has been increasing, which naturally brings up issues such as more consumption of feed and water for animals and a greater environmental burden caused by animal husbandry.

Fermentation may provide one of the solutions for the worsening problem in food security and food safety. Fermentation processes could enrich the human diet by improving product quality through enhancing flavor, aroma, texture, shelf life, and digestibility, and through enrichment of nutrients such as essential amino acids, essential fatty acids, vitamins, and minerals. Moreover, fermentation could also utilize the waste and by-products in agricultural and food processing. In recent years, the advances of breeding, microbial technology, and biotechnology make the technology of solid-state fermentation more powerful than traditional fermentation technology in the sustainable development of the food industry.

10.2 Solid-State Food Fermentation and Global Food Security

10.2.1 Challenges Associated with Global Food Security

Access to affordable and high-quality food has remained as one of the main problems in developing countries. The World Food Summit defined food security as: "Food security exists when all people, at all times, have physical and economic access to sufficient, safe, and nutritious food to meet their dietary needs and food preferences for an active and healthy life." Generally, food security is built on three pillars: food availability, access to food, and use of food, which reflects the global food situation and the trend of research and policy of government in an individual country. Therefore, even if enough food is available for the needs, food insecurity may persist either because of inadequate access or failure to meet the nutritious needs and healthy diets (Sen, 1981). Among the issues associated with food security, food shortage, food nutrition and food sustainable development seem to be particularly urgent.

10.2.1.1 Food Shortage According to the Food and Agriculture Organization (FAO) of the United Nations, there are about 20 million undernourished people who need to achieve food security annually. However, during the first half of the 1990s, only 8 million people out of these undernourished people obtained food security annually. The highly desired acceleration in the speed of hunger reduction will require a stronger effort and higher priority on the part of governments, society, and the private sector.

It is believed that in the next 10 years, the increase in the supply of agricultural products could not meet the expected increase in world population and food shortage, especially in developing countries. In many developing countries, efforts have been made to increase the yield of crops by genetic modification, which is seen as a promising solution to tackle the issues of food shortage. From the perspective of food preservation and value-addition, fermentation has become a powerful approach to conquer global food shortages and generate desirable changes in food products with the addition of microorganisms. For many years people in the Sudan, the largest country in Africa, spent tremendous efforts in tackling its food shortage problem. With the adaptation of food fermentation, the indigenous self-help famine relief has been successfully achieved. Most fermented foods appear to have been developed to ensure the nutrition of the family in the dry season or as survival foods for the drought years, which inevitably occur in the Sudan.

10.2.1.2 Food Nutrition   In the developing countries, the diet of approximately 2.5 billion people contains insufficient amounts of minerals and vitamins. Data shows that 2 million young children die and almost 300,000 children go blind due to a shortage of vitamin A each year (Mann et al., 1998). Research has shown that during the processing of fermentation, raw food is turned into a product that may have enhanced nutritional and/or organoleptic characteristics, contributing to supplementation with components such as amino acids, vitamins (in particular, K and B), and minerals. At the same time, fermentation also has the advantage of eliminating undesirable components that are present in raw materials. Therefore, fermentation could turn vegetable proteins, sometimes an inferior protein source to animal protein, into food resources with improved nutritive value or even substitutes for animal proteins.

In developed countries, although there is little concern for issues such as food shortages, threatens to the health of people from junk food remains high. In America one-third of the food people eat is junk food, which contains a variety of harmful substances to human health. Long-term consumption of large quantities of junk food will cause obesity and increase risk of disease such as high blood pressure, cardiovascular disease, type II diabetes, osteoporosis, and cancer. As most raw materials used in fermented food is low-fat, and fermentation consumes the energy of carbohydrates, fermented food has the characteristics of low-fat and low calorie. Therefore, an increase in the consumption of fermented food may help solve the problem caused by unhealthy diets in developed countries.

Besides the above beneficial effects of fermented foods on human health, incorporation of probiotics into fermented food will bring extra credits to the products. Probiotics are described as "live microorganisms which, when administered in adequate amounts, confer a health benefit on the host." Many foods containing probiotics (such as fermented milks, yogurts, and cheese) fall within the functional food category, which includes any fresh or processed food with health-promoting and/or disease-preventing properties besides the basic nutritional function.

10.2.1.3 Food Sustainable Development   It is well known that there are enormous residues left in the manufacturing of agricultural products, which become not only a waste issue but also create severe environmental problems. As agricultural wastes or biological wastes contain reusable substances such as soluble sugars, fiber, vitamin and mineral substances, solid-state fermentation for specific value-added products using biowastes will generate great economical and environment-friendly advantages.

For example, although soybean residue is rich in nutrition, the high water content, large grained fiber, and strong flavor restrict its easy application in the production of food. Solid-state fermentation has been shown to improve the organoleptic properties of the fermented soybean residues, which have a pleasant taste without the sensation of a beany flavor. With fermentation, the content of vitamin B_2 of the product increased from 0.99 mg/100 g to 1.7 mg/100 g, and the content of vitamin B_{12} increased from zero to 0.56 mg/100 g (Wang, 2007).

In past decades, biofuel production from grains has drawn a great deal of attention. Scientists in multiple nations have conducted research on the production of biofuel with various types of grains. However, thorough consideration of all aspects is necessary to ensure sustainable development.

10.2.2 Approach of Enhancement of Storage Stability
       through Solid-State Fermentation

Microorganisms have long played important roles in the production of food (dairy, fish, and meat products) and alcoholic beverages. In addition, several products of microbial fermentation also act as additives and supplements in food (flavors, colorants, antioxidants, preservatives, sweeteners). There is great interest in the development and use of natural food and additives derived from microorganisms, since they are more desirable than the synthetic ones produced by chemical processes. Ancient people practiced various approaches to food fermentation, and spread salt on meat or soak the vegetables in salt water prior to fermentation to obtain a broad range of microorganisms for fermentation.

Although submerged fermentation remained as the main technique throughout history, solid-state fermentation (SSF) of enzymes, spices, and organic acids possess several biotechnological advantages, such as higher fermentation productivity, higher product stability, lower catabolic repression, mixed cultivation of various microbes, and last but not least, higher end-concentration of products and lower water activity to avoid decay (Holker et al., 2004). SSF is performed on a nonsoluble material that acts both as a physical support and source of nutrients in the absence of free-flowing liquid. Viniegra et al. showed that higher biomass, high enzyme production, and lower protein breakdown contributed to the better production in SSF. A generation of biologically active substances such as organic acids through SSF can not only hold back food materials deterioration, but also extend the shelf life of food.

10.2.2.1 Organic Acids   Organic acids are the primary metabolic products of microorganisms, and organic acid serves as food additives and preservatives to prevent food spoilage and extend the storage period of perishable foods. Organic acids such as lactic acid and citric acid are acidity regulators, which can maintain or change the pH of food substances, improve the flavor, and enhance antioxidants to prevent food spoilage (Rodriguez Couto and Sanroman, 2006).

Lactic acid has two enantiomers, L (+) and D (−), due to the presence of L-lactate dehydrogenase, and L (+)-lactic acid has been preferred for food. Production of lactic acid with SSF has been practiced with various raw materials including wheat bran, bagasse, and sorghum. Investigations have been done on the L (+)-lactic acid production by *Lactobacillus amylophilus* GV6 using wheat bran as both the substrate and carrier. Soccol (Soccol et al., 1994) used bagasse as a carrier to produce L (+)-lactic acid using *Rhizopus oryzae*. Richter (Suryanarayan, 2003b) utilized *Lactobacillus paracasei* to produce L (+)-lactic acid with sorghum as the carrier in the solid-state conditions.

Production of fermented vegetables, dairy products, and meat products has been practiced throughout human history to provide foods with better storage stability than the raw materials. Vacuum packaging of *shibazuke*, traditional Japanese-style salted and fermented pickles, increases the shelf life of refrigerated pickles up to one year, whereas the harvested vegetables are only fresh for a few days. Pasteurized pickles only have a shelf life of around a week when stored in the refrigeration. Therefore, fermentation prolongs the shelf life of shibazuke extensively. Another example is yogurt, the fermented product of milk, which could have a shelf life of 14–21 days. In the process of milk fermentation, lactic acid bacteria break down the milk protein and lactose, which will not only be easy to digest, but also effectively inhibit the growth of undesirable bacteria in milk.

10.2.2.2 Enzymes Enzymatic preservation technology is a new food preservation technology which utilizes the high catalytic efficiency of enzymes to prevent or eliminate the adverse external factors of food and maintain the food quality and characteristics. It is known that glucose oxidase catalyzes the reaction of glucose and oxygen to produce gluconic acid and hydrogen peroxide. Therefore, when glucose oxidase is added to sealed containers together with food, the enzyme can effectively reduce or eliminate oxygen. Therefore, glucose oxidase is currently produced by SSF and applied in food products to effectively prevent the oxidation of food components and play a role in food preservation.

The traditional production of soy sauce in a *koji* preparation is carried out in an open environment, and it is susceptible to bacteria pollution, which affects the breeding and quality of koji and directly affect the quality and output of soy sauce. The application of modern SSF technology can easily overcome these problems, and soy sauce from SSF even has a longer shelf life than that made by traditional methods. Japan has been using SSF for commercial production of enzymes and the production of soy sauce.

10.2.3 Improvement of Food Nutrition through Solid-State Fermentation

Fermentation can preserve food and has the potential to enhance food safety by controlling a great number of pathogens in foods. Furthermore, fermentation can enrich food nutrition and improve digestibility to gain high-value products.

SSF reproduces the natural microbiological processes like composting and ensiling (Rodriguez Couto and Sanroman, 2006; Singhania, 2009). In industrial applications, this natural process can be utilized in a controlled way to produce a desired and high-value product (Pandey, 1992). The low moisture content of the SSF process means that fermentation can only be carried out by a limited number of microorganisms, mainly fungi and especially yeasts, although some bacteria have also been used (Pandey et al., 2000a).

SSF offers numerous advantages for the production of bulk chemicals and enzymes. The process has been known from ancient times and different fungi have been cultivated in SSF. The most typical example of SSF is the fermentation of rice by *Aspergillus oryzae* to initiate the koji process, which changes the rice to a value-added product. The other example is *Penicillium roquefortii* for cheese production, which enriches the flavor and increases the shelf life, accordingly, with higher values.

In Japan, SSF is used commercially to produce industrial enzymes (Suryanarayan, 2003a). In China, SSF has been used extensively to produce brewed foods (such as Chinese wine, soy sauce, and vinegar) since ancient times (Chen, 1992). Since 1986 in Brazil (Robinson and Nigam, 2003), a series of research projects for the value addition of tropical agricultural products and subproducts by SSF has been developed due to the high amounts of agricultural residues. Thus, the production of bulk chemicals and value-added fine products has been produced from these raw materials by means of the SSF technique. Recently, SSF has attracted more and more interest from researchers, since several studies for ethanol, single-cell protein (SPC), mushroom, organic acids, amino acids, enzymes, flavors, colorants, and other substances of interest to the food industry have shown that SSF can give higher yields or better product characteristics than submerged fermentation. In addition, costs are much lower due to the availability, utilization, and value addition of wastes.

Studies about solid-state fermentation of black soybeans with *Bacillus subtilis* BCRC 14715 showed that fermentation enhanced the total phenolic and flavonoid content as well as antioxidant activity of the black soybean extract (Juan and Chou, 2010). In another study, after solid-state fermentation, the protein content of the soybean was 1.18 times than before, and the content of nitrogen in amino acid was 9.12 times than before. Therefore, SSF enhanced the utilization rate and nutritional value of soybeans successfully.

After making of sugar, sugarcane bagasse is typically treated as an agricultural residue and the major by-product of the sugarcane industry, which generally is used as a fuel (Vandenberghe et al., 2000). The bagasse contains about 50% cellulose, 25% hemicellulose, and 25% lignin (Pandey et al., 2000b). Due to the abundant availability of sugarcane bagasse, it can serve as an ideal substrate for microbial processes for the production of value-added products. Attempts have been made to produce protein-enriched animal feed, enzymes, amino acids, organic acids, and compounds of pharmaceutical importance from a bagasse substrate.

SSF have been applied in industry for several decades and show an enormous potential in bioremediation and biological detoxification of hazardous and toxic compounds (Pandey et al., 2000a). Besides, it also plays an important role in biotransformation of crops and crop residues for nutritional enrichment and production of value-added products, such as biologically active secondary metabolites, including antibiotics, alkaloids, plant growth factors and other drugs, enzymes, organic acids, biopesticides, including mycopesticides and bioherbicides, biosurfactants, biofuel, aroma compounds, etc. SSF systems, once misinterpreted as "low-technology" systems during the last two decades appear to have the potential to enhance the value of production. With the advancement of biotechnological innovations, mainly in the area of enzyme and fermentation technology, many new avenues have opened for the application of SSF (Singhania et al., 2009).

10.3 Solid-State Food Fermentation as a Tool Targeting Malnutrition in Developing Countries

10.3.1 Challenges Associated with Malnutrition in Developing Countries

Malnutrition is the condition that results from a disturbed diet in which certain nutrients are lacking, excessive, or in the wrong proportions.

Many different nutrition disorders may arise, depending on which nutrients are under- or overabundant in the diet. Malnutrition is still a major public health concern in the developing countries, particularly in southern Asia and sub-Saharan Africa (Rice et al., 2000. Diets in those populations are frequently deficient in macronutrients (carbohydrates, protein and fat, leading to protein-energy malnutrition) and micronutrients (minerals, electrolytes, and vitamins, leading to specific micronutrient deficiencies) or both.

The high prevalence of bacterial and parasitic diseases in developing countries contributes greatly to malnutrition as well (Levin et al., 1993; Dickson et al., 2000; Brabin et al., 2003; FAO, 2004; Millward and Jackson, 2004; Stoltzfus et al., 2004). Similarly, malnutrition increases one's susceptibility to infections, and is thus a main component of illness and death from disease (Murray and Lopez, 1997; Rice et al., 2000; Black, 2003; Brabin et al., 2003; FAO, 2004). Malnutrition is the most important risk factor in developing countries (Murray and Lopez, 1997; Nemer et al., 2001). It is the direct cause of about 300,000 deaths per year and is indirectly responsible for about half of all deaths in young children (Black, 2003; Brabin et al., FAO, 2004; Murray and Lopez, 1997; Rice et al., 2000). The risk of death is directly correlated with the degree of malnutrition (Chen et al., 1980; Pelletier et al., 1993; Man et al., 1998; Fernandez et al., 2002; Black, 2003).

Poverty is the major underlying cause of malnutrition and its determinants (WHO, 2001). There are many factors resulting in the various degrees of and distribution of protein-energy malnutrition and micronutrient deficiencies: the political and economic situation, the level of education, the season and climate conditions, food production, cultural and religious food customs, breast-feeding habits, prevalence of infectious diseases, the existence and effectiveness of nutrition programs, and the availability and quality of health services (Levin et al., 1993; Brabin et al. 2003; FAO, 2004; Salama et al., 2004; Young et al., 2004). Here, we classify the malnutrition into two categories: protein-energy malnutrition and micronutrient malnutrition.

10.3.1.1 Protein-Energy Malnutrition (Pemberton, 2006)   Worldwide, about 852 million people were undernourished in 2000–2002, with most (815 million) living in developing countries. The absolute number of cases has changed little over the last decade. China had major reductions in its number of cases of protein-energy malnutrition during this period (Fao., 2004). In children, protein-energy malnutrition (PEM) is defined by measurements that fall below 2 standard deviations under the normal weight for age, height for age, and weight for height. The symptom of protein-energy malnutrition occurs early, in children between 6 months and 2 years of age and is associated with early weaning (Kwena et al., 2003). The current view is that most PEM is the result of inadequate intake or poor consumption of food and energy, not a deficiency of one nutrient and not usually simply a lack of dietary protein. Nutritional marasmus is defined as severe wasting, and is now recognized to be more prevalent than kwashiorkor (Williams, 1935; Brabin et al., 2003).

10.3.1.2 Micronutrient Malnutrition Micronutrient malnutrition is defined as diseases caused by a dietary deficiency of vitamins or minerals. Deficiencies in iron, iodine, vitamin A, and zinc are still major public health problems in developing countries (Levin et al., 1993; Diaz et al., 2003), and at least 2 billion people worldwide are suffering micronutrient deficiencies (FAO, 2004). Randomized controlled trials of supplementation are the most excellent method to study the relation between micronutrient deficiencies and health parameters in human populations (Shankar, 2000). The current situation of micronutrient malnutrition is serious. Worldwide, 740 million people are deficient in iodine, about 300 million with goiter and 20 million with brain damage from maternal iodine deficiency during their fetal development (Levin et al., 1993; Hetzel, 2002; Black et al., 2003). About 2 billion people are deficient in zinc, and 1 billion are suffering with iron deficiency anemia. Vitamin A deficiency affects about 250 million, generally pregnant women and young children in developing countries (Aggett, 1989; Krawinkel et al., 1990; Fawzi et al., 1993; Glasziou and Mackerras, 1993; Ross et al., 1993; Cook et al., 1994; WHO, 1995; Fleming and De Silva, 1996; Black, 1998; Shankar and Prasad, 1998; Sikosana et al., 1998; Stoltzfus and Dreyfuss, 1998; Humphrey and Rice, 2000; Vijayaraghavan, 2000; Yip and Ramakrishnan, 2002; Black, 2003; Prasad, 2003; Hesham et al., 2004; Shali et al., 2004; Zimmermann et al., 2005).

Vitamin A deficiency is widespread in developing countries, but seldom seen in developed countries. Night blindness is one of the first signs of vitamin A deficiency, which also causes xerophthalmia and complete blindness. About 250,000 to 500,000 malnourished children in the developing world go blind each year because of a deficiency of vitamin A. The prevalence of night blindness due to vitamin A deficiency is high among pregnant women in many developing countries. Vitamin A deficiency also leads to maternal mortality and other poor outcomes in pregnancy and lactation (Sommer, 1995). Vitamin A deficiency can also make you ease to access infections. In countries where children are not immunized, infectious diseases like measles have higher fatality rates. According to the World Health Organization (WHO), vitamin A deficiency is common in developing countries. Approximately 250,000–500,000 children in developing countries become blind each year due to vitamin A deficiency, with the highest prevalence in Southeast Africa and Asia. One-third of the children around the world suffering from vitamin A deficiency is estimated to be under the age of five. It is predicted to affect the lives of 670,000 children under five per year (Black et al., 2008).

Thiamine deficiency, or beriberi, refers to the lack of thiamine pyrophosphate, the active form of the vitamin known as thiamine (or thiamin), or vitamin B_1. Thiamine takes part in the formation of glucose by acting as a coenzyme for the transketolase in the pentose monophosphate pathway. Persons may become deficient in thiamine either by not ingesting enough vitamin B_1 through the diet or by excess use, which may occur in hyperthyroidism, pregnancy, lactation, or fever. Prolonged diarrhea may impair the body's ability to absorb vitamin B1, and severe liver disease impairs its use (Beers et al., 2000; Cole and Kamen, 2003). Thiamine is a water-soluble vitamin; the body cannot produce thiamine by itself and can only save up to 30 mg of thiamine in its tissues. Thiamine is mostly concentrated in the skeletal muscles, and other organs such as the brain, heart, liver, and kidneys. It is excreted by the kidney (McCormick, 1988; Fauci, 1994; Rosen et al., 1994). Because of this factor, thiamine is easy to find deficient.

Niacin (also known as vitamin B_3, nicotinic acid, and vitamin PP) is an organic complex with the formula $C_6H_5NO_2$. At present, poverty countries find common the condition of niacin deficiency, malnutrition, and chronic alcoholism, but it is rarely seen in developed countries (Pitsavas et al., 2004). Niacin deficiency tends to occur in the areas where people mainly eat maize (corn, the only grain low in niacin). Mild niacin deficiency has been shown to slow metabolism, causing decreased tolerance to cold. Severe deficiency of niacin in the diet causes the disease "pellagra," which is characterized by diarrhea, dermatitis, and dementia, as well as "necklace" lesions on the lower neck, hyperpigmentation, thickening of the skin, inflammation of the mouth and tongue, digestive disturbances, amnesia, delirium, and eventually death, if left untreated. General psychiatric symptoms of niacin deficiency include irritability, poor concentration, anxiety, fatigue, apathy, and depression (Prakash et al., 2008). Studies show that niacin deficiency may be an important factor influencing both the onset and severity of this condition.

Vitamin C (ascorbic acid) plays an important part in the formation and function of collagen, carnitine, hormones, and amino acid. It is essential for wound healing and facilitates recovery from burns. Vitamin C is also an antioxidant, supports immune function, and facilitates the absorption of iron. Lacking in vitamin C can result in scurvy, which leads to the formation of brown spots on the skin, spongy gums, and bleeding from all mucous membranes. In advanced scurvy there are open, suppurating wounds and loss of teeth and, eventually, death. The human body can store only a certain amount of vitamin C, and the body storage will be exhausted if fresh supplies are not consumed. It has been shown that smokers who have diets poor in vitamin C are at a higher risk of lung-borne diseases than those smokers who have higher concentrations of vitamin C in the blood. Nobel Prize winner Linus Pauling and Dr. G. C. Willis have asserted that chronic long-term low blood levels of vitamin C (chronic scurvy) is a cause of atherosclerosis (Rath and Pauling, 1990). Moderately higher blood levels of vitamin C in healthy persons have been found to be prospectively correlated with a decreased risk of cardiovascular disease and ischemic heart disease, and an increase life expectancy. The same study found an inverse relationship between blood vitamin C levels and cancer risk in men, but not in women. An increase in blood level of 20 µmol/L of vitamin C was found epidemiologically to reduce the all-cause risk of mortality, 4 years after measuring it, by about 20% (Khaw et al., 2001). Studies show that with much higher doses of vitamin C, generally between 200 and 6000 mg/day, for the treatment of infections and wounds have shown inconsistent results (Hemilä, 2007). Combinations of antioxidants seem to improve wound healing (Barbosa et al., 2009). Western societies usually consume far more than adequate vitamin C. In 2004, a Canadian Community Health Survey reported that Canadians 19 years and above have intakes of vitamin C from food of 133 mg/d for males and 120 mg/d for females; these are higher than the RDA recommendations. Studies of experimentally-induced scurvy found that all obvious symptoms of scurvy previously induced by an experimental scorbutic diet with extremely low vitamin C content could be completely reversed by additional vitamin C supplementation of only 10 mg a day.

Hypovitaminosis D, a vitamin D deficiency, has many health hazards, which may result in impaired bone mineralization and leads to bone-softening diseases such as rickets in children and osteomalacia and osteoporosis in adults. Vitamin D deficiency may be owed to inadequate intake of vitamin D coupled with insufficient sunlight exposure, disorders limiting vitamin D absorption, and conditions that weaken the conversion from vitamin D to active metabolites including certain liver, kidney, and hereditary disorders.

Vitamin E deficiency can cause neurological problems due to poor nerve conduction. These problems comprise neuromuscular problems such as spine cerebellar ataxia and myopathies. Deficiency can also lead to anemia, owing to oxidative damage to red blood cells. Vitamin E deficiency is rarely in humans and almost never results from a poor diet. It is seen in individuals who cannot absorb dietary fat, has been found in premature, very low birth weight infants, and is seen in persons with rare disorders of fat metabolism (Traber and Sies, 1996). Since some dietary fat is essential for the absorption of vitamin E from the gastrointestinal tract, individuals who cannot metabolize fat may require a supplement of vitamin E. Persons who have been diagnosed with cystic fibrosis, individuals who have had part or all of their stomach removed, and individuals with malabsorptive problems such as Crohn's disease, liver disease, or pancreatic insufficiency may not absorb fat and should consider the need for a vitamin E supplement.

Iodine is an essential microelement, and thyroxine and triiodothyronine, the important thyroid hormones, both contain iodine. In regions where the diet of the inhabitants contain little iodine, typically remote inland areas where there are no marine foods to eat, iodine deficiency gives rise to goiter (endemic goiter), as well as cretinism, which may lead to growth delays and other health problems. While noting recent progress, *The Lancet* (2008) added, that according to WHO, in 2007, nearly 2 billion individuals had insufficient iodine intake, a third being of school age. Thus, iodine deficiency, as the single greatest preventable cause of mental retardation, is an important public health problem.

Iron deficiency anemia (IDA) is a common health concern worldwide that affects more than 3.5 billion people in the developing world, stealing the vitality of both the young and the old and impeding the cognitive development of children. Iron deficiency has a massive, but until recently nearly completely unrecognized, economic cost, which adds more burden on health systems, affects learning and school performance, and reduces the productivity of adults. The World Bank, WHO, and Harvard University have described iron deficiency as having a higher overall cost than any other disease apart from tuberculosis. The consequences of iron deficiency, and particularly iron deficiency anemia are numerous. For infants and children, iron deficiency impedes motor development and coordination, impairs language development and scholastic achievement, psychological and behavioral effects (inattention, fatigue, etc.), and reduces physical activity. Iron deficiency anemia in adults of both sexes leads to decreased physical work and earning capacity and reduced resistance to fatigue. The negative effects of iron deficiency anemia in pregnant women consists of increased maternal morbidity and mortality, increased fetal morbidity and mortality, and increased risk of low birth weight. The prevention and treatment strategy for anemia is based on three strategies: dietary improvements, fortification of staples and condiments with iron, and supplementation.

Zinc deficiency is a lack of sufficient zinc to meet the requirements of biological organisms, which can happen in both plants and animals. The nutrition journals of the 1990s put more emphasis on zinc and zinc deficiency than on PEM. However, zinc deficiency is not declared as a dominate public health concern in any country around the world, and no obvious syndrome is described for zinc deficiency. In Egypt and the Islamic Republic of Iran a condition in males characterized by dwarfism and hypogonadism is related to zinc deficiency. In the United States and elsewhere, zinc deficiency status in children has been correlated with a slowing growth rate, poor appetite, and impaired sense of taste.

Folate deficiency is a lack of folic acid in the diet and the symptoms are usually subtle, and folate deficiency anemia is the medical term given for the condition (Huether and McCance, 2004). When folate deficiency anemia is likely to occur, there are some specific symptoms such as loss of appetite and weight loss. Folate deficiency can also cause weakness, sore tongue, headaches, heart palpitations, irritability, and often behavioral disorders (Haslam and Probert, 1998). Pregnant women with folate deficiency are more likely to give birth to low birth weight and premature infants, and infants with neural tube defects. In adults, anemia (macrocytic, megaloblastic anemia) can be a sign of advanced folate deficiency. In infants and children, folate deficiency can retard growth. Recent studies suggested an involvement in tumorogenesis (especially in colon) by means of demethylation/hypomethylation of rapid replicating tissues. Some of the signs are also due to kinds of medical conditions other than folate deficiency.

Selenium deficiency rarely occurs in healthy well-nourished individuals. Selenium deficiency can cause Keshan disease, which is potentially fatal. It can also lead (together with iodine deficiency) to Kashin-Beck disease. The primary symptom of Keshan disease is myocardial necrosis, resulting in weakening of the heart. Kashin-Beck disease can leads to atrophy, degeneration, and necrosis of cartilage tissue. Keshan disease also causes more susceptibility to illness resulting from other nutritional, biochemical, or infectious diseases. Selenium is also essential for the conversion from the thyroid hormone thyroxine to its more active counterpart, triiodothyronine. At the same time, such a deficiency can cause symptoms of hypothyroidism, including extreme fatigue, mental slowing, goiter, cretinism, and recurrent miscarriage.

10.3.2 A Tool Targeting Malnutrition in Developing Countries

Throughout history, fermentation has been practiced by mankind to boost the nutrition value of food. As a result, various types of fermented foods have been developed and are still produced all over the world. Here, a few examples of traditional fermented food would serve as examples to solve the malnutrition in developing countries.

10.3.2.1 Bean-Based Solid-State Fermented Food   In addition to the high protein content, soybeans contain polyphenols, soluble fiber, and

isoflavones, all of which play various biological activities both *in vivo* and *in vitro*. However, unprocessed soybeans contain antinutritional factors and have a high content of oligosaccharides such as stachyose, raffinose, and melibiose, which are difficult to digest and may result in flatulence. To solve this issue, Asian people have transformed the soybean into more palatable and digestible forms through fermentation since ancient times. During fermentation undesirable components such as allergens and phytates are hydrolyzed, and various fermentation processes generate all kinds of nutritional and unique fermented soybean products such as *doenjiang*, soy sauce, *furu*, and *douchi*.

Doenjang is a soy-based food consumed daily by Koreans. In Korean households, it is a soup base prepared for regular broth-type soup and much thicker stew-type soup, both of which are eaten together with rice. In general, most soybean-based fermented foods are prepared once a year and conserved for 2 to 3 years. With the natural fermentation of *meju* in the late fall, the solid blocks (size varies by household, but approximately $20 \times 15 \times 10$ cm) are made from steamed soybeans. The fermentation goes on for approximately 2 weeks. Then the product is dried over the winter and fermented in brine for 6 to 8 weeks during spring. At the end of the wet fermentation, doenjang (soybean paste, solid material) is separated from *ganjang* (soy sauce, supernatant). The product can be consumed after the next 30 to 50 days of maturation; as a side note, longer maturation is thought to produce a more appealing favor (Han et al., 1998). Doenjang provides abundant flavonoids, and a variety of beneficial vitamins, minerals, and hormones that are considered by many to possess anticarcinogenic properties. The Korean traditional meals typically have fair shares of vegetables and rice, and doenjang is rich in lysine, an essential amino acid that rice lacks. Doenjang also contains linoleic acid and linolenic acid, which play significant roles in normal growth of blood vessels and prevention of blood vessel-related illness. Unlike *miso*, doenjang is often not boiled; after boiling doenjang's efficacy still exists, in dishes such as doenjang *jjigae*.

Natto, a traditional Japanese food, is made from soybeans and fermented with *Bacillus subtilis*. Since natto and the soybean paste miso are rich in protein and beneficial bacteria, they provided a vital source of nutrition in feudal Japan. Natto can be an acquired taste because of its powerful smell, strong flavor, and slippery texture. In Japan natto is widely accepted by the residents in the eastern regions, including Kanto, Tohoku, and Hokkaido. The chemical composition of natto is not greatly different from that of soybeans apart from the vitamin K content. *B. subtilis* (*natto*) produces vitaminK$_2$ (menaquinone-7) (Kudo, 1990; Sumi, 1999). Domestic soybeans, U.S. soybeans, Chinese soybeans, and natto contain 21, 39, 39, and 2148 μg/100 g dry of vitamin K, respectively. During fermentation, *B. subtilis* (*natto*) produces sticky substances, which cover the natto. These sticky substances are made up of polyglutamic acid and levan (fructan) (Claus, 1986). *B. subtilis* (*natto*) also produces proteases and an amylase during fermentation (Ferrari et al. 1993). Peptides or amino acids produced during fermentation compose a part of the natto taste. *B. subtilis* (*natto*) utilizes soybean saccharides and confers to natto a characteristic flavor.

Miso, produced through fermenting rice, barley and/or soybeans, with salt and the fungus *kojikin*, is a traditional Japanese flavoring. The most typical miso is prepared with soy. The product is a thick paste, can be used for sauces and spreads, pickling vegetables or meats, and mixing with *dashi* soup stock to serve as miso soup called *misoshiru*, a Japanese culinary staple. High in protein and rich in vitamins and minerals, miso played an important nutritional part in feudal Japan. Still now miso is widely consumed in Japan, both in traditional and modern cooking, and has been attracting worldwide interest. Miso is typically salty, but its flavor and aroma depend on a variety of factors in the ingredients and fermentation process. There are many kinds of miso available now; miso of different flavors have been described as salty, sweet, earthy, fruity, and savory.

Douchi, the so-called Chinese fermented black bean, is a seasoning most popular in Chinese meals, and is used to make black bean sauce. Douchi is made by fermenting and salting soybeans. The beans become black, soft, and mostly dry during the process. The flavor is sharp, pungent, and spicy in smell, with a taste of salty and somewhat bitter and sweet. Douchi should not be confused with black turtle beans, various common beans that are commonly used in the cuisines of Central America, South America, and the Caribbean. In Japanese, douchi is also referred to by the same name *kanji* and pronounced as "touchi." The process and product are similar to African fermented bean products such as *ogiri* and *iru*.

Tempeh. Tempeh is a traditional soy product originating from Indonesia. It is similar to a very firm vegetarian burger patty, made through a natural culturing and controlled fermentation process that binds soybeans into a cake form. Tempeh is unique among primary traditional soy foods in that it is the only one that isn't originated from East Asia. In the fermentation process the soy protein in tempeh becomes more digestible. In particular, the *Rhizopus* culture can greatly reduce the oligosaccharides that are associated with gas and indigestion. In traditional tempeh-making shops, the starter culture often includes beneficial bacteria that produce vitamins such as vitamin B_{12} (Liem et al., 1977; Truesdell et al., 2006). In western countries, it is more common to by means of a pure culture containing only *Rhizopus oligosporus*, which produces very little vitamin B_{12} and could be missing *Klebsiellapneumoniae* which has been shown to produce significant levels of VB_{12} analogs in tempeh when present. Whether these analogs are true, bioavailable VB_{12}, hasn't been completely confirmed yet. Compared to raw or boiled soybeans, tempeh provides many other health benefits, including perfect digestibility, increased folate content, increased VB_{12} content, content of bacteriocins, reduced levels of trypsin inhibitors, production of fibrinolyticsubtilisin from *Bacillus subtilis* in tempeh, antioxidative properties, increased content of GABA, HAA, and isoflavonoids, etc.

Furu or *sufu* is a fermented soybean curd (tofu) product originated from China. Regardless of differences in color and flavor, most types of furu contain approximately 58 to 70% moisture, 12 to 17% crude protein, 8 to 12% crude lipid, and 6 to 12% carbohydrates. As a result of high levels of salt, the ash content of furu is about 4 to 9%. The minerals measured in furu (based on 100 g fresh weight) include Ca (100 to 230 mg), P (120 to 300 mg/g), and Fe (7 to 16 mg/g). The vitamin's content are thiamin (0.04 to 0.09 mg), riboflavin (0.13 to 0.36 mg), niacin (0.5 to 1.2 mg), and VB_{12} (1.7 to 22 µg) (Su, 1986; Wang and Du, 1998). In general, 18 amino acids have been identified in furu. Glutamic acid and aspartic acid are the most abundant amino acids found in red furu and gray furu. The ratio of the combined glutamic acid and aspartic acid to the total amino acid content is approximately 30%, resulting in a conspicuous savory taste (Yen, 1986). Furu is popular because of its pleasant creamy taste and intense savory flavor, which are usually used to accent the otherwise bland flavors of rice or bread. Chinese people also consider it as an easily digested and healthy food for children, the elderly and the infirm, possibly due to its high levels of calcium, peptides, free amino acids, and enzymes, which result from the fermentation process. The water-soluble proteins in furu have been determined to be 6 to 7 times higher than that in tofu (Liu and Chou, 1994; Wang, 2002). Many physiologically active components, such as soybean peptides, B-vitamins, nucleic glycosides, and aromatic compounds have been found in furu, compared with unfermented soybeans (Chou and Hwan, 2006). The content of vitamin B12, an essential nutrient for the nervous system, in *choufuru* (strong smell, stinky furu) has been shown to be much higher (9.8 to 18.8 mg/100 g) than that in red furu (0.42 to 0.78 mg/100 g) (Wang et al., 2004), suggesting a high activity of microorganisms during fermentation of the choufuru. The antioxidative and antihypertensive effects of furu have also been documented (Wang et al., 2003, 2004).

10.3.2.2 Vegetable-Based Solid-State Fermented Food Kimchi, spelled gimchi, kimchee, or kimchee, is a traditional fermented Korean food, made of vegetables with varied seasonings (Kim and Chun, 2005). There are kinds of kimchi, made within a main vegetable ingredient such as napa cabbage, radish, green onions or cucumber. Kimchi is the most common *banchan* in Korean cuisine. Kimchi is made of various vegetables and contains a high concentration of dietary fiber, while being low in calories. One serving supplies up to 80% of the daily recommended amount of vitamin C and carotene. Most kinds of kimchi contain onions, garlic, and peppers. Kimchi is rich in vitamin A, thiamine (vitamin B_1), riboflavin (vitamin B_2), calcium, and iron, and contains a number of lactic acid bacteria (Yoon et al., 2000; Kim and Chun, 2005; Lee et al., 2005).

The fermentation of cabbage to sauerkraut exhibits a complex system of microbial, chemical, biochemical, enzymatic, and physical processes. Cabbage is composed of crude fibers, lipids, proteins, carbohydrates, and ash in relatively high proportions. The primary change that occurs during fermentation is the bioconversion of carbohydrates to lactic and acetic acids, ethyl alcohol, mannitol, carbon dioxide, and dextrans. The proteins, glucosides, lipids, and other composition of cabbage are affected by the fermentation, resulting in change of the chemical and physical properties of the product. Sauerkraut is considered to be a healthy product as it is an important source of vitamins (especially vitamin C), mineral salts, and dietary fibers.

10.3.2.3 Meat-Based Solid-State Fermented Food Fermented meat such as sausage, ham, salami, chorizo, and pepperoni are consumed all over the world. The physical, microbial, and biochemical changes during fermentation are summarized as follows: growth of lactic acid bacteria and concomitant acidification of the product, reduction of nitrates to nitrites and formation of solubilization, nitrosomyoglobinand gelifcation of myofibrillar and sarcoplasmic proteins, degradation of lipids and proteins, and dehydration. The fat content of meat as consumed is around 2% to 5%, while total fat content differs from age, feeding regimes, and species. The main fatty acids in meat are saturated fatty acids, containing palmitic acid and stearic acid.

About 40% of the fat in meat is monounsaturated, of which oleic acid is one of the key contributors (Enser et al., 1996). Protein of high biological value and micronutrients such as zinc, iron, vitamin B1, niacin equivalents, and vitamin B_{12} significantly contribute to the nutritional value of meat. Mann, 2000 reported that the requirement for iron is one of the most difficult nutritional requirements for humans, because iron deficiency is caused not only by a low intake but also the consequence of low bioavailability. Increased iron requirements may result from physiological or clinical variables. The physical, biochemical, and microbial changes during fermentation are summarized: growth of lactic acid bacteria and associated acidification of the product, reduction of nitrates to nitrites and formation of nitrosomyoglobin, solubilization and gelification of myofibrillar and sarcoplasmic proteins, degradation of proteins and lipids, and dehydration.

10.3.2.4 Fish-Based Solid-State Fermented Food Fermented fish products such as fish paste, fish sauces, or salted fish have been consumed. In Asian countries such as China, the Philippines, Thailand, Malaysia, and Indonesia, fish fermentation has been of great value since earliest times (Vaughn, 1954).

10.3.2.5 Grain-Based Solid-State Fermented Food: Fermented Rice Products
About 80 to 90% of the daily caloric intake of people in China is derived from rice, and China contributes 38% of the world's rice production using 24% of the world's growing area. In addition to the normal forms of rice cereal and rice flour products, many forms of fermented rice products are produced as alternative food sources, due to their nutritive values.

Red mold rice (RMR). RMR consists mainly of cooked nonglutinous rice, red fungi, and secondary metabolites of the fermentation process. Ma et al. reported its ingredients to be: total carbohydrate (73.4%), crude protein (14.7%), moisture (6.0%), fatty acids (2.84%), ash (2.45%), fiber (0.8%), monacolins (0.4%), phosphorus (0.4%), pigments (0.3%), vitamin C (0.03%), organic phosphorus (0.02%), and vitamin A < 70 IU/100g. The trace elements in RMR were: Ca, Mg, Na, Al, Fe, Mn, Cu, Zn, and Se. Of these trace elements, magnesium, and sodium were the most abundant metal elements (Ma et al., 2000).

Molds, yeasts, and bacteria involved in the production eventually lyse. Chinese rice vinegar has great levels of amino acids, making it not only nutritious, but also overcoming the severe taste from acetic acid, hence, making it a balancing food. Vinegar can improve dietary mineral intake and absorption, which may help prevent iron and calcium deficiencies and their related illnesses. The vinegar acid not only helps by releasing iron from cookware into the food, but also improves iron absorption, as shown in studies using intestinal cell models (Rosen et al., 1994; Ujike et al., 2003).

10.3.2.6 Other Solid-State Fermented Food Cheese is recognized as having a high nutritional value because of its normally high content in protein, calcium, vitamins A, riboflavin, and D (Dillon, 2000). Generally, cheese provides number of proteins, calcium, fat and phosphorus. A 30-gram serving of Cheddar cheese comprises about 7 grams of protein and 200 milligrams of calcium. Nutritionally, cheese is essentially concentrated milk: It takes about 200 grams of milk to provide that much protein, and 150 grams to equal the calcium (nutritional data from CNN interactive). Cheese potentially shares other nutritional properties of milk. This seeming discrepancy is called the "French paradox"; the higher rates of consumption of red wine in these countries are often invoked as at least a partial explanation. Some studies claim that cheddar, mozzarella, Swiss, and American cheeses can help to prevent tooth decay (Maidment and Williams, 2000).

Table olives, a traditional fermented functional food, are an essential source of linoleic acid and monounsaturated fatty acids. The benefit of the olive products is the absence of saturated fatty acids. On top of monounsaturated fatty acids, chlorophylls, polyphenols, and carotenoids contribute to the nutritional benefits and biological functions of table olives. The phenolic fraction of table olives is very complex (Boskou et al., 2006) and can vary both in the quality and quantity of phenolic compounds (Uccella, 2000).

Kombucha is a kind of fermented tea. Kombucha contains various species of yeast and bacteria, as well as the organic acids, amino acids, active enzymes, and polyphenols. Finished kombucha contains some of the following components, depending on the source and diversity of the culture: B-vitamins; acetic acid, which is slightly antibacterial; butyric acid; gluconic acid; lactic acid; malic acid; oxalic acid; and usnic acid (Velićanski Aleksandra et al., 2007).

10.4 Solid-State Food Fermentation as a Tool for Food Safety and Hygiene in Developing Countries

10.4.1 Challenges Associated with Food Safety and Hygiene in Developing Countries

10.4.1.1 Current Status of Food Safety and Hygiene in Developing Countries

In the past decade attention on food safety and hygiene has been gradually increasing. Consumers in developing countries are now aware that safety of food is a basic aspect of food quality. The concept of food safety is very broad and implies the absence or acceptable and safe levels of contaminants, adulterants, naturally occurring toxins or any other substance that may make food injurious to health on an acute or chronic basis.

Challenges to food safety continue to arise in unpredictable ways, largely due to changes in food production and supply, environment leading to food contamination, new and emerging germs, toxins, and antibiotic resistance. According to the related reports by World Health Organization (WHO), in constituting the public health of international concern events from 2003 to 2006, the events happening in developing countries occupied the main part. In many Asian countries, for example, dangerous chemicals such as boric acid are used in noodle preparation.

Each year, the world's population grows quickly, and most of this increase occurs in the developing countries. In 2020, the world population will most likely reach 7.6 billion, with an increase of 31% over the mid-1996 population of 5.8 billion. Approximately 98% of the projected population growth over this period will take place in developing countries. It has also been estimated that between the years 1995 and 2020 the developing world's urban population will double, reaching 3.4 billion (Battcock and Azam-Ali, 1998). Population growth is accompanied by many socio-economic problems in these countries, and many people view industrialization as a solution. However, this overall increase in population and in the urban population in particular, poses great challenges to food systems.

With the development of economics, there are more and more small-scale food producers and cottage industries, even outnumbering large-scale commercial producers and processors. Lack of adequate facilities and knowledge to produce foods that are safe and of an acceptable quality, small-scale producers and cottage industries have at times been sources of environmental pollution and unsafe food. Rapid urbanization has given urban services rise to be stretched beyond their limits, resulting in insufficient supplies of potable water, sewage disposal, and other necessary services.

Governments all over the world are improving the status of food safety in individual countries, which are in response to an increasing number of food safety issues and rising consumer concerns. Nowadays consumer awareness is provoked by local and international issues regarding food safety such as the detection of pesticide residues in food, and these cautious consumers then raise the issue in the quest for a chemical-free environment and safe food to eat. In developing countries, governments are facing the issue of protecting their populations against unsafe, adulterated, and poor quality food. There are several reasons for this unsatisfactory situation in developing countries which should be paid attention to at the levels of both the food industry and the government including:

10.4.1.1.1 Foodborne Illnesses   As a widespread and growing public health problem, foodborne illnesses are defined by WHO as diseases, either infectious or toxic in nature, caused by agents that enter the body through the ingestion of food. Both developed and developing countries are facing the threat from foodborne diseases, in particular in developing countries. It was reported that in industrialized countries, the percentage of the population suffering from foodborne diseases each year has been reported to be up to 30%. While less well documented, the food safety issues in developing countries are due to the presence of a wide range of foodborne diseases, including those caused by parasites. The high prevalence of diarrheal diseases in many developing countries suggests major underlying food safety problems. It has been reported that in 2005 alone 1.8 million people died from diarrheal diseases, a great proportion of which can be attributed to contamination of food and drinking water. Additionally, diarrhea is a major cause of malnutrition in infants and young children.

10.4.1.1.1.1 Salmonellosis   This disease is caused by the *Salmonella* bacteria and symptoms are mainly fever, headache, nausea, vomiting, and abdominal pain. *Salmonella* is widely spread in the environment and universally recognized as an important cause of foodborne infections. Examples of foods involved in outbreaks of salmonellosis include eggs, poultry and other meats, raw milk, and chocolate. Fermented foods derived from these materials, such as salami and cheeses, have occasionally been associated with outbreaks of illness.

10.4.1.1.1.2 Campylobacteriosis   Caused by certain species of the *Campylobacter* bacteria, in some cases the infection may lead to chronic health problems, including reactive neurological disorders and arthritis. Acute health effects of campylobacteriosis include severe abdominal pain, nausea, fever, and diarrhea. *C. jejuni, C. coll,* and *C. lari* are responsible for more than 90% of human illness caused by *Campylobacters*, which are normally characterized by profuse diarrhea and abdominal pain (Beumer, 2001). Poorly cooked poultry products and raw milk are most frequently involved in outbreaks, and cross-contamination from raw poultry to ready-to-eat products may also be a significant factor. Although the infective dose of *Campylobacter* is low (approximately 500 cfu), *Campylobacter* infection by fermented foods is probably not important. They cannot grow below 30°C and are sensitive to freezing, low pH, drying and sodium chloride, and so would not survive well in fermented foods. This has been confirmed by artificial contamination of fermenting products such as salami, weaning foods, and yogurt, which resulted in a rapid decrease in numbers (Northolt, 1983; Kingamkono et al., 1996; Morioka et al., 1996).

10.4.1.1.1.3 Enterohaemorrhagic and Listeriosis   These are important foodborne diseases that have emerged over the last decades. Although their incidence is relatively low, they have serious and sometimes lethal health consequences among people, particularly infants, children, and the elderly. Some members of the Enterobacteriaceae family are known pathogens, such as *E. coli* (especially *E. coli* 0 157), *Shigella*, and *Yersinia*, and a great part of the family, including the foodborne pathogens, are components of the excreta of animals. As a result, generally speaking, the whole family or parts of it (coliforms and *E. coli*) are usually used as hygiene indicators for processed foods. Species *of Listeria* are widespread and potential food contaminants, but only *L. monocytogenes* is a human and animal pathogen. Most other cases of listeriosis occur in immune-compromised persons, with a death rate varying from 30% to 50% (Roberts et al., 1996). Because of its widespread existence around the environment, this pathogen can exist in low numbers by practically all raw food products, and processed foods can be polluted from the production environment. Fermented products such as soft mold-ripened cheeses and sausages have been related to listeriosis. The organism is not particularly acid-resisting, but where mold ripening occurs in a fermented product, the rise in pH can allow surviving *Listeria* to resume growth, resulting in problems that occur.

10.4.1.1.1.4 Cholera   A major public health problem in developing countries, it also results in huge economic losses. Although not in water, contaminated foods including vegetables, rice, millet gruel, and a variety of seafood have been implicated in outbreaks of cholera; the disease is caused by the bacterium *Vibrio cholerae*. Vibrios are gram-negative, facultative anaerobic bacteria belonging to the family of the *Vibrionaceae,* and are usually found in aquatic environments. The organism is commonly isolated from river waters and coastal marine waters and, that is to say, from shellfish and other marine animals, but its transmission by fermented fish products has not been reported (West, 1989; Roberts et al., 1996). Refrigerated or frozen storage of contaminated food will not ensure safety because of the good survival of vibrios at low temperatures.

10.4.1.1.2 Toxins Naturally occurring toxins, such as marine biotoxins and mycotoxins, periodically cause severe intoxications. Mycotoxins can be found at measurable levels in many foods, and mycotoxin biosynthesis may be related to the preharvest stage of crop production by fungi that are obligate plant pathogens, or members of the flora responsible for the decay of senescent plant material (Moss, 1996). However, many of these toxic metabolites are produced by fungi growing on postharvest commodities that are stored under inappropriate conditions.

10.4.1.1.3 Persistent Organic Pollutants and Metals Persistent organic pollutants (POP) are compounds that accumulate in the environment and the human body such as polychlorinated biphenyls and dioxins. Dioxins are unwanted coproducts of some industrial processes and waste incineration. Exposure to POPs may result in a wide variety of adverse effects in humans. Lead and mercury can cause neurological damage in infants and children, and exposure to cadmium can also cause kidney damage, usually seen in the elderly.

10.4.1.1.4 Unconventional Agents Unconventional agents such as the agent causing bovine spongiform encephalopathy (BSE, or "mad cow disease") are associated with variant Creutzfeldt-Jakob disease in humans. Consumption of bovine products containing brain tissue is the most likely route for transmission of the agent to humans.

In developing countries, food systems are highly predominated by small-scale producers, which have many disadvantages. As a lot of food passes through a number of food handlers extending the food production, processing, storage and distribution chain, there is a higher risk of exposing food to contamination and adulteration. Furthermore, absence of resources and infrastructure for handling, processing, and storage leads to the severe decreasing of quality and avoidable contamination. For example, in the absence of facilities for home refrigeration in developing countries, fermentation became one of the main options for extension of the shelf life of foods.

Interaction and cooperation on food control is often lacking between industry and government in developing countries. Producing and selling safe and good quality food is the basic responsibility of industry, and it is the duty of the government to ensure compliance by industry to national food quality and safety requirements. Quality and safety of food have to be ensured throughout the food production, processing, storage, and distribution chain. However, national food control systems often suffer from serious inadequacies including insufficient involvement of scientific expertise from academia, industry, and consumers to strengthen the scientific basis for food control decision-making processes. Lack of resources such as trained inspectorate and laboratory staff, and inflexibility of the food control systems, make it difficult to cope with developments in food science and technology.

10.4.2 A Tool for Food Safety and Hygiene in Developing Countries

10.4.2.1 Why Solid-State Food Fermentation Can Be a Tool for Food Safety and Hygiene   Although the current high standard of quality and hygiene of fermented foods has mainly attributed to modern food technology, the principles of the age-old processes have barely changed. Fermentation can increase the security of foods by eliminating their natural toxic components, or by preventing the growth of pathogenic microbes.

10.4.2.1.1 Fermentation Can Remove Natural Toxic Components in Raw Materials   There are various hazards that can be associated with food materials, which can be broadly classified into three categories: extrinsic, intrinsic, and processing/biogenic. Some of these hazards can be severe, such as the botulinum toxin and *bongkrekic* acid poisoning, but the risk they pose can be reduced to acceptable levels by effective and hygienic processing and preparation. The processing of fermented foods starts with potentially hazardous raw materials and transfers them into products with better keeping qualities and reduced safety risk.

10.4.2.1.1.1 Physical Processing Prior to Fermentation A variety of physical unit operations associated with fermented food production may have effects on product safety. Inspection and sorting can remove damaged or obviously infected raw material such as infected meat, damaged or moldy fruit, or mold-infected grains. Interior tissues are generally sterile, and the majority of the surface-associated microflora can be removed by washing, peeling or trimming, substantially improving overall microbiological quality provided hygienical measures are adapted to avoid contamination of the freshly exposed surfaces. Chemical hazards can also be reduced by the physical separation of food components. Adjusting the water activity (Aw) of a material may have a significant effect on microbial hazards. Generally a decrease in Aw is achieved by drying or the addition of solutes to reduce the potential for microbial growth.

10.4.2.1.1.2 Solid-State Fermentation In industrial applications solid-state fermentation can be utilized in a controlled way and offer potential advantages in biological detoxification of hazardous and toxic compounds. Application of *Pleurotus* sp. was applied (Fan et al., 1999) for bioremediation of caffeinated residues. It is envisaged that in some cases the growth of microbes during SSF may utilize and reduce the undesirable antinutrients/toxicants present in the substrates (Reddy and Pierson, 1994). The method has been proved successful in diminishing gossypol content by 78% in cotton seed meal by SSF using *Geotrichum candidum* (Sun et al., 2008). Joshi et al. (2011) carried out the SSF of *J. curcas* seed cake using *Pseudomonas aeruginosa*, and assessed the effect of fermentation on the phorbol esters, the main toxicant of *J. curcas* seed cake. Under optimized SSF conditions, *P. aeruginosa* PseA completely degraded phorbol esters in nine days.

Sometimes food raw materials may contain toxic compounds such as hydrogen cyanide, tannins, etc., and measures are needed to reduce these toxic compounds prior to consumption. Food commodities stored under inappropriate conditions may have the potential problem of contamination with mycotoxins, and SSF has been applied as a tool to detoxify cassava peels and coffee pulp (Ofuya, 1994). Fermentation for 96 h caused a drastic reduction in the HCN levels and soluble tannins of the peels. Essers et al. (1995) studied the impact of six individual and dominant microflora strains in solid substrate fermentation of cassava on cyanogen levels, which included *Geotrichum candidum*, *Mucor racemosus*, *Rhizopus oryzae*, *Neurospora sitophila*, *Rhizopus stolonifer*, and *Bacillus* sp. Reduction in the potential toxic components of cassava during SSF was observed, and the effectiveness varied considerably between the species of microorganisms applied.

Peltonen et al. (2001) assessed the binding activity of aflatoxin B_1 (AFB_1) of 20 strains of lactic acid bacteria and bifidobacteria including *Lactobacillus*, *Bifidobacterium*, and *Lactococcus* strains. Two *Lactobacillus amylovorus* strains and 1 *Lactobacillus rhamnosus* strain removed more than 50% AFB_1, and more than 50% AFB_1 was bound throughout a 72-h incubation period.

Porres et al. (1993) studied the degradation of caffeine and polyphenols in coffee pulp, and the results showed that a reduction of 13%–63%, 28%–70%, and 51%–81% in caffeine, total polyphenols, and condensed tannins, respectively, was achieved. Roussos et al. (1994) studied caffeine degradation by *P. verrucosum* in SSF of coffee pulp, and found the addition of nitrogenous compounds rather inhibited caffeine degradation. Hakil et al. (1998) achieved substantial degradation of caffeine utilizing several strains of filamentous fungi.

Development of bioprocesses for bioremediation and biological detoxification of food commodities has been the main focus of SSF research in current years. SSF has been used in eliminating saponin from whole soybean tempeh by *Rhizopus oligosporus* (Fenwick and Oakenfull, 1983), hydrogen cyanide from cassava peel by *Rhizopus* sp. (Ofuya, 1994), and β-N-oxalyl-α, β-diaminopropionic acid (β-ODAP) from grass-pea by *Aspergillus oryzae* and *Rhizopus oligosporus* (Yigzaw et al., 2004).

10.4.2.1.2 Fermentation Can Prevent Growth of Disease-Causing Microbes
As already noted, the production of fermented foods can contribute to a number of desirable properties such as increased food safety, extended shelf life, and unique flavor or texture. Contrary to the unwanted spoilage or toxin production by disease-causing microbes, fermentation is regarded as a desirable effect of microbial activity in foods. The microbes that may be involved in fermentation include yeasts, molds, and bacteria. As a result of their growth and metabolism, substances of microbial origin are found in the fermented food,

including organic acids, alcohols, esters, aldehydes, which have a profound effect on the quality of the fermented product.

During fermentation lactic acid bacteria (LAB) is one of the main factors for the improvements in food safety arising from microbial activity, which are a group of organisms that predominate in a number of fermented foods. For example, lactic and acetic acids produced by LAB have an inhibitory effect on spoilage bacteria in fermented foods such as yogurt and sourdough bread. The growth and metabolism of LAB inhibit the normal spoilage flora of the food material and bacterial pathogens. LAB is known to produce a variety of antimicrobial factors, which play important roles in the improvement of food safety of fermented food.

The presence of organic acids lowers the pH of food systems and is one of the principal components of the inhibition of microorganisms by fermentation. Bacteriocins showed unique inhibitory activity against microorganisms, for example nisin is inhibitory to most gram-positive bacteria, and has been used as a food preservative with "generally regarded as safe" (GRAS) status in the United States. Heterofermentative LAB produces ethanol of lower concentrations, which is a well-established antimicrobial. Moreover, LAB produces hydrogen peroxide through the activity of flavoprotein oxidases in the presence of oxygen.

10.4.2.2 Potential of Solid-State Food Fermentation for Developing Countries
Fermentation is an applicable technique to target food safety and hygiene issues in developing countries because of the characteristics of low cost and ease of technology. The availability of modern food preservation techniques, such as an efficient cold chain, has reduced the significance of fermentation as a food preservation technology in developed countries today. However, fermentation remains as an important measure of food preservation in developing countries, particularly in tropical developing regions.

Over the years, widespread failure of the survival of medium- to large-scale food-processing enterprises in developing countries has resulted in a growing need to foster the development of the small-scale food industry.

Small-scale fermentation contributes substantially to food security and nutrition, especially in areas that are vulnerable to food shortages.

Moreover, agricultural production in most developing countries is seasonal and variable. Raw materials are usually produced on a small scale for subsistence, and it is difficult to have consistent supplies of raw agricultural materials for processing. Therefore, small-scale fermentation can ease the dependency of populations on food imports, and an appropriate food preservation technology contributes to sustainable development in developing countries. In a word, solid-state fermentation is highly in accordance with the situation in developing countries, where small-scale food producers and cottage industries usually outnumber large-scale commercial producers and processors.

Food sustainable development should always be paramount in modern food processing, and the significant developments in solid-state fermentation technology in the past few years in both industrialized and developing countries have proved the great potential of this technology. It is right to believe that SSF will continue to make a significant contribution to human nutrition, especially in developing countries, where economic problems pose a major barrier to ensuring food security and safety.

References

Aggett, P. (1989). Severe zinc deficiency. *Zinc in Human Biology*. New York: Springer-Verlag, pp. 259–279.

Ashok, P., Carlos R.S., and David, M. (2000a). New developments in solid state fermentation I-bioprocesses and products. *Process Biochemistry*. **35**, 1153–1169.

Ashok, P., Carlos, R.S., Poonam, N., and Soccol, V.T. (2000b). Biotechnological potential of agro-industrial residues. I: Sugarcane bagasse. *Bioresource Technology*. **74**, 69–80.

Barbosa, E., Faintuch, J., Moreira, E.A.M., Da Silva, V.R.G., Pereima, M.J.L., Fagundes, R.L.M., and Wilhelm Filho, D. (2009). Supplementation of vitamin E, vitamin C, and zinc attenuates oxidative stress in burned children: A randomized, double-blind, placebo-controlled pilot study. *Journal of Burn Care and Research*. **30**(5), 859–866.

Battcock, M. and Azam-Ali, S. (1998). *Fermented Fruits and Vegetables: A Global Perspective*. Rome: Food and Agriculture Organization of the United Nations Rome.

Beers, M., Berkow, R., and Bogin, R. (2000). In *The Merck Manual of Geriatrics*. eds. W. Station. NJ: Merck & Co., Inc.

Beumer, R. (2001). Microbiological hazards and their control: Bacteria. *Fermentation and Food Safety*. 141–157.

Black, R. (2003). Micronutrient deficiency: An underlying cause of morbidity and mortality. *Bulletin of the World Health Organization.* **81**(2), 79–79.

Black, R.E. (1998). Therapeutic and preventive effects of zinc on serious childhood infectious diseases in developing countries. *The American Journal of Clinical Nutrition.* **68**(2), 476S–479S.

Black, R.E., Allen, L.H., Bhutta, Z.A., Caulfield, L.E., De Onis, M., Ezzati, M., Mathers, C., and Rivera, J. (2008). Maternal and child undernutrition: Global and regional exposures and health consequences. *The Lancet.* **371**(9608), 243–260.

Black, R.E., Morris, S.S., and Bryce, J. (2003). Where and why are 10 million children dying every year? *The Lancet.* **361**(9376), 2226–2234.

Boskou, G., Salta, F.N., Chrysostomou, S., Mylona, A., Chiou, A., and Andrikopoulos, N.K. (2006). Antioxidant capacity and phenolic profile of table olives from the Greek market. *Food Chemistry.* **94**(4), 558–564.

Brabin, B., Coulter, J., Brabin, B., and Coulter, J. (2003). *Nutrition-Associated Disease.* London: Saunders.

Buchanan and Gibbons (1974). In *Bergey's Manual of Determinative Bacteriology*eds, P.H.A. Sneath, Mair, N.S., Sharpe, M.E., and Holt, J.G., Baltimore, MD: Williams & Willkins, pp. 15–36.

Chen, H.Z. (1992). Advances in solid-state fermentation. *Research and Application of Microbiology.* **3**, 7–10.

Chen, L.C., Chowdhury, A., and Huffman, S.L. (1980). Anthropometric assessment of energy-protein malnutrition and subsequent risk of mortality among preschool aged children. *The American Journal of Clinical Nutrition.* **33**(8), 1836–1845.

Chou, C.C. and Hwan, C.H. (2006). Effect of ethanol on the hydrolysis of protein and lipid during the ageing of a Chinese fermented soya bean curd—sufu. *Journal of the Science of Food and Agriculture.* **66**(3), 393–398.

Cole, P.D. and Kamen, B.A. (2003). "Beriberi" interesting! *Journal of Pediatric Hematology/Oncology.* **25**(12), 924–925.

Cook, J.D., Skikne, B.S., and Baynes, R.D. (1994). Iron deficiency: The global perspective. *Advances in Experimental Medicine and Biology.* **356**, 219–228.

Diaz, J., De Las Cagigas, A., and Rodriguez, R. (2003). Micronutrient deficiencies in developing and affluent countries. *European Journal of Clinical Nutrition.* **57**, S70–S72.

Dickson, R., Awasthi, S., Williamson, P., Demellweek, C., and Garner, P. (2000). Effects of treatment for intestinal helminth infection on growth and cognitive performance in children: Systematic review of randomised trials. *British Medical Journal.* **320**(7251), 1697–1701.

Eck and Gillis (2000). *Cheesemaking: from science to quality assurance.* Paris: Lavoisier Publishing.

Enser, M., Hallett, K., Hewitt, B., Fursey, G., and Wood, J. (1996). Fatty acid content and composition of English beef, lamb and pork at retail. *Meat Science.* **42**(4), 443–456.

Essers, A.J., Jurgens, C.M., and Nout, M.J. (1995). Contribution of selected fungi to the reduction of cyanogen levels during solid substrate fermentation of cassava. *International Journal of Food Microbiology.* **26**(2), 251–257.

Fan, L.F., P.A., Vandenberghe, L.P.S., and Soccol, C.R. (1999). Effect of caffeine on *Pleurotus* sp. and bioremediation of caffeinated residues. *IX European Congress on Biotechnology*. Brussels: Belgium: 2664.

FAO. (2004). Undernourishment around the world. In *The State of Food Insecurity in the World 2004*. Rome: The Organization.

Fauci, A. (1994). *Harrison's Principles of Internal Medicine*. New York: McGraw-Hill.

Fawzi, W.W., Chalmers, T.C., Herrera, M.G., and Mosteller, F. (1993). Vitamin A supplementation and child mortality. *JAMA: The Journal of the American Medical Association*. **269**(7), 898–903.

Fenwick, D.E. and Oakenfull, D. (1983). Saponin content of soybeans and some prepared foods. *Journal of the Science of Food and Agriculture*. **32**, 273–278.

Fernandez, I.D., Himes, J.H., and De Onis, M. (2002). Prevalence of nutritional wasting in populations: Building explanatory models using secondary data. *Bulletin of the World Health Organization*. **80**(4), 282–291.

Ferrari, E., Jarnagin, A.S., and Schmidt, B.F. (1993). Commercial production of extracellular enzymes, pp. 917–937. In A.L. Sonenshein, J.A. Hoch, and R. Losick (ed.), *Bacillus subtilis* and other Gram-positive bacteria. Washington, D.C., American Society for Microbiology.

Fleming, A.F. and De Silva, P.S. (1996). Haematological diseases in the tropics. In *Manson's Tropical Diseases*. pp. 169–243.

Glasziou, P. and Mackerras, D. (1993). Vitamin A supplementation in infectious diseases: A meta-analysis. *British Medical Journal*. **306**(6874), 366.

Hakil, M., Denis, S., Viniegra-Gonzalez, G., and Augur, C. (1998). Degradation and product analysis of caffeine and related dimethylxanthines by filamentous fungi. *Enzyme and Microbial Technology*. **22**(5), 355–359.

Han, B.J., Han, B.R., and Whang, H.S. (1998). Kanjang and doenjang. In *One Hundred Korean Foods that Koreans Must Know*. Seoul: Hyun Am Sa.

Haslam, N. and Probert, C. (1998). An audit of the investigation and treatment of folic acid deficiency. *Journal of the Royal Society of Medicine*. **91**(2), 72.

Hemilä, H. (2007). The role of vitamin C in the treatment of the common cold. *American Family Physician*. **76**(8), 1111–1115.

Hesham, M., Edariah, A., and Norhayati, M. (2004). Intestinal parasitic infections and micronutrient deficiency: A review. *The Medical Journal of Malaysia*. **59**(2), 284–293.

Hetzel, B.S. (2002). Eliminating iodine deficiency disorders: The role of the international council in the global partnership. *Bulletin of the World Health Organization*. **80**(5), 410–412.

Holker, U., Hofer, M., and Lenz, J. (2004). Biotechnological advantages of laboratory-scale solid-state fermentation with fungi. *Applied Microbiology and Biotechnology*. **64**, 175–186.

Huether, S. and McCance, K. (2004). *Understanding Pathophysiology*. St. Louis: Mosby-Year Book Inc.

Humphrey, J.H. and Rice, A.L. (2000). Vitamin A supplementation of young infants. *Lancet*. **356**(9227), 422–424.

Joshi, C., Mathur, P., and Khare, S.K. (2011). Degradation of phorbol esters by *Pseudomonas aeruginosa* (PseA) during solid-state fermentation of deoiled Jatropha curcas seed cake. *Bioresource Technology.* **102**, 4815–4819.

Khaw, K.T., Bingham, S., Welch, A., Luben, R., Wareham, N., Oakes, S., and Day, N. (2001). Relation between plasma ascorbic acid and mortality in men and women in epic-Norfolk prospective study: A prospective population study. European prospective investigation into cancer and nutrition. *Lancet.* **357**(9257), 657–663.

Kim, M. and Chun, J. (2005). Bacterial community structure in kimchi, a Korean fermented vegetable food, as revealed by 16s rRNA gene analysis. *International Journal of Food Microbiology.* **103**(1), 91–96.

Kingamkono, R., Sjoegren, E., Svanberg, U., and Kaijser, B. (1996). Inhibition of different strains of enteropathogens in a lactic-fermenting cereal gruel. *World Journal of Microbiology and Biotechnology.* **11**, 299–303.

Krawinkel, M., Bethge, M., El Karib, A., Ahmet, H., and Mirghani, O. (1990). Maternal ferritin values and foetal iron stores. *Acta Paediatrica.* **79**(4), 467–467.

Kudo, T. (1990). Warfarin antagonism of natto and increase in serum vitamin K by intake of natto. *Artery.* **17**(4), 189–201.

Kwena, A.M., Terlouw, D.J., De Vlas, S.J., Phillips-Howard, P.A., Hawley, W.A., Friedman, J.F., Vulule, J.M., Nahlen, B.L., Sauerwein, R.W., and Ter Kuile, F.O. (2003). Prevalence and severity of malnutrition in pre-school children in a rural area of western Kenya. *The American Journal of Tropical Medicine and Hygiene.* **68**(4 Suppl.), 94–99.

Lee, J.S., Heo, G.Y., Lee, J.W., Oh, Y.J., Park, J.A., Park, Y.H., Pyun, Y.R., and Ahn, J.S. (2005). Analysis of kimchi microflora using denaturing gradient gel electrophoresis. *International Journal of Food Microbiology.* **102**(2), 143–150.

Levin, H.M., Pollitt, E., Galloway, R., and Mcguire, J. (1993). Micronutrient deficiency disorders. *Disease Control Priorities in Developing Countries.* 421–454.

Liem, I., Steinkraus, K. and Cronk, T. (1977). Production of vitamin B-12 in tempeh, a fermented soybean food. *Applied and Environmental Microbiology.* **34**(6), 773–776.

Liu, Y. and Chou, C. (1994). Contents of various types of proteins and water soluble peptides in sufu during aging and the amino acid composition of tasty oligopeptides. *Journal of the Chinese Agricultural Chemical Society.* **32**, 276–276.

Ma, J., Li, Y., Ye, Q., Li, J., Hua, Y., Ju, D., Zhang, D., Cooper, R., and Chang, M. (2000). Constituents of red yeast rice, a traditional Chinese food and medicine. *Journal of Agricultural and Food Chemistry.* **48**(11), 5220–5225.

Maidment, I. and Williams, C. (2000). Drug-induced eosinophilia. *Pharmaceutical Journal.* **264**(7078), 71–76.

Man, W., Weber, M., Palmer, A., Schneider, G., Wadda, R., Jaffar, S., Mulholland, E., and Greenwood, B. (1998). Nutritional status of children admitted to hospital with different diseases and its relationship to outcome in the Gambia, West Africa. *Tropical Medicine and International Health.* **3**(8), 678–686.

Mann, C.C. (1999). Crop scientists seek a new revolution. *Science.* **283**(5400), 310–314.

Mann, N. (2000). Dietary lean red meat and human evolution. *European Journal of Nutrition.* **39**(2), 71–79.

McCormick, D. (1988). Thiamin. In *Modern Nutrition in Health and Disease.* PA: Lea & Febiger.

Millward, D.J. and Jackson, A.A. (2004). Protein/energy ratios of current diets in developed and developing countries compared with a safe protein/energy ratio: Implications for recommended protein and amino acid intakes. *Public Health Nutrition.* **7**(03), 387–405.

Ming-Yen, J. and Cheng-Chun, C. (2010). Enhancement of antioxidant activity, total phenolic and flavonoid content of black soybeans by solid state fermentation with *Bacillus subtilis* BCRC 14715. *Food Microbiology.* **27**, 586–591.

Morioka, Y., Nohara, H., Araki, M., Suzuki, M., and Numata, M. (1996). Studies on the fermentation of soft salami sausage by starter culture. *Animal Sd Technol.* **67**, 204–210.

Moss, M.O. (1996). Mycotoxins. *Mycological Research.* **100**, 513–523.

Murray, C.J.L. and Lopez, A.D. (1997). Global mortality, disability, and the contribution of risk factors: Global burden of disease study. *Lancet.* **349**(9063), 1436–1442.

Nemer, L., Gelband, H., Jha, P., and Duncan, T. (2001). The evidence base for interventions to reduce malnutrition in children under five and school-age children in low and middle-income countries. *Commission on Macroeconomics and Health (CMH) Working Paper WG5.* **11**, 2–11.

Northolt, M.D. (1983). Pathogenic micro-organisms in fermented dairy products. *Netherlands Milk and Dairy Journal.* **37**, 247–248.

Ofuya Co, O.S. (1994). The effects of solid state fermentation on the toxic components of cassava peel. *Process Biochem.* **29**, 25–28.

Pandey, A. (1992). Recent process developments in solid-state fermentation. *Process Biochemistry.* **27**(2), 109–117.

Pelletier, D.L., Frongillo, E.A. Jr., and Habicht, J.P. (1993). Epidemiologic evidence for a potentiating effect of malnutrition on child mortality. *American Journal of Public Health.* **83**(8), 1130–1133.

Peltonen, K., El-Nezami, H., Haskard, C., Ahokas, J., and Salminen, S. (2001). Aflatoxin B1 binding by dairy strains of lactic acid bacteria and bifidobacteria. *Journal of Dairy Science.* **84**, 2152–2156.

Pitsavas, S., Andreou, C., Bascialla, F., Bozikas, V.P., and Karavatos, A. (2004). Pellagra encephalopathy following B-complex vitamin treatment without niacin. *The International Journal of Psychiatry in Medicine.* **34**(1), 91–95.

Porres, C., Alvarez, D., and Calzada, J. (1993). *Biotechnology Advances.* **11**, 519–522.

Prakash, R., Gandotra, S., Singh, L.K., Das, B., and Lakra, A. (2008). Rapid resolution of delusional parasitosis in pellagra with niacin augmentation therapy. *General Hospital Psychiatry.* **30**(6), 581–584.

Prasad, A.S. (2003). Zinc deficiency. *British Medical Journal.* **326**(7386), 409–410.

Rath, M. and Pauling, L. (1990). Immunological evidence for the accumulation of lipoprotein (a) in the atherosclerotic lesion of the hypoascorbemic guinea pig. *Proceedings of the National Academy of Sciences.* **87**(23), 9388–9390.

Reddy, N.R. and Pierson, M.D. (1994). Reduction in antinutritional and toxic components in plant foods by fermentation. *Food Research International.* **27**, 281–290.

Reeta, R.S., Anil, K.P., Carlos, R.S., and Ashok, P. (2009). Recent advances in solid-state fermentation. *Biochemical Engineering Journal.* **44**, 13–18.

Rice, A.L., Sacco, L., Hyder, A., and Black, R.E. (2000). Malnutrition as an underlying cause of childhood deaths associated with infectious diseases in developing countries. *Bulletin of the World Health Organization.* **78**(10), 1207–1221.

Roberts, T.A., Baird-Parker, A.C., and Tompkin, R.B., Eds. (1996). *Microbiological Specifications of Food Pathogens: Microorganisms in Foods.* **5**. London, Blackie Academic & Professional, pp 458–478.

Robinson, T. and Nigam, P. (2003). Bioreactor design for protein enrichment of agricultural residues by solid state fermentation. *Biochemical Engineering Journal.* **13**, 197–203.

Rosen, H., Salemme, H., Zeind, A., Moses, A., Shapiro, A., and Greenspan, S. (1994). Chicken soup revisited: Calcium content of soup increases with duration of cooking. *Calcified Tissue International.* **54**(6), 486–488.

Ross, D.A., Dollimore, N., Smith, P., Kirkwood, B., Arthur, P., Morris, S., Addy, H., Binka, F.N., Gyapong, J., and Tomkins, A. (1993). Vitamin A supplementation in northern Ghana: Effects on clinic attendances, hospital admissions, and child mortality. *The Lancet.* **342**(8862), 7–12.

Roussos, S., Angeles-Aquiahuatl, M.D.L, Trejo-Hernandez, M.D.R., and Marakis, S. (1994). *Journal of Food Science and Technology.* **31**, 316–319.

Rodrıguez Couto, S. and Sanroman, M.A. (2006). Application of solid-state fermentation to food industry: A review. *Journal of Food Engineering.* **76**, 291–302.

Salama, P., Spiegel, P., Talley, L., and Waldman, R. (2004). Lessons learned from complex emergencies over the past decade. *The Lancet.* **364**(9447), 1801–1813.

Sen, A. (1981). In *Poverty and Famines: An Essay on Entitlement and Famines.* Oxford: Clarendon Press.

Shali, T., Singh, C., and Goindi, G. (2004). Prevalence of anemia amongst pregnant mothers and children in Delhi. *Indian Journal of Pediatrics.* **71**(10), 946–946.

Shankar, A.H. (2000). Nutritional modulation of malaria morbidity and mortality. *Journal of Infectious Diseases.* **182**(Suppl. 1), S37–S53.

Shankar, A.H. and Prasad, A.S. (1998). Zinc and immune function: The biological basis of altered resistance to infection. *The American Journal of Clinical Nutrition.* **68**(2), 447S–463S.

Sikosana, P., Bhebhe, S., and Katuli, S. (1998). A prevalence survey of iron deficiency and iron deficiency anaemia in pregnant and lactating women, adult males and pre-school children in Zimbabwe. *Central African Journal of Medicine.* **44**(12), 297–304.

Singhaniaa, R.R. (2009). Recent advances in solid-state fermentation. *Biochemical Engineering Journal.* **44**, 13–18.

Soccol, C.R., Marin, B., and Rimbauh, M. (1994). Potential of solid state fermentation for production of l(+)lactic acid by *Rhizopus oryzae. Applied Micmbiology and Biotechnology.* **41**, 286–290.

Sommer, A. (1995). *Vitamin a Deficiency and Its Consequences: A Field Guide to Detection and Control.* Geneva, WHO.

Stoltzfus, R.J., Chway, H.M., Montresor, A., Tielsch, J.M., Jape, J.K., Albonico, M., and Savioli, L. (2004). Low dose daily iron supplementation improves iron status and appetite but not anemia, whereas quarterly anthelminthic treatment improves growth, appetite and anemia in Zanzibari preschool children. *The Journal of Nutrition.* **134**(2), 348–356.

Stoltzfus, R.J. and Dreyfuss, M.L. (1998). In *Guidelines for the Use of Iron Supplements to Prevent and Treat Iron Deficiency Anemia.* Washington, DC: ILSI Press.

Su, Y.C. (1986). In *Sufu.* Boca Raton, FL: CRC Press.

Sumi, H. (1999). Determination of the vitamin K (menaquinone-7) content in fermented soybean natto and in the plasma of natto-ingesting subjects. *Journal of Home Economics of Japan.* **50**, 309–312.

Sun, Z.T., Liu, C., and Du, J.H. (2008). Optimisation of fermentation medium for the detoxification of free gossypol in cottonseed powder by geotrichum candidum g07 in solid-state fermentation with response surface methodology. *Annals of Microbiology.* **58**, 683–690.

Suryanarayan, S. (2003). Industrial practice in solid state fermentations for secondary metabolite production: The Biocon India experience. *Biochemical Engineering Journal.* **13**, 189–195.

Traber, M.G. and Sies, H. (1996). Vitamin E in humans: Demand and delivery. *Annual Review of Nutrition.* **16**(1), 317–321.

Truesdell, D.D., Green, N.R., and Acosta, P.B. (2006). Vitamin B12 activity in miso and tempeh. *Journal of Food Science.* **52**(2), 493–494.

Uccella, N. (2000). Olive biophenols: Novel ethnic and technological approach. *Trends in Food Science and Technology.* **11**(9), 328–339.

Ujike, S., Shimoji, Y., Nishikawa, Y., Taniguchi, M., Nanda, K., Ito, R., Kawabata, D., and Uenakai, K. (2003). Effect of vinegar on absorption of iron in rats. *Journal of Japanese Society of Nutrition and Food Science.* **56**(6), 371–374.

Vandenberghe, L.P.S., Soccol, C.R., Pandey, A., and Lebeeult, J.M. (2000). Solid state fermentation for the synthesis of citric acid by *Aspergillus niger. Bioresource Technology.* **74**, 175–178.

Vaughn, R.H. (1954). Lactic acid fermentation of cucumbers, sauerkraut and olives. *Industrial Fermentations.* **2**, 417–478.

Velićanski Aleksandra, S., Cvetković Dragoljub, D., Markov Siniša, L., Tumbas Vesna, T., and Savatović Slađana, M. (2007). Antimicrobial and antioxidant activity of lemon balm Kombucha. *Acta Periodica Technologica.* **2007**(38), 165–172.

Vijayaraghavan, K. (2000). Vitamin A deficiency. *The Lancet.* **356**, S41.

Wang, F. (2007). *The Research on Solid-State Fermentation of Soybean Residue.* Beijing: China Agricultural University, pp 48–50.

Wang, J. (2002). The nutrition and health function of sufu: A fermented bean curd. *China Brew.* **4**, 4–6.

Wang, L., Li, L., Fan, J., Saito, M., and Tatsumi, E. (2004). Radical-scavenging activity and isoflavone content of sufu (fermented tofu) extracts from various regions in China. *Food Science and Technology Research.* **10**(3), 324–327.

Wang, L., Saito, M., Tatsumi, E., and Li, L. (2003). Antioxidative and angiotensin I-converting enzyme inhibitory activities of sufu (fermented tofu) extracts. *Japan Agricultural Research Quarterly.* **37**(2), 129–132.

Wang, R. and Du, X. (1998). In *The Production of Sufu in China.* Beijing, China: Light Industry Press, pp. 1–4.

West, P.A. (1989). The human pathogenic vibrios: A public health update with environmental perspectives. *Epidemiollnfm.* 1–34.

WHO (1995). *Global Prevalence of Vitamin A Deficiency.* Geneva: The Organization.

WHO (2001). Commission on macroeconomics and health. *Macroeconomics and Health: Investing in Health for Economic Development.* Ginebra: WHO.

Williams, C. (1935). Kwashiorkor: A nutritional disease of children associated with a maize diet. *The Lancet.* **226**(5855), 1151–1152.

Yen, G.C. (1986). Studies on biogenic amines in foods. I. Determination of biogenic amines in fermented soybean foods by HPLC. *Journal of the Chinese Agricultural Chemical Society.* **24**(2), 211–227.

Yigzaw, Y., Gorton, L., Solomon, T., and Akalu, G. (2004). Fermentation of seeds of teff (*eragrostis teff*), grass-pea (*lathyrus sativus*), and their mixtures: Aspects of nutrition and food safety. *Journal of Agricultural and Food Chemistry.* **52**(5), 1163–1169.

Yip, R. and Ramakrishnan, U. (2002). Experiences and challenges in developing countries. *The Journal of Nutrition.* **132**(4), 827S–830S.

Yoon, J.H., Kang, S.S., Mheen, T.I., Ahn, J.S., Lee, H.J., Kim, T.K., Park, C.S., Kho, Y.H., Kang, K.H., and Park, Y.H. (2000). *Lactobacillus kimchii* sp. Nov., a new species from Kimchi. *International Journal of Systematic and Evolutionary Microbiology.* **50**(5), 1789–1795.

Young, H., Borrel, A., Holland, D., and Salama, P. (2004). Public nutrition in complex emergencies. *The Lancet.* **364**(9448), 1899–1909.

Zimmermann, M.B., Chaouki, N., and Hurrell, R.F. (2005). Iron deficiency due to consumption of a habitual diet low in bioavailable iron: A longitudinal cohort study in Moroccan children. *The American Journal of Clinical Nutrition.* **81**(1), 115–121.

Index

Milton Keynes UK
Ingram Content Group UK Ltd.
UKHW030900141024
449569UK00025B/1304

9 781138 199323